Theoretical Foundations of Functional Data Analysis, with an Introduction to Linear Operators

WILEY SERIES IN PROBABILITY AND STATISTICS

Established by WALTER A. SHEWHART and SAMUEL S. WILKS

A complete list of the titles in this series appears at the end of this volume.

Theoretical Foundations of Functional Data Analysis, with an Introduction to Linear Operators

Tailen Hsing
Professor, Department of Statistics
University of Michigan, USA

Randall Eubank
Professor Emeritus, School of Mathematical and Statistical Sciences, Arizona State University, USA

This edition first published 2015

Registered office
John Wiley & Sons Ltd, The Atrium, Southern Gate, Chichester, West Sussex, PO19 8SQ, United Kingdom

For details of our global editorial offices, for customer services and for information about how to apply for permission to reuse the copyright material in this book please see our website at www.wiley.com.

Library of Congress Cataloging-in-Publication Data applied for.

A catalogue record for this book is available from the British Library.

ISBN: 9780470016916

Set in 10/12pt TimesLTStd by Laserwords Private Limited, Chennai, India

To Our Families

Contents

Preface

This book aims to provide a compendium of the key mathematical concepts and results that are relevant for the theoretical development of functional data analysis (fda). As such, it is not intended to provide a general introduction to fda *per se* and, accordingly, we have not attempted to catalog the volumes of fda research work that have flowed at a brisk pace into the statistics literature over the past 15 years or so. Readers might therefore find it helpful to read the present text alongside other books on fda, such as Ramsay and Silverman (2005), which provide more thorough and practical developments of the topic.

This project grew out of our own struggle in acquiring the theoretical foundations for fda research in diverse fields of mathematics and statistics. With that in mind, the book strives to be self-contained. Rigorous proofs are provided for most of the results that we present. Nonetheless, a solid mathematics background at a graduate level is needed to be able to appreciate the content of the text. In particular, the reader is assumed to be familiar with linear algebra and real analysis and to have taken a course in measure theoretic probability. With this proviso, the material in the book would be suitable for a one-semester, special-topics class for advanced graduate students.

Functional data analysis is, from our perspective, the statistical analysis of sample path data observed from continuous time stochastic processes. Thus, we are dealing with random functions whose realizations fall into some suitable (large) collection of functions. This makes an overview of function space theory a natural starting point for our treatment of fda. Accordingly, we begin with that topic in Chapter 2. There we develop essential concepts such as Sobolev and reproducing kernel Hilbert spaces that play pivotal roles in subsequent chapters. We also lay the foundation that is needed to understand the essential mathematical properties of bounded operators on Banach and, in particular, Hilbert spaces.

Our treatment of operator theory is broken into three chapters. The first of these, Chapter 3, deals with basic concepts such as adjoint, inverse, and projection operators. Then, Chapter 4 investigates the spectral theory that underlies compact operators in some detail. Here, we present both the typical eigenvalue/eigenvector expansion for self-adjoint operators and the somewhat less common singular value expansion that applies in the non-self-adjoint case or, more generally, for operators between two different Hilbert spaces. These expansions make it possible to develop the concepts of Hilbert–Schmidt and trace class operators at a level of generality that makes them useful in subsequent aspects of the text.

The treatment of principal components analysis in Chapter 9 requires some understanding of perturbation theory for compact operators. This material is therefore developed in Chapter 5. As was the case for Chapter 4, we do this for both the self-adjoint and the-non self-adjoint scenarios. The latter instance therefore provides an introduction to the less well-documented perturbation theory for singular values and vectors that, for example, can be employed to investigate the properties of canonical correlation estimators.

The fact that sample paths must be digitized for storage entails that data smoothing of some kind often becomes necessary. Smoothing and regularization problems also arise naturally from the approximate solution of operator equations, functional regression, and various other problems that are endemic to the fda setting. Chapter 6 examines a general abstract smoothing or regularization problem that corresponds to what we call a functional linear model. An explicit form is derived for the associated estimator of the underlying functional parameter. The problems of computation and regularization parameter selection are considered for the case of real valued, scalar response data. A special case of our abstract smoothing scenario leads us back to ordinary smoothing splines and we spend some time studying their associated properties as nonparametric regression estimators.

Chapter 7 aims to establish the probabilistic underpinnings of fda. The mean element, covariance operator, and cross-covariance operators are rigorously defined here for random elements of a Hilbert space. The fda case where a random element has a representation as a continuous time stochastic process is given special treatment that, among other factors, clarifies the relationship between its covariance operator and covariance kernel. A brief foray into representation theory produces congruence relationships that prove useful in Chapter 10. The chapter then concludes with selected aspects of the large sample theory for Hilbert space valued random elements that includes both a strong law and central limit theorem.

The large sample behavior of the sample mean element and covariance operator are studied in Chapter 8. This is relevant for cases where functional data is completely observed. When meaningful discretization occurs, smoothing becomes necessary and we look into the large sample performance of two estimation schema that can be used for that purpose: namely, local linear and penalized least-squares smoothing. Chapter 9 is the principal components counterpart of Chapter 8 in that it investigates the properties of eigenvalues and eigenfunctions associated with the covariance operator estimators that were derived in that chapter.

Chapters 10 and 11 both address bivariate situations. In Chapter 10, the focus is canonical correlation and this concept is used to study a variety of fda problems including functional factor analysis and discriminant analysis. Then, Chapter 11 deals with the problem of functional regression with a scalar response and functional predictor. An asymptotically, optimal penalized least-squares estimator is investigated in this setting.

We have been fortunate to have talented coworkers and students that have generously shared their ideas and expertise with us on many occasions. A nonexhaustive list of such important contributors includes Toshiya Hoshikawa, Ana Kupresanin, Yehua Li, Heng Lian, Yolanda Munoz-Maldanado, Rosie Renaut, Hyejin Shin, and Jack Spielberg. We sincerely appreciate the invaluable help they have provided in bringing this book to fruition. The inspiration for much of the development in Chapter 10 can be traced to the serendipitous path through academics that brought us into contact with Anant Kshirsagar and Emanuel Parzen. We gratefully acknowledge the profound influence these two great scholars have had on this as well as many other aspects of our writing. TH also wishes to thank Ross Leadbetter for introducing him to the world of research and Ray Carroll for his support which has opened doors to many possibilities, including this book.

1

Introduction

Briefly stated, a *stochastic process* is an indexed collection of random variables all of which are defined on a common probability space $(\Omega, \mathscr{F}, \mathbb{P})$. If we denote the index set by E, then this can be described mathematically as

$$\{X(t, \omega) : t \in E, \omega \in \Omega\},$$

where $X(t, \cdot)$ is a $\mathscr{F}$-measurable function on the sample space Ω. The ω argument will generally be suppressed and $X(t, \omega)$ will typically be shortened to just $X(t)$.

Once the $X(t)$ have been observed for every $t \in E$, the process has been realized and the resulting collection of real numbers is called a *sample path* for the process. Functional data analysis (fda), in the sense of this text, is concerned with the development of methodology for statistical analysis of data that represent sample paths of processes for which the index set is some (closed) interval of the real line; without loss, the interval can be taken as $[0, 1]$. This translates into observations that are functions on $[0, 1]$ and data sets that consist of a collection of such random curves.

From a practical perspective, one cannot actually observe a functional data set in its entirety; at some point, digitization must occur. Thus, analysis might be predicated on data of the form

$$x_i(j/r), j = 1, \ldots, r, i = 1, \ldots, n,$$

involving n sample paths $x_1(\cdot), \ldots, x_n(\cdot)$ for some stochastic process with each sample path only being evaluated at r points in $[0, 1]$. When viewed from this perspective, the data is inherently finite dimensional and the temptation is to treat it as one would data in a multivariate analysis (mva) context.

Theoretical Foundations of Functional Data Analysis, with an Introduction to Linear Operators,
First Edition. Tailen Hsing and Randall Eubank.

However, for truly functional data, there will be many more "variables" than observations; that is, $r \gg n$. This leads to drastic ill conditioning of the linear systems that are commonplace in mva which has consequences that can be quite profound. For example, Bickel and Levina (2004) showed that a naive application of multivariate discriminant analysis to functional data can result in a rule that always classifies by essentially flipping a fair coin regardless of the underlying population structure.

Rote application of mva methodology is simply not the avenue one should follow for fda. On the other hand, the basic mva techniques are still meaningful in a certain sense. Data analysis tools such as canonical correlation analysis, discriminant analysis, factor analysis, multivariate analysis of variance (MANOVA), and principal components analysis exist because they provide useful ways to summarize complex data sets as well as carry out inference about the underlying parent population. In that sense, they remain conceptually valid in the fda setting even if the specific details for extracting the relevant information from data require a bit of adjustment. With that in mind, it is useful to begin by cataloging some of the multivariate methods and their associated mathematical foundations, thereby providing a roadmap of interesting avenues for study. This is the subject of the following section.

1.1 Multivariate analysis in a nutshell

mva is a mature area of statistics with a rich history. As a result, we cannot (and will not attempt to) give an in-depth overview of mva in this text. Instead, this section contains a terse, mathematical sketch of a few of the methods that are commonly employed in mva. This will, hopefully, provide the reader with some intuition concerning the form and structure of analogs of mva techniques that are used in fda as well as an appreciation for both the similarities and the differences between the two fields of study. Introductions to the theory and practice of mva can be found in a myriad of texts including Anderson (2003), Gittins (1985), Izenman (2008), Jolliffe (2004), and Johnson and Wichern (2007).

Let us begin with the basic set up where we have a p-dimensional random vector $X = (X_1, \dots, X_p)^T$ having (variance-)covariance matrix

$$\mathscr{K} = \mathbb{E}\left[(X - m)(X - m)^T\right] \tag{1.1}$$

with

$$m = \mathbb{E}X \tag{1.2}$$

the mean vector for X. Here, $\mathbb{E}$ corresponds to mathematical expectation and v^T indicates the transpose of a vector v. The matrix $\mathscr{K}$ admits an eigenvalue–eigenvector decomposition of the form

$$\mathscr{K} = \sum_{j=1}^{p} \lambda_j e_j e_j^T \tag{1.3}$$

for eigenvalues $\lambda_1 \geq \cdots \geq \lambda_p \geq 0$ and associated orthonormal eigenvectors $e_j = \left(e_{1j}, \ldots, e_{pj}\right)^T, j = 1, \ldots, p$ that satisfy

$$e_i^T \mathscr{K} e_j = \lambda_j \delta_{ij},$$

where δ_{ij} is 1 or 0 depending on whether or not i and j coincide. This provides a basis for principal components analysis (pca).

We can use the eigenvectors in (1.3) to define new variables $Z_j = e_j^T(X - m)$, which are referred to as principal components. These are linear combinations of the original variables with the weight or loadings e_{ij} that is applied to X_i in the jth component indicating its importance to Z_j; more precisely,

$$\mathrm{Cov}(Z_j, X_i) = \lambda_j e_{ij}.$$

In fact,

$$X = m + \sum_{j=1}^{p} Z_j e_j \tag{1.4}$$

as, if $\mathscr{K}$ is full rank, $e_1, \ldots, e_p$ provide an orthonormal basis for $\mathbb{R}^p$; this is even true when $\mathscr{K}$ has less than full rank as $e_j^T X$ is zero with probability one when $\lambda_j = 0$. The implication of (1.4) is that X can be represented as a weighted sum of the eigenvectors of $\mathscr{K}$ with the weights/coefficients being uncorrelated random variables having variances that are the eigenvalues of $\mathscr{K}$.

In practice, one typically retains only some number $q < p$ of the components and views them as providing a summary of the (covariance) relationship between the variables in X. As with any type of summarization, this results in a loss of information. The extent of this loss can be gauged by the proportion of the total X variance $V := \mathrm{trace}(\mathscr{K})$ that is recovered by the principal components that are retained. In this regard, we know that

$$V = \sum_{j=1}^{p} \lambda_j$$

while the variance of the jth component is

$$\begin{aligned} \mathrm{Var}(Z_j) &= e_j^T \mathscr{K} e_j \\ &= \lambda_j. \end{aligned}$$

Thus, the jth component accounts for $100\lambda_j/V$ percentage of the total variance and $100\left(1 - \sum_{k=1}^{j} \lambda_k/V\right)$ is the percentage of variability that is not accounted for by $Z_1, \ldots, Z_j$.

Principal components possess various optimality features such as the one catalogued in Theorem 1.1.1.

Theorem 1.1.1 $\mathrm{Var}(Z_j) = \max_{\{e^T e=1, e^T \mathscr{K} e_i=0, i=1,\ldots,j-1\}} \mathrm{Var}(e^T X)$.

The proof of this result is, e.g., a consequence of developments in Section 4.2. It can be interpreted as saying that the jth principle component is the linear combination of X that accounts for the maximum amount of the remaining total variance after removing the portion that was explained by $Z_1, \ldots, Z_{j-1}$.

The discussion to this point has been concerned with only the population aspects of pca. Given a random sample $x_1, \ldots, x_n$ of observations on X, we estimate $\mathscr{K}$ by the sample covariance matrix

$$\mathscr{K}_n = (n-1)^{-1} \sum_{i=1}^{n} \left(x_i - \bar{x}_n\right)\left(x_i - \bar{x}_n\right)^T \tag{1.5}$$

with

$$\bar{x}_n = n^{-1} \sum_{i=1}^{n} x_i \tag{1.6}$$

the sample mean vector. As $\mathscr{K}_n$ is positive semidefinite, it has the eigenvalue–eigenvector representation

$$\mathscr{K}_n = \sum_{j=1}^{p} \lambda_{jn} e_{jn} e_{jn}^T, \tag{1.7}$$

where the e_{in} are orthonormal and satisfy

$$e_{in}^T \mathscr{K}_n e_{jn} = \lambda_{jn} \delta_{ij}.$$

This produces the sample principle components $z_{jn} = e_{jn}^T(x - \bar{x}_n)$ for $j = 1, \ldots, p$ with $x = (x_1, \ldots, x_p)^T$ and the associated scores $e_{jn}^T(x_i - \bar{x}_n), i = 1, \ldots, n$ that provide sample information concerning the Z_j.

Theorems 9.1.1 and 9.1.2 of Chapter 9 can be used to deduce the large sample behavior of the sample eigenvalue–eigenvector pairs, $(\lambda_{jn}, e_{jn}), j = 1, \ldots, r$. The limiting distributions of $\sqrt{n}(\lambda_{jn} - \lambda_j)$ and $\sqrt{n}(e_{jn} - e_j)$ are found to be normal which provides a foundation for hypothesis testing and interval estimation.

The next step it to assume that X consists of two subsets of variables that we indicate by writing $X = (X_1^T, X_2^T)^T$, where $X_1 = (X_{11}, \ldots, X_{1p})^T$ and

$X_2 = (X_{21}, \ldots, X_{2q})^T$. Questions of interest now concern the relationships that may exist between X_1 and X_2. Our focus will be on those that are manifested in their covariance structure. For this purpose, we partition the covariance matrix $\mathscr{K}$ for X from (1.1) as

$$\mathscr{K} = \begin{bmatrix} \mathscr{K}_1 & \mathscr{K}_{12} \\ \mathscr{K}_{21} & \mathscr{K}_2 \end{bmatrix}. \tag{1.8}$$

Here, $\mathscr{K}_1, \mathscr{K}_2$ are the covariance matrices for X_1, X_2, respectively, and $\mathscr{K}_{12} = \mathscr{K}_{21}^T$ is sometimes called the cross-covariance matrix.

The goal is now to summarize the (cross-)covariance properties of X_1 and X_2. Analogous to the pca approach, this will be accomplished using linear combinations of the two random vectors. Specifically, we seek vectors $a_1 \in \mathbb{R}^p$ and $a_2 \in \mathbb{R}^q$ that maximize

$$\rho^2(a_1, a_2) = \frac{\mathrm{Cov}^2(a_1^T X_1, a_2^T X_2)}{\mathrm{Var}\left(a_1^T X_1\right) \mathrm{Var}\left(a_2^T X_2\right)}. \tag{1.9}$$

This optimization problem can be readily solved with the help of the singular value decomposition: e.g., Corollary 4.3.2. Assuming that X_1, X_2 contain no redundant variables, both $\mathscr{K}_1$ and $\mathscr{K}_2$ will be positive-definite with nonsingular square roots $\mathscr{K}_i^{1/2}, i = 1, 2$. This allows us to write

$$\rho^2(a_1, a_2) = \frac{\left(\tilde{a}_1^T \mathscr{R}_{12} \tilde{a}_2\right)^2}{\tilde{a}_1^T \tilde{a}_1 \tilde{a}_2^T \tilde{a}_2}, \tag{1.10}$$

where

$$\mathscr{R}_{12} = \mathscr{K}_1^{-1/2} \mathscr{K}_{12} \mathscr{K}_2^{-1/2}, \tag{1.11}$$

$\tilde{a}_1 = \mathscr{K}_1^{1/2} a_1$ and $\tilde{a}_2 = \mathscr{K}_2^{1/2} a_2$. The matrix $\mathscr{R}_{12}$ can be viewed as a multivariate analog of the linear correlation coefficient between two variables. Using the singular value decomposition in Corollary 4.3.2, we can see that (1.10) is maximized by choosing $\tilde{a}_1, \tilde{a}_2$ to be the pair of singular vectors $\tilde{a}_{11}, \tilde{a}_{21}$ that correspond to its largest singular value ρ_1. The optimal linear combinations of X_1 and X_2 are therefore provided by the vectors $a_{11} = \mathscr{K}_1^{-1/2} \tilde{a}_{11}$ and $a_{21} = \mathscr{K}_2^{-1/2} \tilde{a}_{21}$. The corresponding random variables $U_{11} = a_{11}^T X_1$ and $U_{21} = a_{21}^T X_2$ are called the first canonical variables of the X_1 and X_2 spaces, respectively. They each have unit variance and correlation $\rho_1$that is referred to as the first canonical correlation.

The summarization process need not stop after the first canonical variables. If $\mathscr{K}_{12}$ has rank r, then there are actually $r - 1$ additional canonical variables that can be found: namely, for $j = 2, \ldots, r$, we have

$$U_{1j} = a_{1j}^T X_1 \tag{1.12}$$

and

$$U_{2j} = a_{2j}^T X_2, \tag{1.13}$$

where $a_{1j} = \mathscr{K}_1^{-1/2} \tilde{a}_{1j}$, $a_{2j} = \mathscr{K}_2^{-1/2} \tilde{a}_{2j}$ with $\tilde{a}_{1j}, \tilde{a}_{2j}$, the other singular vector pairs from $\mathscr{R}_{12}$ that correspond to its remaining nonzero singular values $\rho_2 \geq \cdots \geq \rho_r > 0$. For each choice of the index j, the random variable pair (U_{1j}, U_{2j}) is uncorrelated with all the other canonical variable pairs and has corresponding canonical correlation ρ_j. When all this is put Together, it gives us

$$\mathscr{R}_{12} = \mathscr{K}_1^{1/2} \mathscr{A}_1 \mathscr{D} \mathscr{A}_2^T \mathscr{K}_2^{1/2} \tag{1.14}$$

with

$$\mathscr{A}_i = [a_{i1}, \ldots, a_{ir}]$$

the matrix of canonical weight vectors for $X_i, i = 1, 2$, and

$$\mathscr{D} = \operatorname{diag}(\rho_1, \ldots, \rho_r)$$

a diagonal matrix containing the corresponding canonical correlations.

There are various other ways of characterizing the canonical correlations and vectors. As they stem from the singular values and vectors of $\mathscr{R}_{12}$, they are the eigenvalue and eigenvectors obtained from $\mathscr{K}_1^{-1/2} \mathscr{K}_{12} \mathscr{K}_2^{-1} \mathscr{K}_{21} \mathscr{K}_1^{-1/2}$ and $\mathscr{K}_2^{-1/2} \mathscr{K}_{21} \mathscr{K}_1^{-1} \mathscr{K}_{12} \mathscr{K}_2^{-1/2}$. For example, the squared canonical correlations and canonical vectors of the X_1 space can be derived from the linear system

$$\mathscr{K}_1^{-1} \mathscr{K}_{12} \mathscr{K}_2^{-1} \mathscr{K}_{21} a_1 = \rho^2 a_1. \tag{1.15}$$

The population canonical correlations and associated canonical vectors can be estimated from the sample covariance matrix (1.5). For this purpose, one partitions $\mathscr{K}_n$ analogous to $\mathscr{K}$ and carries out the same form of singular value decomposition except using sample entities in place of $\mathscr{K}_1, \mathscr{K}_2$, and $\mathscr{K}_{12}$. The resulting sample canonical correlations have a limiting multivariate normal distribution under various conditions as detailed in Muirhead and Waternaux (1980).

The vectors X_1 and X_2 are linearly independent when $\mathscr{K}_{12}$ is a matrix of all zeros which we now recognize as being equivalent to $\rho_1 = \cdots = \rho_{\min(p,q)} = 0$. Test statistics for this and other related hypotheses can be developed from the sample canonical correlations.

Canonical correlation occupies a pervasive role in classical multivariate analysis. One place it arises naturally is in the (linear) prediction of X_1 from X_2. A best linear unbiased predictor(BLUP) is provided by the vector $\beta_0 + \beta_1 X_2$, where β_0, β_1 are, respectively, the $p \times 1$ vector and $p \times q$ matrix that minimize

$$\mathbb{E}(X_1 - b_0 - b_1 X_2)^T (X_1 - b_0 - b_1 X_2)$$

as a function of b_0, b_1. The minimizers are easily seen to be

$$\beta_0 = m_1 - \beta_1 m_2$$
$$\beta_1 = \mathcal{K}_{12}\mathcal{K}_2^{-1},$$

wherein $m_j = \mathbb{E}X_j, j = 1, 2$. From this, we recognize β_1 as the least-squares regression coefficients for the regression of X_1 on X_2. Now, premultiply (1.14) by $\mathcal{K}_1^{1/2}$ and postmultiply by $\mathcal{K}_2^{-1/2}$ to obtain

$$\beta_1 = \sum_{j=1}^{r} \rho_j \mathcal{K}_1 a_{1j} a_{2j}^T,$$

where, again, r is the rank of $\mathcal{K}_{12}$. This establishes a fundamental relationship between the best linear predictor and the canonical variables: namely,

$$\beta_0 + \beta_1 X_2 = \beta_0 + \sum_{j=1}^{r} \rho_j \mathcal{K}_1 a_{1j} U_{2j}. \tag{1.16}$$

Thus, canonical correlation lies at the heart of the linear prediction of X_1 from X_2. The converse is seen to be true as well by simply interchanging the roles of X_1 and X_2 in the above-mentioned discussion.

In finding our linear predictor for X_1, we chose a priori to restrict attention to only those that were linear functions of X_2. A related, but distinct, variation on this theme is to presume the existence of a linear model that relates X_1 to X_2 by an expression such as

$$X_1 = \beta_0 + \beta_1 X_2 + \varepsilon \tag{1.17}$$

with β_0 and β_1 of dimension $p \times 1$ and $p \times q$ as before and ε a $p \times 1$ random vector that is uncorrelated with X_2 while having zero mean and covariance matrix

$$\mathcal{K}_\varepsilon = \mathbb{E}\varepsilon\varepsilon^T.$$

In this event,

$$\mathcal{K}_{12} = \beta_1 \mathcal{K}_2, \tag{1.18}$$
$$\mathcal{K}_1 = \beta_1 \mathcal{K}_2 \beta_1^T + \mathcal{K}_\varepsilon$$
$$= \mathcal{K}_{12}\mathcal{K}_2^{-1}\mathcal{K}_{21} + \mathcal{K}_\varepsilon \tag{1.19}$$

and the canonical correlations now satisfy the relationship

$$\mathcal{A}_1 \mathcal{D} \mathcal{A}_2^T = (\beta_1 \mathcal{K}_2 \beta_1^T + \mathcal{K}_\varepsilon)^{-1} \beta_1. \tag{1.20}$$

Weyl's inequality (e.g., Thompson and Freede, 1971) tells us that the eigenvalues of $\left(\beta_1 \mathcal{K}_2 \beta_1^T + \mathcal{K}_\varepsilon\right)$ are at least as large as those for $\mathcal{K}_\varepsilon$. Thus, β_1 is null if and only if $\mathcal{D}$ in (1.20) is the zero matrix.

Factor analysis is another multivariate analysis method that aims to examine the relationship between the two sets of variables. However, in this setting, only the values of the response variable X_1 are observed while X_2 is viewed as a collection of latent variables whose values represent the object of the analysis. The basic premise is that X_1 and X_2 are linearly related in that

$$X_1 = X_2 + \varepsilon \tag{1.21}$$

with ε a vector of zero mean random errors with variance–covariance matrix $\mathscr{K}_\varepsilon$ as before. The goal is now to use X_1 to predict the unobserved values of X_2.

Canonical correlation again provides the tool that allows us to make a modicum of progress toward solving the prediction problem posed by model (1.21). In this regard, we can look for a linear combination $a_1^T X_1$ that is maximally correlated with a linear combination $a_2^T X_2$ of X_2. There are some simplifications in this instance in that $\mathscr{K}_{12} = \mathscr{K}_2$ and $\mathscr{K}_1 = \mathscr{K}_2 + \mathscr{K}_\varepsilon$ lead us to consideration of the objective function

$$\begin{aligned}\mathrm{Corr}^2(a_1^T X_1, a_2^T X_2) &= \frac{\left(a_1^T \mathscr{K}_2 a_2\right)^2}{a_1^T \mathscr{K}_1 a_1 a_2^T \mathscr{K}_2 a_2} \\ &= \frac{\left(\tilde{a}_1^T \mathscr{K}_1^{-1/2} \mathscr{K}_2^{1/2} \tilde{a}_2\right)^2}{\tilde{a}_1^T \tilde{a}_1 \tilde{a}_2^T \tilde{a}_2}\end{aligned}$$

with $\tilde{a}_1 = \mathscr{K}_1^{1/2} a_1, \tilde{a}_2 = \mathscr{K}_2^{1/2} a_2$. Thus, for example, the optimal correlations ρ and choices for $\tilde{a}_1$ can be obtained as solutions of the eigenvalue problem

$$\mathscr{K}_1^{-1/2} \mathscr{K}_2 \mathscr{K}_1^{-1/2} \tilde{a}_1 = \rho^2 \tilde{a}_1.$$

A little algebra then reveals this to be equivalent to finding solutions of

$$\mathscr{K}_1 a_1 = \frac{1}{1-\rho^2} \mathscr{K}_\varepsilon a_1. \tag{1.22}$$

To proceed further, we need to impose some additional structure on X_2. The standard approach is to assume that

$$X_2 = \Phi Z \tag{1.23}$$

for some unknown $p \times r$ matrix

$$\Phi = \{\phi_{ij}\}_{i=1:p, j=1:r} = \left[\phi_1, \ldots, \phi_r\right] \tag{1.24}$$

and $Z = (Z_1, \ldots, Z_r)^T$ a vector of zero mean random variables with

$$\mathbb{E}\left[ZZ^T\right] = I.$$

The elements of Z are referred to as *factors* while the elements of Φ are called *factor loadings*. A typical identifiability constraint arising from maximum likelihood estimation is to have

$$\phi_i^T \mathscr{K}_\varepsilon^{-1} \phi_j = 0, i \neq j. \tag{1.25}$$

This has the consequence of making $\Phi^T \mathscr{K}_\varepsilon^{-1} \Phi$ a diagonal matrix, which we will subsequently presume to be the case.

When (1.23) holds

$$\begin{aligned}\mathscr{K}_1 &= \Phi\Phi^T + \mathscr{K}_\varepsilon \\ &= \mathscr{K}_\varepsilon^{1/2} \left(\mathscr{K}_\varepsilon^{-1/2} \Phi\Phi^T \mathscr{K}_\varepsilon^{-1/2} + I \right) \mathscr{K}_\varepsilon^{1/2}.\end{aligned}$$

It is readily verified that under (1.25) the matrix $\mathscr{K}_\varepsilon^{-1/2} \Phi\Phi^T \mathscr{K}_\varepsilon^{-1/2}$ has eigenvalues $\gamma_j = \phi_j^T \mathscr{K}_\varepsilon^{-1} \phi_j$ with associated eigenvectors $\mathscr{K}_\varepsilon^{-1/2} \phi_j$. However, this means that $\mathscr{K}_\varepsilon^{-1/2} \mathscr{K}_1 \mathscr{K}_\varepsilon^{-1/2}$ has eigenvalues $1 + \gamma_j$ associated with this same set of eigenvectors; i.e.,

$$\left[\mathscr{K}_\varepsilon^{-1/2} \mathscr{K}_1 \mathscr{K}_\varepsilon^{-1/2} - \left(1 + \gamma_j\right) I \right] \mathscr{K}_\varepsilon^{-1/2} \phi_j = 0 \tag{1.26}$$

or

$$\left[\mathscr{K}_1 - (1 + \gamma_j) \mathscr{K}_\varepsilon \right] \mathscr{K}_\varepsilon^{-1} \phi_j = 0. \tag{1.27}$$

Comparing this with (1.22) leads to the conclusion that $\mathscr{K}_\varepsilon^{-1} \phi_j$ is a canonical weight vectors for the X_1 space with $\gamma_j = \rho_j^2 / (1 - \rho_j^2)$ obtained from its corresponding canonical correlation.

To see where these developments might take us consider the unrealistic scenario where we knew $\mathscr{K}_1$ and $\mathscr{K}_\varepsilon$ but not Φ. If that were the case, the coefficient matrix could be recovered from the canonical weight functions for the X_1 space as $\Phi = \mathscr{K}_\varepsilon \left[a_{11}, \ldots, a_{1r} \right]$. We could then predict Z via the best linear unbiased predictor: namely, the linear transformation of X_1 that minimizes the prediction error $\mathbb{E}(Z - \mathscr{L}X_1)^T (Z - \mathscr{L}X_1)$ over all possible choices for the $r \times p$ matrix $\mathscr{L}$. The minimum is attained with $\mathscr{L} = \mathrm{Cov}(Z, X_1) \mathscr{K}_1^{-1}$ giving

$$\hat{Z} = \Phi^T \left(\Phi\Phi^T + \mathscr{K}_\varepsilon \right)^{-1} X_1 \tag{1.28}$$

as the optimal predictor.

Of course, in practice, we will not know either of $\mathscr{K}_1$ or $\mathscr{K}_\varepsilon$. However, given a random sample of values from X_1, the first of these two quantities is easy to estimate using the sample variance–covariance matrix. While estimation of $\mathscr{K}_\varepsilon$ is more problematic, there are various ways this can be accomplished that produce at least acceptable initial estimators. Such estimators can be substituted into (1.27) to obtain an estimator of Φ. This, in turn, provides an update of $\mathscr{K}_\varepsilon = \mathscr{K}_1 - \Phi\Phi^T$. The result is one possible iterative estimation algorithm that is employed in the factor analysis genre.

Our particular development of factor analysis is due to Rao (1955). A detailed treatments of this and many other factor analysis-related topics can be found in Basilevsky (1994).

A case of particular interest that can be treated with a linear model is MANOVA and its predictive analog known as discriminant analysis. To develop these ideas, we begin with the model

$$X_1 = \overline{m} + \left[m_1 - \overline{m}, \ldots, m_{q+1} - \overline{m}\right] \tilde{X}_2 + \varepsilon,$$

where $\tilde{X}_2 = \left(\tilde{X}_{21}, \ldots, \tilde{X}_{2(q+1)}\right)^T$ has a multinomial distribution with $\sum_{j=1}^{q+1} \tilde{X}_{2j} = 1$ and success probabilities $\pi_1, \ldots, \pi_{q+1}$, $m_1, \ldots, m_{q+1}$ are the X_1 mean vectors for the $q+1$ different populations, $\overline{m} = \sum_{j=1}^{q+1} \pi_j m_j$ is the grand mean and ε is a p-variate random vector with mean zero and covariance matrix $\mathscr{K}_\varepsilon$. As, $\pi_{q+1} = 1 - \sum_{j=1}^{q} \pi_j$ and $\tilde{X}_{2(q+1)} = 1 - \sum_{j=1}^{q} \tilde{X}_{2j}$, the previous model can be equivalently expressed as

$$X_1 = m_{q+1} + \left[m_1 - m_{q+1}, \ldots, m_q - m_{q+1}\right] X_2 + \varepsilon \tag{1.29}$$

with

$$X_2 := \left(\tilde{X}_{21}, \ldots, \tilde{X}_{2q}\right)^T. \tag{1.30}$$

This corresponds to (1.17) with $\beta_1 = \left[(m_1 - m_{q+1}), \ldots, (m_q - m_{q+1})\right]$ and $\beta_0 = m_{q+1}$.

To apply formula (1.20), we must calculate $\mathscr{K}_1, \mathscr{K}_{12}$ and $\mathscr{K}_2$. In this regard, first note that

$$\begin{aligned} \mathbb{E}X_2 &= (\pi_1, \ldots, \pi_q)^T \\ &=: \pi \end{aligned}$$

and

$$\begin{aligned} \mathscr{K}_2 &= \mathrm{Var}(X_2) \\ &= \mathrm{diag}\left(\pi_1, \ldots, \pi_q\right) - \pi\pi^T. \end{aligned}$$

Then, one may check that

$$\mathscr{K}_2^{-1} = \mathrm{diag}\left(\pi_1^{-1}, \ldots, \pi_q^{-1}\right) + \pi_{q+1}^{-1} 1 1^T$$

for a q-vector 1 of all unit elements. From these identities, we obtain

$$\mathscr{K}_{12} = \left[\pi_1(m_1 - \overline{m}), \ldots, \pi_q(m_q - \overline{m})\right]$$

and

$$\mathscr{K}_1 = \mathscr{K}_B + \mathscr{K}_\varepsilon$$

with

$$\mathcal{K}_B = \sum_{j=1}^{q+1} \pi_j (m_j - \overline{m})(m_j - \overline{m})^T = \mathcal{K}_{12}\mathcal{K}_2^{-1}\mathcal{K}_{21}.$$

Relations (1.18) and (1.15) can now be used to see that in this instance the canonical correlations and canonical vectors of the X_1 space are characterized by

$$\mathcal{K}_\varepsilon^{-1}\mathcal{K}_B a = \frac{\rho^2}{1-\rho^2} a. \tag{1.31}$$

Thus, all the canonical correlations are zero if and only if $\mathcal{K}_B$ is the zero matrix, which, in turn, is equivalent to the standard MANOVA null hypothesis that all of the $q+1$ populations have the same mean vector. Statistical tests for this null model can then be constructed from the sample canonical correlations.

If the mean vectors are the same for all of our $q+1$ populations, it will not generally be possible to distinguish between them on the basis of location. However, if the MANOVA null model is rejected, one can expect to have at least some success in categorizing incoming observations according to population membership. The process of doing so is often referred to as *discriminant analysis*. While there are many discrimination methods that appear in the literature, our focus here will be limited to Fisher's classical proposal. The idea is to find a linear combination, or discriminant function, $h^T X_1$ that provides the maximum separation between the populations in the sense of maximizing the ratio

$$\frac{h^T \mathcal{K}_B h}{h^T \mathcal{K}_\varepsilon h}. \tag{1.32}$$

However, this is just a variation of a problem we already encountered with pca. The solution is the largest eigenvalue λ_1 of $\mathcal{K}_\varepsilon^{-1/2}\mathcal{K}_B\mathcal{K}_\varepsilon^{-1/2}$ with the optimal discriminant function weight vector $h_1 = \mathcal{K}_\varepsilon^{-1/2} u_1$ obtained from the eigenvector u_1 that corresponds to λ_1. These quantities are characterized by

$$\mathcal{K}_\varepsilon^{-1/2}\mathcal{K}_B\mathcal{K}_\varepsilon^{-1/2} u_1 = \lambda_1 u_1$$

or, equivalently, by

$$\mathcal{K}_\varepsilon^{-1}\mathcal{K}_B h_1 = \lambda_1 h_1.$$

So, $\lambda_1 = \rho_1^2/(1-\rho_1^2)$ with ρ_1 the first canonical correlation.

Additional discriminant functions are provided by maximizing (1.32) conditional on the resulting linear combinations of variables being uncorrelated with the discriminant functions that have already been determined from this iterative process. The resulting eigenvalues will, of course, enjoy the same relation to their corresponding canonical correlations. If we now opt to retain $r \le \min(p, q)$ discriminant functions having weight vectors $h_1, \dots h_r$, a new observation x is classified as being from the population whose index minimizes

$$\sum_{j=1}^{r} (h_j^T x - h_j^T m_i)^2 \tag{1.33}$$

over $i = 1, \dots, q+1$.

As one might expect, the relationship between Fisher's discriminant functions and canonical variables goes much deeper than just the eigenvalue connection. The equivalence of Fisher's discriminant analysis and canonical correlation is revealed by observing that

$$\begin{aligned}\delta_{ij} &= \frac{a_{1i}^T \mathscr{K}_B a_{1j}}{\rho_i^2} \\ &= \frac{h_i^T \mathscr{K}_B h_j}{\lambda_i}.\end{aligned}$$

This shows that the h_i and a_{1i} are eigenvector of $\mathscr{K}_B$ that have been adjusted to have norms λ_i and ρ_i^2. So, $a_{1i} = \rho_i h_i / \sqrt{\lambda_i}$. In particular, this means that (1.33) will produce the same classification as would be obtained using canonical variables with the criterion

$$\sum_{j=1}^{r} \frac{(a_j^T x - a_j^T m_i)^2}{1 - \rho_j^2}.$$

A typical situation with data would have us observing p dimensional random vectors

$$X_{ij}, i = 1, \dots, q+1, j = 1, \dots, n_i$$

with $n = \sum_{i=1}^{q+1} n_i$. Then, estimators of $\mathscr{K}_\varepsilon, \mathscr{K}_B$ are

$$n^{-1} \sum_{i=1}^{q+1} \sum_{j=1}^{n_i} \left(X_{ij} - \overline{X}_i\right) \left(X_{ij} - \overline{X}_i\right)^T$$

and

$$n^{-1} \sum_{i=1}^{q+1} n_i \left(\overline{X}_i - \overline{X}\right) \left(\overline{X}_i - \overline{X}\right)^T,$$

respectively, with $\overline{X}_i = n_i^{-1} \sum_{j=1}^{n_i} X_{ij}, \overline{X} = n^{-1} \sum_{i=1}^{q+1} n_i \overline{X}_i$. We then carry out classification by replacing $\mathscr{K}_\varepsilon, \mathscr{K}_B$ by their sample analogs in the previous formulation.

1.2 The path that lies ahead

The basic concepts described in Section 1.1 remain conceptually valid in the context of fda. The first challenge faced by researchers in the area lies in developing them fully and rigorously from a mathematical perspective. This is the essential precursor to the growth and maturation of inferential methodology in general and for fda in particular. One cannot estimate a "parameter" if it is undefined and it is easy to take a misstep in fda formulations that lead to exactly such a conundrum.

The theory of multivariate analysis is inextricably interwoven with matrix theory. If the observations in fda are viewed as vectors of infinite length, then we would anticipate that the infinite dimensional analog of matrices would represent the tools of the trade for advancing our understanding of this emerging field. Such entities are called linear operators and, in particular, compact operators give us the infinite dimensional extension of matrices that arise naturally in fda. Just as one cannot venture far into multivariate analysis without understanding matrix theory, one cannot expect to appreciate the mathematical aspects of fda without a thorough background in compact operators and their Hilbert–Schmidt and trace class variants.

After developing the necessary background on linear spaces in Chapter 2, we accumulate some of the essential ingredients of functional analysis and operator theory in Chapters 3–5. Chapter 3 provides a general overview that introduces linear operators and linear functionals as well as fundamental concepts such as the inverse and adjoint of an operator, nonnegative and projection operators. Compact operators are then treated in Chapter 4 where we develop both eigenvalue and singular value expansions and treat the special cases of Hilbert–Schmidt and trace class operators. This work plays an important role throughout the remainder of this text. Chapter 5 deals with perturbation theory for compact operators and provides key tools for the treatment of functional pca in Chapter 9.

Data smoothing methods tend to be a prominent aspect of most fda inferential methodology making a foray into the mathematical aspects of smoothing somewhat de rigueur for this particular treatise. Our treatment of the topic focuses on an abstract penalized smoothing problem that can be specialized to recover the spline smoothers that are most common in the fda literature. Penalized smoothing methods recur in Chapters 8 and 11 where we study their performance for estimation of certain functional parameters.

Classical statistics is concerned with inference about the distribution of a basic random variable that we are able to sample repeatedly. The same can be said for fda except that the meaning of the "random variable" phrase requires a bit of reinterpretation before it becomes relevant to that setting. What is needed is the concept of a random element of a Hilbert space. That idea along with its associated probabilistic machinery is developed in Chapter 7.

Here, we extend the concepts of a mean vector and covariance matrix to the infinite dimensional scenario and develop some asymptotic theory that is relevant for random samples of Hilbert space valued random elements.

Chapters 8–11 will provide extended, detailed illustrations of how the mathematical machinery in Chapters 2–6 can be used to address problems that arise in the fda environment. In Chapter 8, this takes the form of analysis of the large sample properties of three types of estimators of the mean element and covariance function: ones that are based on the sample mean element and covariance operator for completely observed functional data and local linear and penalized least-squares estimators for the discretely observed case. This is followed in Chapter 9 with an investigation of the asymptotic behavior of the principle components estimators that are produced by the covariance estimators introduced in Chapter 8.

Chapter 10 deals with the bivariate case where, for example, one has two stochastic processes and wishes to analyze their dependence structure. Somewhat more generally, it provides a development of abstract canonical correlation for two Hilbert space valued random elements. By specializing this theory to the fda stochastic processes context, we are then able to obtain parallels of results in Section 1.1 for functional analogs of linear prediction, regression, factor analysis, MANOVA, and discriminant analysis.

Finally, Chapter 11 deals with the important case of bivariate data having both a scalar and functional (i.e., stochastic process) response. The large sample properties of a particular penalized least-squares estimator of the regression coefficient function are investigated in this setting.

2

Vector and function spaces

In Chapter, we loosely defined functional data to be a collection of sample paths for a stochastic process (or processes) with the index set $[0, 1]$. Thus, the data are functions that may have various properties such as being continuous or square integrable with probability one. Characteristics such as these signify a commonality that becomes amenable to treatment through the study of function spaces. An understanding of function spaces is an essential first step in dealing with functional data. Beyond that, the properties of linear functionals and operators on function spaces that are studied in Chapters 3 and 4 lie at the heart of the functional analogs of the basic concepts from multivariate analysis.

The purpose of this chapter is to present the function space theory that we perceive to be most relevant for fda. Clearly, it will be impossible to give a comprehensive treatment of each topic that we touch upon here. Certain function space concepts like reproducing kernel Hilbert spaces and Sobolev spaces are treated in some detail in view of their relative obscurity and relevance for our targeted audience. However, many topics such as vector, metric, Banach, and Hilbert spaces are included for completeness and to serve as a review of material that the reader will have hopefully seen elsewhere. Thorough expositions of these concepts can be found in, e.g., Luenberger (1969) and Rynne and Youngson (2001). It is also presumed that the reader is comfortable with standard measure and integration theory with Billingsley (1995) and Royden and Fitzpatrick (2010) being two of the standard sources that provide introductions to this area.

Theoretical Foundations of Functional Data Analysis, with an Introduction to Linear Operators,
First Edition. Tailen Hsing and Randall Eubank.

2.1 Metric spaces

One of the signature features of the p-dimensional Euclidean space $\mathbb{R}^p$ is the ease with which one can define measures of distance or closeness. This can be accomplished in a variety of ways. However, all such measures must possess certain properties; namely, they must all be metrics in the sense defined in the following definition.

Definition 2.1.1 *A metric on a set* $\mathbb{M}$ *is a function* $d : \mathbb{M} \times \mathbb{M} \to \mathbb{R}$ *that satisfies*

1. $d(x_1, x_2) \geq 0$,
2. $d(x_1, x_2) = 0$ *if* $x_1 = x_2$,
3. $d(x_1, x_2) = d(x_2, x_1)$ *and*
4. $d(x_1, x_3) \leq d(x_1, x_2) + d(x_2, x_3)$

for $x_1, x_2, x_3 \in \mathbb{M}$. *We refer to the pair* $(\mathbb{M}, d)$ *as a metric space.*

Example 2.1.2 *Let* $x_i = (x_{i1}, \ldots, x_{ip}) \in \mathbb{R}^p, i = 1, 2$. *Metrics for* $\mathbb{R}^p$ *include the Euclidean distance*

$$d(x_1, x_2) = \sqrt{\sum_{j=1}^{p} (x_{1j} - x_{2j})^2}$$

and

$$d(x_1, x_2) = \max_{1 \leq j \leq p} |x_{1j} - x_{2j}|.$$

On the other hand, squared Euclidean distance is not a metric.

Example 2.1.3 *A somewhat more exotic example of a metric space is* $C[0, 1]$*: the set of continuous functions on* $[0, 1]$. *Recall that continuous functions defined on a closed intervals are uniformly continuous. The sup metric is then well defined as*

$$d(f, g) = \sup\{|f(t) - g(t)| : t \in [0, 1]\}$$

for $f, g \in C[0, 1]$. *Conditions 1–4 in Definition 2.1.1 can be easily verified for this case.*

Once we have a way to assess distances, it becomes possible to define parallels of many of the familiar point-set concepts associated with numbers on the real line. For example, the idea of open and closed sets takes the expected form with absolute values being replaced by an abstract metric.

Definition 2.1.4 *Let* $(\mathbb{M}, d)$ *be a metric space with* $E \subset \mathbb{M}$*. Then,* E *is said to be open if for every* $e \in E$ *there exists an* $\epsilon > 0$ *such that* $\{x \in \mathbb{M} : d(x, e) < \epsilon\} \subset E$*. The subset* E *is closed if* $\{x \in \mathbb{M} : x \notin E\}$ *is open.*

Related concepts such as denseness, separability, etc., now translate virtually unchanged to the metric space setting.

Definition 2.1.5 *The closure* $\overline{E}$ *of* $E \subset \mathbb{M}$ *is the smallest closed set in* $\mathbb{M}$ *that contains* E*.*

Definition 2.1.6 *A set* $E \subset \mathbb{M}$ *is dense in* $\mathbb{M}$ *if* $\overline{E} = \mathbb{M}$*.*

Definition 2.1.7 *A metric space* $\mathbb{M}$ *is separable if it has a countable, dense subset.*

Example 2.1.8 *The p-dimensional real space,* $\mathbb{R}^p$*, is separable as the set of vectors with rational components is countable and is dense in* $\mathbb{R}^p$*.*

If we have a function mapping from one metric space to another, then its smoothness (i.e., how the value of the function changes as the argument changes) is an important consideration for functional data. The most basic form of smoothness is continuity.

Definition 2.1.9 *Let* $(\mathbb{M}_1, d_1)$ *and* $(\mathbb{M}_2, d_2)$ *be metric spaces and let* f *be a function defined on* $\mathbb{M}_1$ *that takes values in* $\mathbb{M}_2$*. Then,* f *is continuous at* $x \in \mathbb{M}_1$ *if for every* $\epsilon > 0$ *there is a* $\delta_{x,\epsilon} > 0$ *such that for all* $y \in \mathbb{M}_1$ *satisfying* $d_1(x, y) < \delta_{x,\epsilon}$ *we have* $d_2(f(x), f(y)) < \epsilon$*. The function* f *is uniformly continuous when* $\delta_{x,\epsilon}$ *does not depend on* x*.*

An equivalent definition of continuity can be based on convergence of sequences in the sense defined in the following definition.

Definition 2.1.10 *Let* $(\mathbb{M}, d)$ *be a metric space with* $\{x_n\}$ *a sequence of points in* $\mathbb{M}$*. The sequence converges to* $x \in \mathbb{M}$*, denoted by* $x_n \to x$*, if* $d(x_n, x) \to 0$ *as* $n \to \infty$*.*

With this notation, we can say that f is continuous at x if $f(x_n) \to f(x)$ whenever $x_n \to x$.

The Cauchy criterion arises as a means to determine the convergence of a sequence of real numbers. The result is that such a sequence converges if and only if it is a Cauchy sequence in the sense of our following definition.

Definition 2.1.11 *A sequence of elements $\{x_n\}$ in a metric space $(\mathbb{M}, m)$ is said to be a Cauchy sequence if* $\sup_{m,n\geq N} d(x_m, x_n) \to 0$ *as* $N \to \infty$.

Now, if $x_n \to x$, part 4 of Definition 2.1.1 (also called the triangle inequality) entails that

$$d(x_n, x_m) \leq d(x_n, x) + d(x_m, x) \to 0$$

as $m, n \to \infty$. So, any convergent sequence is necessarily Cauchy. Unfortunately, the converse if not always true and cases where it is true are sufficiently important to merit a special name.

Definition 2.1.12 *A metric space $(\mathbb{M}, m)$ is said to be complete if every Cauchy sequence is convergent.*

Example 2.1.13 *It is easy to show that $\mathbb{R}$ is complete by arguing that for any Cauchy sequence $\{x_n\}$ in $\mathbb{R}$,* $\liminf x_n$ *must equal* $\limsup x_n$ *and be finite. Thus, a sequence being Cauchy is equivalent to convergence and the Cauchy criterion stems from that fact. This is all readily generalized to $\mathbb{R}^p$.*

A somewhat more challenging case than the previous example is provided by $C[0, 1]$. Our task in this instance is to establish

Theorem 2.1.14 *The set of continuous functions on the interval* $[0, 1]$ *equipped with the sup metric is complete and separable.*

Proof: Let $\{f_n\}$ be a Cauchy sequence in $C[0, 1]$: i.e.,

$$\sup_{m,n\geq N} \sup_{t\in[0,1]} |f_m(t) - f_n(t)| \to 0 \tag{2.1}$$

as $N \to \infty$. Then, for each $t \in [0, 1]$, the sequence $\{f_n(t)\}$ is a Cauchy sequence in $\mathbb{R}$ and necessarily converges. Let f be the point-wise limit of f_n. As $\sup_{n\geq N} f_n(t) \downarrow f(t), \inf_{n\geq N} f_n(t) \uparrow f(t)$ as $N \to \infty$,

$$|f_N(t) - f(t)| \leq \sup_{n\geq N} f_n(t) - \inf_{n\geq N} f_n(t) = \sup_{m,n\geq N} |f_m(t) - f_n(t)|$$

which tends to 0 uniformly by (2.1) and thereby establishes completeness.

To show that $C[0, 1]$ is separable, we present a proof that employs some simple tools from probability to establish an elementary version of the Stone–Weierstrass Theorem. Let $f \in C[0, 1]$ and define the Bernstein polynomial of degree n by

$$f_n(x) = \sum_{m=0}^{n} \binom{n}{m} x^m (1 - x)^{n-m} f(m/n).$$

As the collection of all polynomials is countable, it suffices to show that

$$\lim_{n\to\infty} \sup_{x\in[0,1]} |f_n(x) - f(x)| = 0. \tag{2.2}$$

By uniform continuity, for any given $\epsilon > 0$, there exists a $\delta > 0$ such that if $|x - y| < \delta$ then $|f(x) - f(y)| < \epsilon$. Fix $x \in [0, 1]$ and let Y have a binomial distribution with n trials and success probability x. In that case, $f_n(x) = \mathbb{E}[f(Y/n)]$ and

$$\begin{aligned}|f_n(x) - f(x)| &\le \mathbb{E}|f(Y/n) - f(x)| \\ &= \mathbb{E}[|f(Y/n) - f(x)|I(|Y/n - x| \le \delta)] \\ &\quad + \mathbb{E}[|f(Y/n) - f(x)|I(|Y/n - x| > \delta)]\end{aligned}$$

with $I(A)$ the indicator function for the set A. By the choice of δ,

$$\mathbb{E}[|f(Y/n) - f(x)|I(|Y/n - x| \le \delta)] \le \epsilon.$$

On the other hand, Chebyshev's inequality gives

$$\begin{aligned}\mathbb{E}[|f(Y/n) - f(x)|I(|Y/n - x| > \delta)] &\le 2 \sup_y |f(y)| \frac{\mathrm{Var}(Y)}{n^2\delta^2} \\ &= 2 \sup_y |f(y)| \frac{nx(1 - x)}{n^2\delta^2}.\end{aligned}$$

In combination, the two cases lead us to the conclusion that

$$\limsup_{n\to\infty} \sup_{x\in[0,1]} |f_n(x) - f(x)| \le \epsilon$$

for each $\epsilon > 0$. As $\epsilon > 0$ is arbitrary, (2.2) follows. □

Two other important properties of metric spaces are the following.

Definition 2.1.15 *A metric space $\mathbb{M}$ is compact if every sequence of elements in $\mathbb{M}$ contains a subsequence that converges to an element of $\mathbb{M}$. If $\overline{\mathbb{M}}$ is compact, $\mathbb{M}$ is termed relatively compact.*

Definition 2.1.16 *A metric space $(\mathbb{M}, d)$ is totally bounded if for any $\epsilon > 0$ there exists $y_1, \ldots, y_n$ in $\mathbb{M}$ such that $\mathbb{M} \subset \cup_{i=1}^n \{x \in \mathbb{M} : d(x, y_i) < \epsilon\}$ for some finite positive integer n.*

The following result, known as the Heine–Borel theorem, is proved, for example, in Royden and Fitzpatrick (2010).

Theorem 2.1.17 *A metric space is compact if it is complete and totally bounded.*

Example 2.1.18 *The p-dimensional unit square* $[0, 1]^p$ *is a compact metric space under either of the metrics in Example 2.1.2.*

2.2 Vector and normed spaces

Metric spaces represent a natural topological abstraction of $\mathbb{R}^p$ that is suitable for our purposes. What remains is the addition of an appropriate algebraic structure. This is provided by the vector space concept that is introduced in this section.

It is safe to say that data deriving from complex vis-a-vis real valued random variables is a rarity in statistics. This comment carries over to functional data and, for that reason, we will confine attention to vector spaces over $\mathbb{R}$ throughout this text unless explicitly stated to the contrary.

Definition 2.2.1 *A vector space* $\mathbb{V}$ *is a set of elements, referred to as vectors, for which two operations have been defined: addition and scalar multiplication. Given two vectors* v_1, v_2*, addition returns another vector denoted by* $v_1 + v_2$*. Given a vector* v *and* $a \in \mathbb{R}$*, scalar multiplication returns a vector denote by* av*. The addition and multiplication operations are assumed to satisfy*

1. $v_1 + v_2 = v_2 + v_1$,
2. $v_1 + (v_2 + v_3) = (v_1 + v_2) + v_3$,
3. $a_1(a_2v) = (a_1a_2)v$,
4. $a(v_1 + v_2) = av_1 + av_2$, $(a_1 + a_2)v = a_1v + a_2v$, *and*
5. $1v = v$.

In addition, there is a unique element 0 *with the property that* $v + 0 = v$ *for every* $v \in \mathbb{V}$ *and corresponding to each element* v *there is another element* $-v$ *such that* $v + (-v) = 0$.

We will often work with subsets of vector spaces. When such a subset admits the same algebraic operations as its parent vector space we refer to it as a *linear subspace* or merely as a *subspace*. In particular, beginning with a subset A of a vector space $\mathbb{V}$, we can create a subspace by forming a new set that contains all the finite dimensional linear combinations of its elements. This is referred to as the *span* of A and denoted by $\text{span}(A)$. One characterization of $\text{span}(A)$ is as the intersection of all subspaces in $\mathbb{V}$ that contain A. In that sense, $\text{span}(A)$ is the smallest subspace that contains the elements of A.

The linear span concept can be sharpened somewhat by introducing the idea of a basis. First, we need to define what we mean by linear independence.

Definition 2.2.2 *Let $\mathbb{V}$ be a vector space and $B = \{v_1, \ldots, v_k\} \subset \mathbb{V}$ for some finite, positive integer k. The collection B is said to be linearly independent if $\sum_{i=1}^{k} a_i v_i = 0$ entails that $a_i = 0$ for all i.*

Then, we have

Definition 2.2.3 *If $B = \{v_1, \ldots, v_k\}$ is a linearly independent subset of the vector space $\mathbb{V}$ and* span$(B) = \mathbb{V}$, *B is said to be a basis for $\mathbb{V}$.*

Note that it is necessarily the case that if $\{v_1, \ldots, v_k\} \subset \mathbb{V}$ is a basis and v is any element of $\mathbb{V}$ there are coefficients $b_1, \ldots, b_k$ such that $v = \sum_{j=1}^{k} b_j v_j$. Moreover, the integer k is unique in the following sense.

Theorem 2.2.4 *If, for finite k and ℓ, $B_1 = \{v_1, \ldots, v_k\}$ and $B_2 = \{u_1, \ldots, u_\ell\}$ are both bases of the vector space $\mathbb{V}$, then $k = \ell$.*

Proof: Suppose that $\ell > k$. As B_1 is a basis the span of B_1 coincides with $\mathbb{V}$. Thus, any element of $\mathbb{V}$ can be written as a linear combination of the elements of B_1. In particular, this means that there are coefficients $a_{ij}, j = 1, \ldots, k$ such that

$$u_i = \sum_{j=1}^{k} a_{ij} v_j, \quad 1 \le i \le \ell.$$

Let $\mathscr{A} = \{a_{ij}\}_{i=1:\ell, j=1:k}$. Then, by the method of elimination, the linear system $\mathscr{A} x = 0$ can be shown to have a nonzero solution $x = (x_1, \ldots, x_\ell)^T$. Thus,

$$\sum_{i=1}^{\ell} x_i u_i = \sum_{j=1}^{k} \sum_{i=1}^{\ell} x_i a_{ij} v_j = 0$$

which means that B_2 is linearly dependent and cannot be a basis. □

As in the Euclidean case, the number of basis elements provides us with a value that can be viewed as the dimensionality of the space. This works in the expected way provided that the number of basis elements in finite. Specifically, if a vector space $\mathbb{V}$ has a basis consisting of $p < \infty$ elements, it is said to have dimension p which we then denote by $\dim(\mathbb{V}) = p$. However, it turns out that our labor to create the abstract vector space machinery has not moved us very far from our Euclidean starting point as a consequence of the next result.

Theorem 2.2.5 *Any p-dimensional vector space is isomorphic to $\mathbb{R}^p$.*

Proof: Let $\mathbb{V}$ be a vector space with basis $\{v_1, \ldots, v_p\}$. Each x can be uniquely written as $x = \sum_{i=1}^{p} a_i(x)v_i$ and the mapping $x \mapsto (a_1(x), \ldots, a_p(x))$ is one-to-one and onto $\mathbb{R}^p$. □

When a vector space does not possess a finite dimensional basis, it is said to be infinite dimensional. Even the meaning of the term "basis" becomes open for debate in this instance. While the idea can be extended to deal with this eventuality, the process of doing so entails considerable complications (such as convergence of infinite expansions) that are not relevant for this book. Our focus will instead be on the more readily accessible case of orthonormal bases for separable Hilbert spaces that arise in the following section.

We will subsequently have use for the following concept.

Definition 2.2.6 *Let $\mathbb{V}_1, \mathbb{V}_2$ be vector spaces. A transformation T mapping $\mathbb{V}_1$ into $\mathbb{V}_2$ is said to be linear if $T(a_1v_1 + a_2v_2) = a_1T(v_1) + a_2T(v_2)$ for all $a_1, a_2 \in \mathbb{R}$ and all $v_1, v_2 \in \mathbb{V}_1$.*

Example 2.2.7 *The identity mapping $I(x) = x$ is clearly linear.*

Example 2.2.8 *In $\mathbb{R}^p$, linear transformations coincide with matrix multiplication.*

Things become much more interesting when we merge the metric and vector space concepts. This is accomplished by adding a measure of distance called a *norm* to obtain a *normed vector space* or just *normed space*.

Definition 2.2.9 *Let $\mathbb{V}$ be a vector space. A norm on $\mathbb{V}$ is a function $\|\cdot\| : \mathbb{V} \mapsto \mathbb{R}$ such that*

1. $\|v\| \geq 0$,
2. $\|v\| = 0$ *if* $v = 0$,
3. $\|av\| = |a|\|v\|$, *and*
4. $\|v_1 + v_2\| \leq \|v_1\| + \|v_2\|$

for all $v, v_1, v_2 \in \mathbb{V}$ and $a \in \mathbb{R}$.

Property 4 in Definition 2.2.9 is typically referred to as the *triangle inequality*. This is consistent with the usage of this phrase in the previous section because of the following result.

Theorem 2.2.10 *If $\mathbb{V}$ is a vector space with norm $\|\cdot\|$, then $d(x, y) := \|x - y\|$ for $x, y \in \mathbb{V}$ is a metric.*

Example 2.2.11 *Let $\mathbb{V}$ be a finite-dimensional vector space with basis $\{v_1, \ldots, v_p\}$. Then, if $v = \sum_{i=1}^{p} a_i v_i$,*

$$\|v\| = \left(\sum_{i=1}^{p} a_i^2 \right)^{1/2}$$

is a norm. To verify the triangle inequality in this case, let $w = \sum_{i=1}^{p} b_i v_i$ and observe that

$$\begin{aligned}
\|v + w\|^2 &= \sum_{i=1}^{p} (a_i + b_i)^2 \\
&\leq \sum_{i=1}^{p} a_i^2 + 2\left(\sum_{i=1}^{p} a_i^2 \right)^{1/2} \left(\sum_{i=1}^{p} b_i^2 \right)^{1/2} + \sum_{i=1}^{p} b_i^2 \\
&= (\|v\| + \|w\|)^2,
\end{aligned}$$

where the inequality follows from the Cauchy–Schwarz inequality.

Example 2.2.12 *The ℓ^2 space of square summable sequences consists of elements of the form $(x_1, x_2, \ldots)$, where $x_i \in \mathbb{R}$ and $\sum_{i=1}^{\infty} x_i^2 < \infty$. The vector space operations of addition and multiplication are defined by*

$$\begin{aligned}
(x_1, x_2, \ldots) + (y_1, y_2, \ldots) &= (x_1 + y_1, x_2 + y_2, \ldots), \\
a(x_1, x_2, \ldots) &= (ax_1, ax_2, \ldots)
\end{aligned}$$

and the norm is

$$\|(x_1, x_2, \ldots)\| = \left(\sum_{i=1}^{\infty} x_i^2 \right)^{1/2}.$$

As shown subsequently in Example 2.3.6, ℓ^2 is also complete.

With the norm topology in place, we now possess one way to extend the idea of a linear span to allow for infinite dimensions.

Definition 2.2.13 *Let A be a subset of a normed vector space. The closed span of A, denoted by $\overline{\text{span}(A)}$, is the closure of* span(*A*) *with respect to the metric induced by the norm.*

This definition is quite important and arises in various contexts going forward. Thus, it is worthwhile to explore it in a bit more detail. We can think of

$\overline{\text{span}}(A)$ as being built up from all finite dimensional linear combinations of the elements of A. We then append all the limits of sequences of these linear combinations to obtain the final space of interest. This latter step is simplified somewhat when we can draw on completeness so that only Cauchy sequences need consideration. Finite dimensional linear combinations are therefore by construction dense in $\overline{\text{span}}(A)$.

Theorem 2.2.14 *Suppose that A is a subset of a separable, complete normed vector space $\mathbb{X}$. Then, if $\{x_n\}$ is dense in A, $\overline{\text{span}}(A)$ consists of all finite dimensional linear combinations of the x_n and the limits in $\mathbb{X}$ of Cauchy sequences of such linear combinations.*

A particular norm may be difficult to work with for some purposes. In such instances, the following concept can be useful

Definition 2.2.15 *Two norms $\|\cdot\|_1$ and $\|\cdot\|_2$ on a vector space $\mathbb{V}$ are equivalent if there are constants $c, C \in (0, \infty)$ such that for all $x \in \mathbb{V}$,*

$$c\|x\|_1 \leq \|x\|_2 \leq C\|x\|_1.$$

The point here is that if $\|\cdot\|_1$ and $\|\cdot\|_2$ are equivalent, the metric spaces induced by the two norms have the same "topological" properties; for instance, the classes of open sets are the same, if one is complete then the other is also complete, etc. Thus, one can replace a particular norm with another that is equivalent, but possibly more mathematically tractable, if such a thing can be found. This is not so difficult in finite dimensions in view of the following theorem.

Theorem 2.2.16 *Let $\mathbb{V}$ be a finite-dimensional normed space. Then,*

1. *all norms for $\mathbb{V}$ are equivalent and*
2. *$\mathbb{V}$ is a complete and separable metric space in any metric generated by a norm.*

Proof: By Theorem 2.2.5, we can assume without loss of generality that $\mathbb{V} = \mathbb{R}^p$. So, we first show the equivalence of the Euclidean norm $\|\cdot\|_1$ and an arbitrary norm $\|\cdot\|_2$ on $\mathbb{R}^p$. In this regard, let e_i be a vector of all zeros except for a one as its ith component and write $x = (x_1, \ldots, x_p)$ and $y = (y_1, \ldots, y_p)$ as $x = \sum_{i=1}^p x_i e_i$ and $y = \sum_{i=1}^p y_i e_i$, respectively. Let $\epsilon > 0$ and suppose that

$\|x - y\|_1 < \epsilon/(\sum_{i=1}^{p} \|e_i\|_2)$. Then,

$$\begin{aligned} |\,\|x\|_2 - \|y\|_2| &\le \|x - y\|_2 \\ &\le \left(\max_i |x_i - y_i|\right) \sum_{i=1}^{p} \|e_i\|_2 \\ &\le \|x - y\|_1 \sum_{i=1}^{p} \|e_i\|_2 \quad < \epsilon. \end{aligned}$$

So, $f(\cdot) := \|\cdot\|_2$ is a continuous function from $(\mathbb{R}^p, d_1)$ to $(\mathbb{R}^p, d_2)$, where d_i is the metric defined by $\|\cdot\|_i$.

Now, consider the restriction of f to the unit sphere $\|x\|_1 = 1$. The range of this function is some closed interval $[c, C]$ with $c, C \in (0, \infty)$. As a result, when $\|x\|_1 = 1$,

$$c \le \|x\|_2 \le C$$

and, hence,

$$c \le \frac{\|x\|_2}{\|x\|_1} \le C$$

for any $x \neq 0$. We can argue similarly for another norm $\|\cdot\|_3$ to see that

$$c^* \le \frac{\|x\|_3}{\|x\|_1} \le C^*$$

for $0 < c^* \le C^* < \infty$. In combination, the two inequalities lead to

$$\frac{c^*}{C}\|x\|_2 \le \|x\|_3 \le \frac{C^*}{c}\|x\|_2$$

and thereby establish the first part of the theorem.

For part 2, we again need only consider $\mathbb{R}^p$ with the Euclidean norm. The separability of $\mathbb{R}^p$ was discussed in Example 2.1.8. The completeness of $\mathbb{R}^p$ was the subject of Example 2.1.13. □

Both of the conclusions in Theorem 2.2.16 are generally false for infinite dimensions.

Example 2.2.17 *From Theorem 2.1.14, we know that the space $C[0, 1]$ is complete under the sup norm. Another norm is provided by*

$$\|f\| = \left(\int_0^1 f^2(t)dt\right)^{1/2}.$$

But, by taking $E_n = [.5(1 - 1/n), .5(1 + 1/n)]$ *with*

$$f_n(t) = (1 - 2n|.5 - t|)I(t \in E_n),$$

we produce a sequence of functions $\{f_n\}$ *that is Cauchy but does not have a limit in* $C[0, 1]$.

2.3 Banach and $\mathbb{L}^p$ spaces

As seen in Section 2.2, not every normed vector space is complete. However, many useful normed vector spaces are complete and those that exhibit this property receive the following designation.

Definition 2.3.1 *A Banach space is a normed vector space which is complete under the metric associated with the norm.*

As we saw earlier, the spaces $C[0, 1]$ with the sup norm and finite-dimensional normed vector spaces are examples of Banach spaces. For fda, a more important Banach space is the following.

Definition 2.3.2 *Let* $(E, \mathscr{B}, \mu)$ *be a measure space and for* $p \in [1, \infty)$ *denote by* $\mathbb{L}^p(E, \mathscr{B}, \mu)$ *the collection of measurable functions* f *on* E *that satisfy* $\int_E |f|^p d\mu < \infty$. *Define*

$$\|f\|_p = \left(\int_E |f|^p d\mu\right)^{1/p} \tag{2.3}$$

when $f \in \mathbb{L}^p(E, \mathscr{B}, \mu)$.

The collection of measurable functions f *on* E *that are finite a.e.* μ *is denoted by* $\mathbb{L}^\infty(E, \mathscr{B}, \mu)$, *for which we define*

$$\begin{aligned} \|f\|_\infty &= \operatorname{ess\,sup}_{s \in E} |f(s)| \\ &= \inf\{x \in \mathbb{R} : \mu(s : |f(s)| > x) = 0\}. \end{aligned} \tag{2.4}$$

Of course, the notation $\|\cdot\|_p$ suggests that we are working with a norm. Properties 1 and 2 of Definition 2.2.9 can be easily verified for $\|\cdot\|_p$, while property 4 follows from Minkowski's inequality that we give in Theorem 2.3.4. Somewhat more problematic is showing that $\|f\|_p = 0$ means that $f = 0$. Indeed, this is not true and we can only conclude that $f = 0$ a.e. μ in such an instance. To bypass this stumbling block, we must adopt the convention that functions differing only on a set of μ measure 0 are identified as being the same function. That is, we define the equivalence relation $\sim$ by

$$f \sim g \quad \text{if} \quad \mu\{x : f(x) \neq g(x)\} = 0$$

and focus on the quotient space $\mathbb{L}^p(E, \mathscr{B}, \mu)/\sim$ of equivalence classes. For convenience, we continue to denote this space as $\mathbb{L}^p(E, \mathscr{B}, \mu)$ or $\mathbb{L}^p$ if the associated σ-field and measure are clear from the context. With this modification, $\|\cdot\|_p$ defined in (2.3) and (2.4) are norms for their respective spaces that will be referred to as the $\mathbb{L}^p$ norms.

The key to working with the $\mathbb{L}^p$ spaces is the Hölder and Minkowski inequalities. The former of these takes the form

Theorem 2.3.3 *For $1 \leq p, q \leq \infty$ suppose that $f_1 \in \mathbb{L}^p$ and $f_2 \in \mathbb{L}^q$ with $1/p + 1/q = 1$. Then,*

$$\|f_1 f_2\|_1 \leq \|f_1\|_p \|f_2\|_q.$$

Proof: The case of $p = \infty, q = 1$ is straightforward. It follows immediately upon observing that $|fg| \leq \|f\|_\infty |g|$.

Now assume that $p, q \in (0, \infty)$ with $p^{-1} + q^{-1} = 1$. In this event, for any two real number $a_1, a_2 > 0$, we have

$$\begin{aligned}\log\,(a_1 a_2) &= p^{-1} \log\,(a_1^p) + q^{-1} \log\,(a_2^q)\\ &\leq \log\,(p^{-1} a_1^p + q^{-1} a_2^q)\end{aligned}$$

as log is concave. Upon exponentiating both sides, we obtain Young's inequality

$$a_1 a_2 \leq p^{-1} a_1^p + q^{-1} a_2^q.$$

Now replace a_1, a_2 by functions $|f_1|, |f_2|$ for which $\|f_1\|_p = \|f_2\|_q = 1$ to finish the proof. □

With the aid of Hölder's inequality, we can establish Minkowski's inequality. As mentioned earlier, an important application of Minkowski's inequality is establishing the triangle inequality for $\|\cdot\|_p$.

Theorem 2.3.4 *Suppose that $p \geq 1$ and $f_1, f_2 \in \mathbb{L}^p$. Then $\|f_1 + f_2\|_p \leq \|f_1\|_p + \|f_2\|_p$.*

Proof: The cases $p = 1$ and ∞ are straightforward. For $p \in (1, \infty)$, first note that the function x^p is convex for $x \geq 0$. Thus, $|f_1/2 + f_2/2|^p \leq (|f_1|^p + |f_2|^p)/2$ and, hence, $|f_1 + f_2|^p \leq 2^{p-1}(|f_1|^p + |f_2|^p)$. Consequently, if f_1 and f_2 are both in $\mathbb{L}^p$ so is their sum.

We can treat $\|f_1 + f_2\|_p^p$ as being finite and employ the triangle inequality to obtain

$$\|f_1 + f_2\|_p^p \leq \int_E |f_1||f_1 + f_2|^{p-1} d\mu + \int_E |f_2||f_1 + f_2|^{p-1} d\mu.$$

Now apply Hölder's inequality to the two integrals using $q = p/(p-1)$ to finish the proof. □

The following result is known as the Riesz–Fischer Theorem.

Theorem 2.3.5 *The space $\mathbb{L}^p$ is complete for each $p \geq 1$.*

Proof: We focus on $p \in [1, \infty)$. Let $\{f_n\}$ be a Cauchy sequence in $\mathbb{L}^p$: i.e.,

$$\lim_{N\to\infty} \sup_{m,n\geq N} \|f_m - f_n\|_p = 0. \tag{2.5}$$

Choose an integer subsequence $\{n_k\}$ such that

$$C := \sum_{k=1}^{\infty} \|f_{n_{k+1}} - f_{n_k}\|_p < \infty.$$

By Minkowsky's inequality,

$$\left\| \sum_{k=1}^{\infty} |f_{n_{k+1}} - f_{n_k}| \right\|_p \leq C.$$

Thus,

$$\sum_{k=1}^{\infty} |f_{n_{k+1}}(s) - f_{n_k}(s)| < \infty \tag{2.6}$$

a.e. μ and from the triangle inequality

$$|f_{n_{k_2}}(s) - f_{n_{k_1}}(s)| \leq \sum_{k=k_1}^{k_2-1} |f_{n_{k+1}}(s) - f_{n_k}(s)|,$$

for $s \in E$ and $k_1 < k_2$. This allows us to conclude that $\{f_{n_k}(s)\}$ is Cauchy and hence convergent a.e. μ.

Now define

$$f(s) = \begin{cases} \lim_{k\to\infty} f_{n_k}(s), & \text{if the limit exists and is finite,} \\ 0, & \text{otherwise,} \end{cases}$$

and fix arbitrary $\epsilon > 0$. Using Fatou's Lemma along with (2.5) reveals that for all n sufficiently large

$$\begin{aligned}\int_E |f_n - f|^p d\mu &= \int_E \liminf_{k\to\infty} |f_n - f_{n_k}|^p d\mu \\ &\leq \liminf_{k\to\infty} \int_E |f_n - f_{n_k}|^p d\mu < \epsilon.\end{aligned}$$

Thus, $f \in \mathbb{L}^p$ and $\|f_n - f\|_p \to 0$. □

The sequence spaces ℓ^p are an important special case of the $\mathbb{L}^p$ spaces.

Example 2.3.6 *The ℓ^p space for $p \in [1, \infty]$ consists of elements of the form $x = (x_1, x_2, \ldots)$, where $x_i \in \mathbb{R}$ and $\sum_{i=1}^{\infty} |x_i|^p < \infty$ if $p \in [1, \infty)$ and $\max_i |x_i| < \infty$ if $p = \infty$. The norm for the space is*

$$\|x\|_p = \begin{cases} \left(\sum_{i=1}^{\infty} |x_i|^p\right)^{1/p}, & p \in [1, \infty), \\ \max_i |x_i|, & p = \infty. \end{cases}$$

We can view this from the perspective of the $\mathbb{L}^p(E, \mathscr{B}, \mu)$ spaces with $E = \mathbb{Z}^+$ and μ chosen to be the counting measure. Thus, the ℓ^p spaces are also Banach spaces.

The Banach space of random variables with finite pth moments will appear in a number of places going forward.

Example 2.3.7 *Let $(\Omega, \mathscr{F}, \mathbb{P})$ be a probability space. Then, the space $\mathbb{L}^p(\Omega, \mathscr{F}, \mathbb{P})$ contains random variables defined on $(\Omega, \mathscr{F}, \mathbb{P})$ with*

$$\|X\|_p = \left(\int_\Omega |X|^p d\mathbb{P}\right)^{1/p} = (\mathbb{E}|X|^p)^{1/p} < \infty,$$

where $\mathbb{E}$ denotes expected value.

Next, we turn to the case where $E = [0, 1]$, $\mathscr{B}$ is the Borel σ-field of $[0, 1]$ and μ is Lebesgue measure. For convenience, we will refer to these $\mathbb{L}^p$ spaces simply as $\mathbb{L}^p[0, 1]$. As $\mathbb{L}^p$ contains equivalence classes, it is meaningless to speak of a function's value $f(t)$ at any one specific t for $f \in \mathbb{L}^p[0, 1]$ because singletons in $[0, 1]$ have Lebesgue measure zero. One might argue from this that smoothness concepts such as continuity are not relevant in $\mathbb{L}^p[0, 1]$, which makes the following result especially interesting.

Theorem 2.3.8 *The set of equivalence classes that correspond to functions in $C[0, 1]$ is dense in $\mathbb{L}^p[0, 1]$.*

Proof: It suffices to show that for any uniformly bounded measurable function f on $[0,1]$ there exists a sequence of continuous functions g_n on $[0,1]$ such that $\int_0^1 |f - g_n|^2 d\mu \to 0$. Toward this goal first recall (cf. Billingsley, 1995) that for any Borel set B of $[0,1]$ and any $\epsilon > 0$, there exist intervals (a_i, b_i), $1 \le i \le k$, such that $\mu(B\Delta \cup_{i=1}^k (a_i, b_i)) < \epsilon/2$ where Δ stands for symmetric difference; it is therefore possible to construct a continuous function f such that

$$\int_0^1 |I_B - f|^p d\mu \le \int_0^1 |I_B - I_{\cup_{i=1}^k (a_i,b_i)}|^p d\mu + \int_0^1 |I_{\cup_{i=1}^k (a_i,b_i)} - f|^p d\mu \le \epsilon,$$

where we have now used I_A to represent the indicator function for a set A. Generalizing this from indicator to simple functions, we arrive at the conclusion that for any simple function f and any $\epsilon > 0$ there exists a continuous function g such that $\int_0^1 |f - g|^p d\mu < \epsilon$.

Recall that if f is a uniformly bounded, nonnegative measurable function then there exist a sequence of nonnegative simple functions f_n such that $f_n \uparrow f$ uniformly and $\int_0^1 |f - f_n|^p d\mu \downarrow 0$. Combining this fact with what we already know about approximating simple functions and applying Minkowsky's inequality, we conclude that for any uniformly bounded, nonnegative measurable function f there exist a sequence of continuous functions g_n such that $\int_0^1 |f - g_n|^p d\mu \to 0$. For a general uniformly bounded measurable function, it suffices to consider the positive parts and negative parts separately. □

Our last example in this section is concerned with functions having bounded variation on $[0,1]$. As we will see in Section 3.2, such functions play an interesting role in characterizing a certain key property of the Banach space $C[0,1]$.

Example 2.3.9 *The total variation of a function f on $[0,1]$ is*

$$TV(f) = \sup \sum_{i=1}^{n} |f(t_i) - f(t_{i-1})|, \tag{2.7}$$

where the supremum is taken over all partitions $0 = t_0 < t_1 < \cdots < t_n = 1$ of the interval $[0,1]$. If $TV(f)$ is finite, f is said to be of bounded variation. A function on $[0,1]$ is of bounded variation if and only if it can be expressed as the difference of two finite nondecreasing functions.

Let the space $BV[0,1]$ be the collection of all functions of bounded variation on $[0,1]$ equipped with the norm

$$\|f\| = |f(0)| + TV(f). \tag{2.8}$$

It can be verified that $\|f\|$ is a norm and that $BV[0,1]$ is a Banach space.

2.4 Inner Product and Hilbert spaces

Banach spaces provide one natural extension of finite dimensional, normed vector spaces. However, they do so without an immediate extension of the orthogonality concept that is so important in the Euclidean case. In some instances, this feature can be recovered by introducing an abstract analog of the dot or inner product.

Definition 2.4.1 *A function* $\langle \cdot, \cdot \rangle$ *on a vector space* $\mathbb{V}$ *is called an inner product if it satisfies*

1. $\langle v, v \rangle \geq 0,$
2. $\langle v, v \rangle = 0$ *if* $v = 0,$
3. $\langle a_1 v_1 + a_2 v_2, v \rangle = a_1 \langle v_1, v \rangle + a_2 \langle v_2, v \rangle$*, and*
4. $\langle v_1, v_2 \rangle = \langle v_1, v_2 \rangle$

for every $v, v_1, v_2 \in \mathbb{V}$ *and* $a_1, a_2 \in \mathbb{R}$.

A vector space with an associated inner product is called an inner-product space. The following result establishes the connection between such spaces and the normed spaces of Section 2.3.

Theorem 2.4.2 *An inner product* $\langle \cdot, \cdot \rangle$ *on a vector space* $\mathbb{V}$ *produces a norm* $\| \cdot \|$ *defined by* $\|v\| = \langle v, v \rangle^{1/2}$ *for* $v \in \mathbb{V}$*. The inner product then satisfies the Cauchy–Schwarz inequality*

$$|\langle v_1, v_2 \rangle| \leq \|v_1\| \|v_2\| \tag{2.9}$$

for $v_1, v_2 \in \mathbb{V}$ *with equality in (2.9) if* $v_1 = a_1 + a_2 v_2$ *for some* $a_1, a_2 \in \mathbb{R}$.

Proof: Let $v_3 = v_1 - \left(\langle v_1, v_2 \rangle / \|v_2\|^2 \right) v_2$ and observe that $\langle v_3, v_2 \rangle = 0$. Thus,

$$\begin{aligned}
\|v_1\|^2 &= \|v_3 + \left(\langle v_1, v_2 \rangle / \|v_2\|^2 \right) v_2\|^2 \\
&= \|v_3\|^2 + \| \left(\langle v_1, v_2 \rangle / \|v_2\|^2 \right) v_2\|^2 \\
&\geq \langle v_1, v_2 \rangle^2 / \|v_2\|^2.
\end{aligned}$$

The only way that equality can be achieved in this last expression is when $\|v_3\| = 0$.

To verify that $\|\cdot\|$ is a norm, we need only check the triangle inequality. This is accomplished with

$$\begin{aligned}\|v_1 + v_2\|^2 &= \|v_1\|^2 + 2\langle v_1, v_2\rangle + \|v_2\|^2 \\ &\leq \|v_1\|^2 + 2|\langle v_1, v_2\rangle| + \|v_2\|^2 \\ &\leq \|v_1\|^2 + 2\|v_1\|\|v_2\| + \|v_2\| \\ &= (\|v_1\| + \|v_2\|)^2. \end{aligned}$$ □

Although any inner product naturally defines a metric, the following example shows that not every metric space exhibits the structure that is necessary to also be an inner-product space.

Example 2.4.3 *The norm* $\|f\| = \sup\{|f(x)| : x \in [0, 1]\}$ *on* $C[0, 1]$ *is not induced by an inner product. To see this consider the functions* $f, g \in C[0, 1]$ *defined by*

$$f(x) \equiv 1, \quad g(x) = x, \quad x \in [0, 1].$$

Then,

$$\begin{aligned}\|f + g\|^2 + \|f - g\|^2 &= 4 + 1 = 5, \\ 2(\|f\|^2 + \|g\|^2) &= 2(1 + 1) = 4.\end{aligned}$$

This choice for f and g fails to satisfy the parallelogram rule

$$\|f + g\|^2 + \|f - g\|^2 = \langle f + g, f + g\rangle + \langle f - g, f - g\rangle = 2(\|f\|^2 + \|g\|^2)$$

that must hold for any two function with a norm-induced inner product.

The standard norm for an inner-product space is the one defined through its inner product as in Theorem 2.4.2. Topological properties then derive from those of the corresponding metric space whose metric is the standard norm. With that in mind, we are lead to conclude that inner products are continuous functions under the norm induced topology.

Theorem 2.4.4 *Let* $\{v_{1n}\}, \{v_{2n}\}$ *be sequences and* v_1, v_2 *elements in an inner product space* $\mathbb{V}$ *with inner product and norm* $\langle\cdot,\cdot\rangle$ *and* $\|\cdot\|$. *If* $\|v_i - v_{in}\| \to 0, i = 1, 2$ *then* $\langle v_{1n}, v_{2n}\rangle \to \langle v_1, v_2\rangle$.

Proof: The proof follows from

$$\begin{aligned}|\langle v_{1n}, v_{2n}\rangle - \langle v_1, v_2\rangle| &\leq |\langle v_{1n} - v_1, v_{2n}\rangle| + |\langle v_1, v_{2n} - v_2\rangle| \\ &\leq \|v_{1n} - v_1\|\|v_{2n}\| + \|v_1\|\|v_{2n} - v_2\|.\end{aligned}$$ □

Definition 2.4.5 *A complete inner-product space is called a Hilbert space.*

Example 2.4.6 *From Theorem 2.2.16, if follows that any finite-dimensional inner-product space is a Hilbert space.*

Example 2.4.7 *The ℓ^2 space from Example 2.3.6 is a Hilbert space. The inner product of elements $v_i = (v_{i1}, v_{i2}, \ldots), i = 1, 2$ is*

$$\langle v_1, v_2 \rangle = \sum_{j=1}^{\infty} v_{1j} v_{2j}.$$

Inner product spaces provide the framework that is needed to extend the concepts of perpendicular vectors and subspaces to abstract settings.

Definition 2.4.8 *Elements x_1, x_2 of an inner-product space $\mathbb{X}$ are said to be orthogonal if $\langle x_1, x_2 \rangle = 0$. A countable collection of elements $\{e_1, e_2, \ldots\}$ is said to be an orthonormal sequence if $\|e_j\| = 1$ for all j and the e_j are pairwise orthogonal.*

The inner products of elements from an orthonormal sequence with the elements of its parent Hilbert space are of paramount interest in the study of inner-product spaces. To be precise, let $\{e_1, e_2, \ldots\}$ be an orthonormal sequence in an inner product space $\mathbb{X}$ with associated inner product $\langle \cdot, \cdot \rangle$. The corresponding generalized Fourier coefficients for $x \in \mathbb{X}$ are $\langle x, e_j \rangle, j = 1, \ldots$ It is customary to drop the "generalized" from their name and refer to them simply as Fourier coefficients and we will adhere to that practice throughout the remainder of this text. The Fourier coefficients of an element are square summable due to the following result known as Besssel's inequality.

Theorem 2.4.9 *Let $\{e_1, e_2, \ldots\}$ be an orthonormal sequence in an inner-product space $\mathbb{X}$. For any $x \in \mathbb{X}$, $\sum_{i=1}^{\infty} \langle x, e_i \rangle^2 \leq \|x\|^2$ and, therefore, $\sum_{i=1}^{\infty} \langle x, e_i \rangle e_i$ converges in $\mathbb{X}$.*

Proof: The result is an immediate consequence of the fact that

$$0 \leq \left\| x - \sum_{i=1}^{n} \langle x, e_i \rangle e_i \right\|^2 = \|x\|^2 - \sum_{i=1}^{n} \langle x, e_i \rangle^2$$

for all n. □

The standard approach to creating an orthonormal sequence is the Gram–Schmidt algorithm described in the following theorem. The result is easily verified by induction.

Theorem 2.4.10 *Let $\{x_n\}$ be a countable collection of elements in a Hilbert space such that every finite subcollection of $\{x_n\}$ is linearly independent. Define $e_1 = x_1/\|x_1\|$ and $e_i = v_i/\|v_i\|$ for*

$$v_i = x_i - \sum_{j=1}^{i-1} \langle x_i, e_j \rangle e_j.$$

Then, $\{e_n\}$ is an orthonormal sequence and $\overline{\text{span}\{x_n\}} = \overline{\text{span}\{e_n\}}$.

We would now like to expand the basis concept to the infinite dimensional Hilbert space context. The first step is to come to some agreement on precisely what the term "basis" might mean in this instance. With that in mind, the next definition provides one possible starting point.

Definition 2.4.11 *An orthonormal sequence $\{e_n\}$ in a Hilbert space $\mathbb{H}$ is called an orthonormal basis or a* complete orthonormal system *(CONS) if $\overline{\text{span}\{e_n\}} = \mathbb{H}$.*

A simple application of the continuity of the inner product gives us one way to check that Definition 2.4.11 is applicable.

Theorem 2.4.12 *An orthonormal sequence $\{e_n\}$ in a Hilbert space $\mathbb{H}$ is a CONS if $\langle x, e_n \rangle = 0$ for all n implies that $x = 0$.*

This allows us to see that a CONS $\{e_n\}$ provides a representation for the elements of a Hilbert space as linear combinations of the basis elements in an extended sense. Specifically, let $\{e_n\}$ be a CONS for a Hilbert space $\mathbb{H}$ and let x be any element of $\mathbb{H}$. Now take $\tilde{x} = \sum_{j=1}^{\infty} \langle x, e_j \rangle e_j$ which is well defined as a result of Bessel's inequality. As, $\langle x - \tilde{x}, e_j \rangle = 0$ for every j, we reach the conclusion.

Theorem 2.4.13 *Every element x of a Hilbert space $\mathbb{H}$ with CONS $\{e_j\}$ can be expressed in terms of the Fourier expansion*

$$x = \sum_{j=1}^{\infty} \langle x, e_j \rangle e_j \tag{2.10}$$

and

$$\|x\|^2 = \sum_{j=1}^{\infty} \langle x, e_j \rangle^2. \tag{2.11}$$

The identity (2.11) strengthens Bessel's inequality and is known as Parseval's relation.

The CONS concept also provides a way to characterize separability.

Theorem 2.4.14 *A Hilbert space is separable if it has an orthonormal basis.*

Proof: Assume that $\{e_j\}$ is a CONS for the Hilbert space $\mathbb{H}$. By Theorem 2.4.13, it is easy to show that the countable subset of elements x with $\langle x, e_j \rangle$ in the set of rationals for all j is dense. Thus, $\mathbb{H}$ is separable. Conversely, any countable dense subset of elements can be transformed into an orthonormal sequence via the Gram–Schmidt algorithm of Theorem 2.4.10. □

The next topic for consideration plays an important role in the representation of stochastic processes studied in Section 7.6.

Definition 2.4.15 *Two metric spaces* $(\mathbb{M}_1, d_1)$ *and* $(\mathbb{M}_2, d_2)$ *are said to be isometrically isomorphic or* congruent *if there exists a bijective function* $\Psi : \mathbb{M}_2 \mapsto \mathbb{M}_1$ *such that* $d_2(x_1, x_2) = d_1(\Psi(x_1), \Psi(x_2))$ *for all* $x_1, x_2 \in \mathbb{M}_2$.

The case of most interest to us in when the two metric spaces in the definition are Hilbert spaces. In that case, we have the following.

Theorem 2.4.16 *Let* $\mathbb{H}_i, i = 1, 2,$ *be Hilbert spaces with inner products* $\langle \cdot, \cdot \rangle_i, i = 1, 2$. *Suppose that for some index set E there are collections of vectors* $\mathbb{U}_i = \{u_{it} : t \in E\}$ *such that* $\overline{\text{span}(\mathbb{U}_i)} = \mathbb{H}_i, i = 1, 2$. *If for every* $s, t \in E$

$$\langle u_{1s}, u_{1t} \rangle_1 = \langle u_{2s}, u_{2t} \rangle_2, \tag{2.12}$$

$\mathbb{H}_1$ *and* $\mathbb{H}_2$ *are congruent.*

Proof: The proof is an immediate consequence of the denseness of the sets $\mathbb{U}_1$ and $\mathbb{U}_2$ and the continuity of the inner product in a Hilbert space. □

An application of Theorem 2.4.16 gives us

Theorem 2.4.17 *Any infinite-dimensional separable Hilbert space is congruent to* ℓ^2.

Proof: Let $\{e_j\}_{j=1}^{\infty}$ be an orthonormal basis for an infinite-dimensional separable Hilbert space $\mathbb{H}$. In ℓ^2, define the orthonormal basis $\{\phi_j\}_{j=1}^{\infty}$ with ϕ_j being a sequence of all zeros except for a 1 as its jth entry. Now apply Theorem 2.4.16 with $E = \mathbb{Z}^+$. □

We defined the Banach space $\mathbb{L}^p(E, \mathscr{B}, \mu)$ in Section 2.3. In this collection of Banach spaces, the only one that is also a Hilbert space is $\mathbb{L}^2(E, \mathscr{B}, \mu)$ for which the inner product is defined as

$$\langle f_1, f_2 \rangle = \int_E f_1 f_2 d\mu$$

for $f_1, f_2 \in \mathbb{L}^2(E, \mathscr{B}, \mu)$.

In the fda, literature attention has focused on $\mathbb{L}^2[0, 1]$: namely, the $\mathbb{L}^2$ space with $E = [0, 1]$, $\mathscr{B}$ the Borel σ-field of $[0, 1]$ and μ Lebesgue measure. The following result catalogs several Fourier bases for $\mathbb{L}^2[0, 1]$. We note in passing that they all consist of continuous and, in fact, infinitely differentiable functions.

Theorem 2.4.18 *The following sets of functions*

$$B_1 = \{f_0(x) = 1, f_n(x) = \sqrt{2}\cos(n\pi x), n \geq 1\},$$
$$B_2 = \{g_n(x) = \sqrt{2}\sin(n\pi x), n \geq 1\}$$

and

$$B_3 = \{h_0(x) = 1, h_{2n-1}(x) = \sqrt{2}\sin(2n\pi x), h_{2n}(x) = \sqrt{2}\cos(2n\pi x), n \geq 1\}$$

are all orthonormal bases for $\mathbb{L}^2[0, 1]$.

Proof: It is clear that B_1, B_2, and B_3 are orthonormal. Hence, we need only show that they are bases. We begin with B_1 and pick an arbitrary $f \in \mathbb{L}^2[0, 1]$ for which we wish to show that for each $\epsilon > 0$ there exist an integer m_ϵ and real coefficients $a_{1\epsilon}, a_{2\epsilon}, \ldots, a_{m_\epsilon \epsilon}$ with $\|f - f_\epsilon\| < \epsilon$ and $f_\epsilon = \sum_{i=0}^{m_\epsilon} a_{i\epsilon} f_i$.

From Theorem 2.3.8, for any $f \in \mathbb{L}^2[0, 1]$ and $\epsilon > 0$, there exist $g \in C[0, 1]$ such that $\|f - g\| < \epsilon/2$. Now observe that the function $\cos^{-1}$ is a continuous bijection, so that we can define the continuous function $h(s) = g((1/\pi)\cos^{-1} s)$ for $s \in [-1, 1]$. Then, it follows that (cf. Theorem 2.1.14) there is a polynomial p such that $|h(s) - p(s)| < \epsilon/2$ uniformly in $s \in [-1, 1]$. Hence, writing $k(x) = p(\cos \pi x), x \in [0, 1]$, we have $|g(x) - k(x)| = |h(\cos \pi x) - p(\cos \pi x)| < \epsilon/2$ for $x \in [0, 1]$ so that $\|g - k\| < \epsilon/2$ and $\|f - k\| < \epsilon$. As powers of cosine functions can be expressed as linear combinations of cosine function, the result has been shown.

Now consider B_2. For $f \in \mathbb{L}^2[0, 1]$ and any $\delta > 0$, define $f_\delta(x) = f(x) I_{(\delta, 1]}(x)$. Given $\epsilon > 0$, there obviously exists $\delta > 0$ such that $\|f - f_\delta\| < \epsilon/2$.

Now $h(x) = f_\delta(x)/\sin x \in \mathbb{L}^2[0,1]$ and so there is a function $k(x) = \sum_{i=0}^{m} a_i \cos(i\pi x)$ such that $\|h - k\| < \epsilon/2$. However,

$$\begin{aligned}
\|h - k\|^2 &= \int_0^\delta k^2(x)dx + \int_\delta^1 \left[\frac{f_\delta(x)}{\sin^2(\pi x)} - k(x)\right]^2 dx \\
&\geq \int_0^\delta k^2(x)\sin^2(\pi x)dx + \int_\delta^1 [f_\delta(x) - k(x)\sin(\pi x)]^2 dx \\
&= \int_0^1 [f_\delta(x) - k(x)\sin(\pi x)]^2 dx = \|f_\delta(\cdot) - k(\cdot)\sin(\pi\cdot)\|^2
\end{aligned}$$

and $k(\cdot)\sin(\pi\cdot)$ can be expressed as a linear combination of sine functions.

Finally, consider B_3 and suppose that it is not a basis. In that Case, there exists a nonzero function $f \in \mathbb{L}^2[0,1]$ such that $\langle f, h_0\rangle = \langle f, h_n\rangle = 0, n \geq 1$. However, this means that

$$\begin{aligned}
0 &= \int_0^1 f(x)dx = \int_{-1}^1 f(x/2 + 1/2)dx \\
&= \int_0^1 [f(x/2 + 1/2) + f(-x/2 + 1/2)]dx, \\
0 &= \int_0^1 f(x)\cos(2n\pi x)dx = \int_{-1}^1 f(x/2 + 1/2)\cos(n\pi x + n\pi)dx \\
&= (-1)^n \int_{-1}^1 f(x/2 + 1/2)\cos(n\pi x)dx \\
&= (-1)^n \int_0^1 [f(x/2 + 1/2) + f(-x/2 + 1/2)]\cos(n\pi x)dx
\end{aligned}$$

and, similarly, that

$$0 = (-1)^n \int_0^1 [f(x/2 + 1/2) - f(-x/2 + 1/2)]\sin(n\pi x)dx$$

for each $n \geq 1$. In view of what was already proved for B_1 and B_2, we conclude that for any $x \in [0, 1]$

$$f(x/2 + 1/2) - f(-x/2 + 1/2) = f(x/2 + 1/2) + f(-x/2 + 1/2) = 0$$

which, in turn, implies that $f(x/2 + 1/2) = f(-x/2 + 1/2) = 0$: i.e., $f(x) = 0$ for all $x \in [0, 1]$. □

2.5 The projection theorem and orthogonal decomposition

It is safe to say that almost every statistical problem eventually leads to some type of optimization problem with the classical Gauss–Markov Theorem providing an important case in point. Optimization in vector and function spaces becomes much more tractable when there is an inherent geometry that can be exploited to aid in the characterization of extrema. This is undoubtedly why Hilbert spaces have occupied such a central role in statistics.

The following result is fundamental in optimization theory.

Theorem 2.5.1 *Let $\mathbb{M}$ be a closed convex set in a Hilbert space $\mathbb{H}$. For every $x \in \mathbb{H}$, $\|x - y\|$ has a unique minimizer $\hat{x}$ in $\mathbb{M}$ that satisfies*

$$\langle x - \hat{x}, y - \hat{x} \rangle \leq 0 \tag{2.13}$$

for all $y \in \mathbb{M}$.

Proof: Our proof follows along the lines of that in Luenberger (1969). First we observe that by the definition of infimum, we can obtain a sequence $y_n \in \mathbb{M}$ such that

$$\lim_{n\to\infty} \|x - y_n\|^2 = \inf_{y\in\mathbb{M}} \|x - y\|^2.$$

The first step is to show that this sequence is Cauchy thereby insuring that it converges to some element in $\mathbb{M}$. For this purpose, we use the identity

$$\|y_n - y_m\|^2 = 2\|x - y_n\|^2 + 2\|x - y_m\|^2 - 4\left\|x - \frac{y_n + y_m}{2}\right\|^2.$$

From convexity, $(y_n + y_m)/2 \in \mathbb{M}$ and, hence

$$\|y_n - y_m\|^2 \leq 2\|x - y_m\|^2 + 2\|x - y_m\|^2 - 4 \inf_{y\in\mathbb{M}} \|x - y\|^2.$$

Thus, $\{y_n\}$ is Cauchy and must have a limit $\hat{x}$ that attains the infimum.

If there were another element $\tilde{x}$ in $\mathbb{M}$ that also attained the infimum, we could create the alternating series having $y_n = \hat{x}$ for n even and $y_n = \tilde{x}$ for n odd. It is now trivially true that $\lim \|x - y_n\|^2 = \inf_{y\in\mathbb{M}} \|x - y\|^2$ and our previous argument can be used to see that the sequence is Cauchy. So, it must have a limit and that can only be true if $\hat{x} = \tilde{x}$.

Suppose now that there is a $y \in \mathbb{M}$ for which $\langle x - \hat{x}, y - \hat{x} \rangle > 0$. Then, if we let $x(a) = ay + (1 - a)\hat{x} \in \mathbb{M}$ for $a \in (0, 1)$, the derivative of $\|x - x(a)\|^2$ at $a = 0$ is negative. Thus, there is a choice of a that makes $\|x - x(a)\|^2$

smaller than $\|x-\hat{x}\|^2$ which is a contradiction. On the other hand, if $\langle x-\hat{x}, y-\hat{x}\rangle \le 0$,

$$\|x-y\|^2 = \|x-\hat{x}\|^2 + 2\langle x-\hat{x}, \hat{x}-y\rangle + \|\hat{x}-y\|^2 \ge \|x-\hat{x}\|^2.$$

□

An important application of the previous result is the *projection theorem.*

Theorem 2.5.2 *Let* $\mathbb{M}$ *be a closed linear subspace of a Hilbert space* $\mathbb{H}$. *For any element* $x \in \mathbb{H}$, *there exists a unique element of* $\mathbb{M}$ *that minimizes* $\|x-y\|$ *over* $y \in \mathbb{M}$. *The minimizer* $\hat{x}$ *is uniquely determined by the condition*

$$\langle \hat{x}, y\rangle = \langle x, y\rangle \tag{2.14}$$

for all $y \in \mathbb{M}$.

Proof: In view of Theorem 2.5.1, all we need to verify is (2.14). As $\mathbb{M}$ is a subspace, we can choose $y = 0$ and $y = 2\hat{x}$ in (2.13) to see that $\langle x-\hat{x}, \hat{x}\rangle = 0$. Thus, (2.13) gives

$$\langle x-\hat{x}, y\rangle \le 0$$

for all $y \in \mathbb{M}$. Replacing y by $-y$ in this inequality finishes the proof. □

The element $\hat{x}$ in Theorem 2.5.2 is called the *projection* of x onto $\mathbb{M}$. It is uniquely determined by the fact that the residual or error term $x-\hat{x}$ is orthogonal to the entire $\mathbb{M}$ subspace. The collection of all elements that have this orthogonality property is of independent interest.

Definition 2.5.3 *Let* $\mathbb{X}$ *be an inner-product space with* $\mathbb{M} \subset \mathbb{X}$. *The orthogonal complement of* $\mathbb{M}$ *is the set*

$$\mathbb{M}^\perp = \{x \in \mathbb{X} : \langle x, y\rangle = 0 \textit{ for all } y \in \mathbb{M}\}.$$

Theorem 2.5.2 can now be seen to have the consequence that when $\mathbb{M}$ is a closed subspace every element x in $\mathbb{H}$ can be uniquely expressed as

$$x = x_1 + x_2 \tag{2.15}$$

for $x_1 \in \mathbb{M}$ and $x_2 \in \mathbb{M}^\perp$.

Definition 2.5.4 *Let* $\mathbb{M}_1$ *and* $\mathbb{M}_2$ *be orthogonal subspaces of* $\mathbb{X}$; *i.e.,* $x_1 \perp x_2$ *for all* $x_i \in \mathbb{M}_i$. *Then, the collection* $\{x_1 + x_2 : x_i \in \mathbb{M}_i, i = 1, 2\}$ *is denoted by* $\mathbb{M}_1 \oplus \mathbb{M}_2$ *and is referred to as the orthogonal direct sum of* $\mathbb{M}_1$ *and* $\mathbb{M}_2$. *If* $\mathbb{M}_1$ *and* $\mathbb{M}_2$ *are not orthogonal but satisfy* $\mathbb{M}_1 \cap \mathbb{M}_2 = \{0\}$, *then* $\{x_1 + x_2 :$

$x_i \in \mathbb{M}_i, i = 1,2\}$ *is denoted by* $\mathbb{M}_1 + \mathbb{M}_2$ *and is referred to as the algebraic direct sum of* $\mathbb{M}_1$ *and* $\mathbb{M}_2$.

Using the orthogonal direct sum notation, the decomposition in (2.15) can be restated as

Theorem 2.5.5 *Let* $\mathbb{M}$ *be a closed subspace of a Hilbert space* $\mathbb{H}$. *Then*

$$\mathbb{H} = \mathbb{M} \oplus \mathbb{M}^{\perp}. \tag{2.16}$$

Some additional properties of the orthogonal complement are catalogued in our following theorem.

Theorem 2.5.6 *Let* $\mathbb{H}$ *be a Hilbert space with* $\mathbb{M}$ *a subset of* $\mathbb{H}$. *Then,*

1. $\mathbb{M}^{\perp}$ *is a closed subspace,*
2. $\mathbb{M} \subset (\mathbb{M}^{\perp})^{\perp}$, *and*
3. $(\mathbb{M}^{\perp})^{\perp} = \overline{\mathbb{M}}$ *if* $\mathbb{M}$ *is a subspace.*

Proof: First, it is clear that $\mathbb{M}^{\perp}$ is a linear space. By the Cauchy–Schwarz inequality, the mapping $x \mapsto \langle x, a\rangle$ is continuous for any $a \in \mathbb{X}$. If $x_1, x_2, \ldots$ are in $\mathbb{M}^{\perp}$ and $x_n \to x$ in $\mathbb{X}$, then $\langle x, a\rangle = \lim_{n\to\infty}\langle x_n, a\rangle = 0$ for all $a \in \mathbb{M}$. Thus, $\mathbb{M}^{\perp}$ is closed.

Part 2 of the theorem comes from observing that if $x \in \mathbb{M}$ it must be orthogonal to every $y \in \mathbb{M}^{\perp}$; i.e., it is in $(\mathbb{M}^{\perp})^{\perp}$. Finally, from parts 1 and 2, we conclude that $\overline{\mathbb{M}} \subset (\mathbb{M}^{\perp})^{\perp}$. By Theorem 2.5.5, any element $x \in (\mathbb{M}^{\perp})^{\perp}$ can be uniquely written as $x = y + z$ for some $y \in \overline{\mathbb{M}}$ and $z \in \overline{\mathbb{M}}^{\perp} \cap (\mathbb{M}^{\perp})^{\perp}$. However,

$$\overline{\mathbb{M}}^{\perp} \cap (\mathbb{M}^{\perp})^{\perp} \subset \mathbb{M}^{\perp} \cap (\mathbb{M}^{\perp})^{\perp} = \{0\}.$$

□

2.6 Vector integrals

Suppose now that we have a function f on a measure space $(E, \mathscr{B}, \mu)$ that takes on values in a Banach space $\mathbb{X}$. In the case of $\mathbb{X} = \mathbb{R}$, we are familiar with the Lebesgue integral $\int f d\mu$ and may recall its definition as the limit of integrals of simple functions that take on only finitely many values. Here we wish to extend this idea to the case of a general Banach space. The resulting abstract notion of an integral is called a vector integral and there are various ways such integrals can be constructed.

The objective of this section is to give a concise, but rigorous and self-contained, treatise concerning vector integrals that is appropriate for the target audience of this book. In particular, one place in fda where these integrals are relevant is the definition of the mean element and covariance operator for a random element of a Hilbert space (Section 7.2). Readers interested in additional details on vector integrals may consult sources such as Diestel and Uhl (1977), Dunford and Schwarz (1988), and Yosida (1971).

The integral we will focus on is due to Bochner (1933) and, accordingly, is referred to as the Bochner integral. The construction of these integrals parallels that of the Lebesgue integral in real analysis beginning with a definition of simple functions and their integrals.

Definition 2.6.1 *A function $f : E \to \mathbb{X}$ is called simple if it can be represented as*

$$f(\omega) = \sum_{i=1}^{k} I_{E_i}(\omega) g_i \tag{2.17}$$

for some finite k, $E_i \in \mathscr{B}$ and $g_i \in \mathbb{X}$.

Definition 2.6.2 *Any simple function $f(\omega) = \sum_{i=1}^{k} I_{E_i}(\omega) g_i$ with $\mu(E_i) < \infty$ for all i is said to be integrable and its Bochner integral is defined as*

$$\int_E f d\mu = \sum_{i=1}^{k} \mu(E_i) g_i. \tag{2.18}$$

It is not difficult to verify that this definition does not depend on the particular representation of f. In particular, the E_i can be chosen without loss to be disjoint. To see this merely observe that if any two sets E_i and E_j in a simple function overlap, which part of the function can be rewritten in an equivalent form using disjoint sets as

$$g_i I_{E_i \cap E_j^C}(\omega) + g_j I_{E_j \cap E_i^C}(\omega) + (g_i + g_j) I_{E_i \cap E_j}(\omega)$$

with A^C indicating the complement of the set A.

The previous definition is extended to a general measurable function from E to $\mathbb{X}$ as follows.

Definition 2.6.3 *A measurable function f is said to be Bochner integrable if there exists a sequence $\{f_n\}$ of simple and Bochner integrable functions such that*

$$\lim_{n \to \infty} \int_E \|f_n - f\| d\mu = 0. \tag{2.19}$$

In this case, the Bochner integral of f is defined as

$$\int_E f d\mu = \lim_{n \to \infty} \int_E f_n d\mu. \tag{2.20}$$

To see that this definition has merit first observe from (2.18) and the triangle inequality that

$$\left\| \int_E f d\mu \right\| \leq \int_E \|f\| d\mu \tag{2.21}$$

for any simple function f. When applied to the simple function $f_n - f_m$ in (2.19) this inequality produces

$$\left\| \int_E f_n d\mu - \int_E f_m d\mu \right\| \leq \int_E \|f_n - f_m\| d\mu.$$

However, this upper bound is just a Lebesgue integral and the triangle inequality assures us that

$$\int_E \|f_n - f_m\| d\mu \leq \int_E \|f - f_n\| d\mu + \int_E \|f - f_m\| d\mu$$

which converges to zero as $m, n \to \infty$ by (2.19). This shows that $\{\int_E f_n d\mu\}$ is a Cauchy sequence and the completeness of $\mathbb{X}$ can now be invoked to conclude that the limit in (2.20) must exist. It is also independent of the approximating sequence since we can combine two approximating sequences into a third that must also be convergent.

The definition of the Bochner integral relies on the existence of an approximating sequence of simple and Bochner integrable functions $\{f_n\}$. Of course, this condition may not be satisfied for any particular function and we would at least like to have a sufficient condition that we could more easily check to see if it were true. The following result serves that purpose for a scenario that is sufficiently general to be applicable in most fda applications.

Theorem 2.6.4 *Let f be a measurable function from E to $\mathbb{X}$ with*

$$\int_E \|f\| d\mu < \infty. \tag{2.22}$$

Suppose that for each n there exists a finite-dimensional subspace $\mathbb{X}_n$ of $\mathbb{X}$ such that

$$\lim_{n\to\infty} \int_E \|f - g_n\| d\mu = 0 \tag{2.23}$$

for some measurable g_n taking value in $\mathbb{X}_n$. Then, there exist simple and Bochner integrable functions f_n such that (2.19) holds.

Proof: Define

$$\tilde{\mathbb{X}}_n = \mathbb{X}_n \cap \{g \in \mathbb{X} : \|g\| \in [n^{-1}, n]\}$$

and

$$E_n = \{\omega \in E : g_n(\omega) \in \tilde{\mathbb{X}}_n\}.$$

Markov's inequality produces

$$\mu(E_n) \le n \int_{E_n} \|g_n\| d\mu \le n \int_E \|g_n\| d\mu < \infty. \quad (2.24)$$

As $\tilde{\mathbb{X}}_n$ is bounded and finite-dimensional, it is totally bounded (cf. Section 2.1). Thus, the Heine–Borel theorem (Theorem 2.1.17) can be invoked to see that there is a finite partition $B_i, 1 \le i \le k$, of $\tilde{\mathbb{X}}_n$ such that each B_i is in the Borel σ-field for $\mathbb{X}$ and has diameter less than $(n\mu(E_n))^{-1}$. For an arbitrary element $b_i \in B_i$, set

$$f_n(\omega) = \sum_{i=1}^{k} b_i I_{\{g_n(\omega) \in B_i\}}.$$

Note that $f_n = 0$ on E_n^c and hence (2.24) entails that f_n is simple and Bochner integrable as a simple function. Also, by construction,

$$\|f_n(\omega) - g_n(\omega)\| \le \max_{1 \le i \le k} \sup_{x \in B_i} \|b_i - x\| \le (n\mu(E_n))^{-1} \quad (2.25)$$

for $\omega \in E_n$. Thus, by the triangle inequality,

$$\begin{aligned} &\int_E \|f_n - f\| d\mu \\ &\le \int_E \|f_n - g_n\| d\mu + \int_E \|g_n - f\| d\mu \\ &= \int_{E_n} \|f_n - g_n\| d\mu + \int_{E_n^c} \|f_n - g_n\| d\mu + \int_E \|g_n - f\| d\mu. \end{aligned}$$

The first and third terms in the last expression tend to zero by (2.25) and (2.23), respectively, while the second term reduces to $\int_{E_n^c} \|g_n\| d\mu$ because $f_n = 0$ on E_n^c. Thus, to verify (2.19), we only need to establish that

$$\lim_{n \to \infty} \int_{E_n^c} \|g_n\| d\mu = 0,$$

or, equivalently, by (2.23) that

$$\lim_{n \to \infty} \int_E \|f\| I(\|g_n\| > n) d\mu = 0 \quad (2.26)$$

and that

$$\lim_{n\to\infty} \int_E \|f\| I(\|g_n\| < n^{-1}) d\mu = 0. \tag{2.27}$$

First, from (2.23) and Markov's inequality,

$$\mu(\|g_n\| > n) \le n^{-1} \int_E \|g_n\| d\mu \to 0.$$

As $\|f\|$ is integrable, (2.26) follows easily (cf. Exercise 5.6 of Resnick, 1998) from this relation. To show (2.27), for $\epsilon > 0$, write

$$\int_E \|f\| I(\|g_n\| < n^{-1}) d\mu = \int_E \|f\| I(\|g_n\| < n^{-1}, \|f\| > \epsilon) d\mu + \int_E \|f\| I(\|g_n\| < n^{-1}, \|f\| \le \epsilon) d\mu$$

so that

$$\limsup_{n\to\infty} \int_E \|f\| I(\|g_n\| < n^{-1}) d\mu \le \limsup_{n\to\infty} \int_E \|f\| I(\|g_n\| < n^{-1}, \|f\| > \epsilon) d\mu + \int_E \|f\| I(\|f\| \le \epsilon) d\mu.$$

The first term on the right of the inequality is zero as

$$\mu(\|g_n\| < n^{-1}, \|f\| > \epsilon) \le \frac{\int_E \|f - g_n\| I(\|g_n\| < n^{-1}, \|f\| > \epsilon) d\mu}{\epsilon - n^{-1}} \le \frac{\int_E \|f - g_n\| d\mu}{\epsilon - n^{-1}} \to 0$$

due to Markov's inequality and (2.23). Hence, (2.27) follows by now letting $\epsilon \to 0$ and applying Lebesgue's dominated convergence theorem. □

The following result shows that the existence of g_n in (2.23) is guaranteed if $\mathbb{X}$ is a separable Hilbert space.

Theorem 2.6.5 *Suppose $\mathbb{X}$ is a separable Hilbert space and f is a measurable function from E to $\mathbb{X}$ with $\int_E \|f\| d\mu < \infty$. Then, f is Bochner integrable.*

Proof: If suffices to verify (2.23) in Theorem 2.6.4. For that purpose, take $\mathbb{X}_n$ to be span$\{e_1, \dots, e_n\}$ for any CONS $\{e_j\}_{j=1}^{\infty}$ of $\mathbb{X}$ and g_n the projection of f on $\mathbb{X}_n$. Then, (2.23) follows from Lebesgue's dominated convergence theorem. □

Another connection with Lebesgue integration is the following Bochner integral version of the dominated convergence theorem.

Theorem 2.6.6 *Let* $\{f_n\}$ *be a sequence of Bochner integrable functions in* $\mathbb{X}$ *that converges to some* $f \in \mathbb{X}$. *If there is a nonnegative Lebesgue integrable function* g *such that* $\|f_n\| \leq g$ *for all* n *a.e.* μ, *then* f *is Bochner integrable and* $\int_E f d\mu = \lim_{n\to\infty} \int_E f_n d\mu$.

Proof: In combination, $\|f - f_n\| \leq 2g$ and $\|f - f_n\| \to 0$ allow us to apply the Lebesgue dominated convergence theorem to obtain $\int_E \|f - f_n\| d\mu \to 0$. As the f_n are Bochner Integrable, we may find simple functions $\tilde{f}_n$ such that $\int_E \|f_n - \tilde{f}_n\| d\mu \to 0$. These new functions satisfy (2.19) because

$$\int_E \|f - \tilde{f}_n\| d\mu \leq \int_E \|f - f_n\| d\mu + \int_E \|f_n - \tilde{f}_n\| d\mu$$

and the theorem has been proved. □

The Bochner integral also has a feature similar to the monotonicity of the Lebesgue integral: namely,

Theorem 2.6.7 *If* f *is Bochner integrable, then* $\|\int_E f d\mu\| \leq \int_E \|f\| d\mu$.

Proof: Let $f_n = \sum_{i=1}^n g_i I_{E_i}(\omega)$ be a simple function with $E_i \cap E_j = \phi$ for $i \neq j$. Then,

$$\begin{aligned}\left\|\int_E f_n(\omega) d\mu\right\| &= \left\|\sum_{i=1}^n g_i \mu(E_i)\right\| \\ &\leq \sum_{i=1}^n \|g_i\| \mu(E_i) = \int_E \|f_n\| d\mu.\end{aligned}$$

So, if $\{f_n\}$ is a sequence of simple Bochner integrable functions that satisfie (2.19),

$$\begin{aligned}\left\|\int_E f d\mu\right\| &\leq \left\|\int_E f d\mu - \int_E f_n d\mu\right\| + \left\|\int_E f_n d\mu\right\| \\ &\leq \left\|\int_E f d\mu - \int_E f_n d\mu\right\| + \int_E \|f_n\| d\mu \\ &\leq \left\|\int_E f d\mu - \int_E f_n d\mu\right\| + \int_E \|f_n - f\| d\mu + \int_E \|f\| d\mu\end{aligned}$$

and the result follows upon taking limits with respect to n. □

2.7 Reproducing kernel Hilbert spaces

The concept of a reproducing kernel Hilbert space (RKHS) owes its origin to the work of Moore (1916) and Aronszajn (1950). An overview of statistical applications for RKHS theory is provided in Berlinet and Thomas-Agnan (2004). We will see RKHSs arise in Chapter 6 and as part of the representation theory for second-order stochastic processes discussed in Section 7.6.

Throughout this section, we restrict our attention to the case where $\mathbb{H}$ is a Hilbert space of real valued functions defined on some set E. With this being understood, we can now define the reproducing kernel concept.

Definition 2.7.1 *A bivariate function K on $E \times E$ is said to be a reproducing kernel (rk) for $\mathbb{H}$ if*

1. *for every $t \in E$, $K(\cdot, t) \in \mathbb{H}$ and*
2. *K satisfies the reproducing property that for every $f \in \mathbb{H}$ and $t \in E$*

$$f(t) = \langle f, K(\cdot, t) \rangle. \tag{2.28}$$

When $\mathbb{H}$ possesses an rk it is said to be an RKHS.

Example 2.7.2 *Let $\mathbb{H}$ be a finite-dimensional Hilbert space with $\{e_1, \dots, e_p\}$ an associated orthonormal basis. Now define*

$$K(s, t) = \sum_{i=1}^{p} e_i(s) e_i(t)$$

for $s, t \in E$. Clearly, $K(\cdot, t) \in \mathbb{H}$. Also, for $1 \leq j \leq p$,

$$\langle e_j, K(\cdot, t) \rangle = \sum_{i=1}^{p} \langle e_j, e_i \rangle e_i(t) = e_j(t)$$

from which the reproducing property follows at once. Thus, $\mathbb{H}$ is an RKHS.

The theory of reproducing kernels is driven by the properties of nonnegative definite functions. We say that a bivariate function on a set $E \times E$ is nonnegative definite, or nonnegative for short when there is no ambiguity, if for any set of real numbers $\{a_j\}_{j=1}^{n}$, any set of elements $t_1, \dots, t_n$ from E, and any $n \in \mathbb{Z}^+$

$$\sum_{i=1}^{n} \sum_{j=1}^{n} a_i a_j K(t_i, t_j) \geq 0. \tag{2.29}$$

Bivariate functions that are nonnegative definite and symmetric in their arguments are sometimes referred to as *kernels*. The following two results are fundamental.

Theorem 2.7.3 *The rk K of an RKHS $\mathbb{H}$ is unique, symmetric with $K(s,t) = K(t,s)$ for all $s,t \in E$ and nonnegative definite.*

Proof: Suppose that there are two kernels K_1 and K_2 for $\mathbb{H}$. Then, by the reproducing property

$$f(t) = \langle f, K_1(\cdot,t)\rangle = \langle f, K_2(\cdot,t)\rangle,$$

for all f and all t which means that

$$\langle f, K_1(\cdot,t) - K_2(\cdot,t)\rangle = 0$$

for all f and all t. Thus, $K_1 = K_2$.

Symmetry is a consequence of the reproducing property as

$$K(s,t) = \langle K(\cdot,t), K(\cdot,s)\rangle = \langle K(\cdot,s), K(\cdot,t)\rangle = K(t,s).$$

However, this also means that

$$\begin{aligned}\sum_{i=1}^{n}\sum_{j=1}^{n} a_i a_j K(t_i,t_j) &= \sum_{i=1}^{n}\sum_{j=1}^{n} a_i a_j \langle K(\cdot,t_i), K(\cdot,t_j)\rangle \\ &= \left\langle \sum_{i=1}^{n} a_i K(\cdot,t_i), \sum_{i=1}^{n} a_i K(\cdot,t_i) \right\rangle \ge 0.\end{aligned}$$

□

The following result is known as the Moore–Aronszajn Theorem.

Theorem 2.7.4 *Suppose that $K(s,t), s,t \in E$, is a symmetric and positive definite function. Then, there is a unique Hilbert space $\mathbb{H}(K)$ of functions on E with K as its rk.*

Proof: Set

$$\mathbb{H}_0 := \mathrm{span}\{K(\cdot,t) : t \in E\} = \left\{ \sum_{i=1}^{n} a_i K(\cdot,t_i) : a_i \in \mathbb{R}, t_i \in E, n \in \mathbb{Z}^+ \right\}.$$

Define the bivariate function $\langle\cdot,\cdot\rangle_0$ on $\mathbb{H}_0$ by

$$\left\langle \sum_{i=1}^{m} a_i K(\cdot,s_i), \sum_{j=1}^{n} b_j K(\cdot,t_j) \right\rangle_0 := \sum_{i=1}^{m}\sum_{j=1}^{n} a_i b_j K(s_i,t_j).$$

Clearly, $\langle\cdot,\cdot\rangle_0$ is bilinear and the assumption that K is nonnegative definite implies that $\langle f,f\rangle_0 \geq 0$ for $f \in \mathbb{H}_0$.

To establish that $\langle\cdot,\cdot\rangle_0$ is an inner product, it suffices to verify that $\langle f,f\rangle_0 = 0$ means that $f = 0$. Note that

$$\langle f, K(\cdot,t)\rangle_0 = f(t)$$

for $f \in \mathbb{H}_0$, $t \in E$ and

$$|\langle f,g\rangle_0|^2 \leq \langle f,f\rangle_0\langle g,g\rangle_0$$

if $f, g \in \mathbb{H}_0$. So, if $\langle f,f\rangle_0 = 0$, for $t \in E$,

$$|f(t)|^2 = |\langle f, K(\cdot,t)\rangle_0|^2 \leq \langle f,f\rangle_0 K(t,t) = 0,$$

showing that $f = 0$. Thus, $\langle\cdot,\cdot\rangle_0$ is indeed an inner product on $\mathbb{H}_0$.

Now we proceed to complete $\mathbb{H}_0$. Suppose that $\{f_n\}_{n=1}^{\infty}$ is a Cauchy sequence in $\mathbb{H}_0$. By the Cauchy–Schwarz inequality, for each $t \in E$ and n_1, n_2,

$$\begin{aligned} |f_{n_1}(t) - f_{n_2}(t)| &= |\langle f_{n_1} - f_{n_2}, K(\cdot,t)\rangle_0| \\ &\leq \|f_{n_1} - f_{n_2}\|_0 K^{1/2}(t,t). \end{aligned}$$

This implies that $\{f_n(t)\}_{n=1}^{\infty}$ is a Cauchy sequence in $\mathbb{R}$ and hence has a limit. Thus, for any Cauchy sequence $\{f_n\}_{n=1}^{\infty}$ in $\mathbb{H}_0$, we can define a function f by taking the pointwise limit of f_n. If the limit f also happens to be in $\mathbb{H}_0$, then $\|f_n - f\|_0 \to 0$. To see why this is true, assume without loss of generality that $f = 0$ and consider the identity

$$\|f_m - f_n\|_0^2 = \|f_n\|_0^2 + \|f_m\|_0^2 - 2\langle f_n,f_m\rangle_0.$$

For fixed n, the pointwise convergence of f_m to 0 implies that $\langle f_n,f_m\rangle_0 \to 0$ as $m \to \infty$, which entails that

$$\limsup_{m\to\infty} \|f_m - f_n\|_0^2 = \|f_n\|_0^2 + \limsup_{m\to\infty} \|f_m\|_0^2.$$

As $\{f_n\}$ is Cauchy, letting $n \to \infty$ on both sides shows that $\|f_n\|_0 \to 0$.

Let $\mathbb{H}$ be the collection of functions on E that are the pointwise limits of all Cauchy sequences in $\mathbb{H}_0$. For $f, g \in \mathbb{H}$, suppose that $\{f_n\}_{n=1}^{\infty}$ and $\{g_n\}_{n=1}^{\infty}$ are arbitrary Cauchy sequences in $\mathbb{H}_0$ that converge to f and g pointwise. It is easy to see that $\langle f_n, g_n\rangle_0$ is also a Cauchy sequence and has a limit. Moreover, if $\{\tilde{f}_n\}_{n=1}^{\infty}$ and $\{\tilde{g}_n\}_{n=1}^{\infty}$ are another pair of Cauchy sequences in $\mathbb{H}_0$ that converge to f and g pointwise, then $f_n - \tilde{f}_n$ and $g_n - \tilde{g}_n$ both converge to 0 pointwise and we know from the previous paragraph that they both converge to 0 in norm in $\mathbb{H}_0$. This implies that the limit of $\langle f_n, g_n \rangle_0$ only depends on f, g. Thus, the bivariate function

$$\langle f, g \rangle := \lim_{n\to\infty} \langle f_n, g_n \rangle_0$$

is well defined. It is straightforward to establish that $\langle \cdot, \cdot \rangle$ is an inner product and $\mathbb{H}$ is an RKHS with rk. K.

Finally, we demonstrate that $\mathbb{H}$ with this choice for its inner product must be the only Hilbert space for which K is the rk. Indeed, suppose to the contrary that there is another Hilbert space $\mathbb{G}$ with inner product $\langle \cdot, \cdot \rangle_{\mathbb{G}}$ for which K is the rk. Then, it is easy to see that $\mathbb{G}$ contains $\mathbb{H}_0$ and hence $\mathbb{H}$ as subspaces. By Theorem 2.5.5, we can write

$$\mathbb{G} = \mathbb{H} \oplus \mathbb{H}^{\perp}.$$

For any $f \in \mathbb{H}^{\perp}$, we have

$$f(t) = \langle f, K(\cdot, t) \rangle_{\mathbb{G}} = 0$$

for $t \in E$ which shows that $\mathbb{H}^{\perp} = \{0\}$. □

The notation $\mathbb{H}(K)$ in Theorem 2.7.4 will denote the RKHS with rk. K throughout the book. A fundamental implication of the proof of Theorem 2.7.4 is that linear combinations of the kernel functions that correspond to finitely many elements from E are dense in $\mathbb{H}(K)$. With that in mind, we can establish the following result.

Theorem 2.7.5 *If E is a separable metric space and K is continuous on $E \times E$, $\mathbb{H}(K)$ is separable and the functions in $\mathbb{H}(K)$ are continuous on E.*

Proof: By assumption, there is a countable set $\{t_1, t_2, \ldots\}$ that is dense in E. If $f \in \mathbb{H}(K)$,

$$|f(t) - f(s)| = |\langle f, K(\cdot, t) - K(\cdot, s) \rangle| \le \|f\| \|K(\cdot, t) - K(\cdot, s)\|.$$

However,

$$\|K(\cdot, t) - K(\cdot, s)\|^2 = K(t, t) - 2K(t, s) + K(s, s) \tag{2.30}$$

which converges to zero as $t \to s$ by the continuity of K.

By (2.30) and the proof of Theorem 2.7.4, the collection of functions

$$\left\{ \sum_{i=1}^{n} a_i K(\cdot, t_i) \, : \, a_i \in \mathbb{Q}, n \in \mathbb{Z}^+ \right\},$$

with $\mathbb{Q}$ denoting the set of rationals, is dense in $\mathbb{H}(K)$. This implies that $\mathbb{H}(K)$ is separable. □

A property of RKHSs that sets them apart from many function spaces is that norm convergence of functions in the RKHS entails point-wise convergence as well.

Theorem 2.7.6 *Let $\mathbb{H}$ be an RKHS containing functions on E. If $f, f_1, f_2, \ldots$ are functions in $\mathbb{H}$ such that* $\lim_{n\to\infty} \|f_n - f\| = 0$, *then* $\lim_{n\to\infty} |f_n(t) - f(t)| = 0$ *for all $t \in E$. The convergence is uniform if* $\sup_{t\in E} K(t,t) < \infty$.

Proof: By the reproducing property and the Cauchy–Schwarz inequality

$$|f_n(t) - f(t)| \leq \|f_n - f\| K^{1/2}(t, t)$$

from which the result follows. □

It turns out that this theorem can be strengthened. Boundedness of the evaluation functionals $\ell_t(t) := f(t), t \in E$ actually characterizes an RKHS as a result of Theorem 3.2.3 in Chapter 3.

The problem that is generally encountered in an fda setting is that we are in possession of a kernel that is a nonnegative definite function, namely, a covariance kernel (cf. Theorem 7.3.1). The interest is then in determining the form of the corresponding RKHS. The following result provides a useful tool for making such determinations. It is typically referred to as the *integral representation theorem*; see, e.g., Parzen (1970).

Theorem 2.7.7 *Let K be a function over the set $E \times E$ which admits the representation*

$$K(t, t') = \int_S g(t, s) g(t', s) d\mu(s) \tag{2.31}$$

for $t, t' \in E$, where $(S, \mathscr{B}, \mu)$ is a measure space and $\{g(t, \cdot) : t \in E\}$ a collection of functions in $\mathbb{L}^2(S, \mathscr{B}, \mu)$. Then, the RKHS $\mathbb{H}$ corresponding to K consists of all functions of the form

$$f(t) = \int_S F(s) g(t, s) d\mu(s) \tag{2.32}$$

for some unique element $F \in \overline{\text{span}\{g(t,\cdot) : t \in E\}} \subset \mathbb{L}^2(S, \mathscr{B}, \mu)$. *The RKHS norm for* $f \in \mathbb{H}$ *is*

$$\|f\| = \|F\|_2 \tag{2.33}$$

with $\|\cdot\|_2$ *and* $\langle \cdot, \cdot \rangle_2$ *the* $L^2(S, \mathscr{B}, \mu)$ *norm and inner products.*

Proof: It suffices to observe that

$$K(t, t') = \langle K(t, \cdot), K(t', \cdot) \rangle = \langle g(t, \cdot), g(t', \cdot) \rangle_2$$

which establishes a congruence relationship between $\mathbb{H}$ and $\overline{\text{span}\{g(t,\cdot) : t \in E\}}$ as a result of Theorem 2.4.16. If $f \in \mathbb{H}$ is the image of $F \in \overline{\text{span}\{g(t,\cdot) : t \in E\}}$ under this congruence, the reproducing property gives

$$f(t) = \langle f, K(t, \cdot) \rangle = \langle F, g(t, \cdot) \rangle_2$$

as was to be shown. □

Example 2.7.8 *A simple, but fundamentally important, example of an RKHS can be obtained when E is finite-dimensional. Thus, let* $E = \{t_1, \ldots, t_p\}$ *in which case the kernel K is equivalent to the matrix* $\mathscr{K} = \{K(t_i, t_j)\}_{i,j=1:p}$.

The RKHS is now found to be the set of functions on E of the form

$$f(\cdot) = \sum_{i=1}^{p} a_i K(\cdot, t_i),$$

where $(a_1, \ldots, a_p)$ *is perpendicular to the null space of* $\mathscr{K}$*: i.e., the set of vectors a for which* $\mathscr{K}a = 0$. *Note that* $f(\cdot)$ *can take on only p values which means that it has a p-vector representation as* $(f(t_1), \ldots, f(t_p))^T = \mathscr{K}a$ *for* $a = (a_1, \ldots, a_p)^T$. *For notational clarity in this instance, we will use* $f(\cdot)$ *to indicate its representation as a function on E and f to denote its vector form. With that convention, the inner product between* $f_1(\cdot), f_2(\cdot) \in \mathbb{H}(K)$ *is*

$$\langle f_1(\cdot), f_2(\cdot) \rangle = f_1^T \mathscr{K}^- f_2 \tag{2.34}$$

with $\mathscr{K}^-$ *any generalized inverse of* $\mathscr{K}$*: i.e., any matrix that satisfies* $\mathscr{K}\mathscr{K}^-\mathscr{K} = \mathscr{K}$. *The Moore–Penrose generalized inverse of* $\mathscr{K}$ *(see, e.g., Section 3.5) is one possible choice for* $\mathscr{K}^-$ *and, of course, we use* $\mathscr{K}^- = \mathscr{K}^{-1}$ *when* $\mathscr{K}$ *is invertible.*

Example 2.7.9 *An elementary illustration of the integral representation theorem arises from the kernel*

$$\begin{aligned} K(s, t) &= \min(s, t) \\ &= \int_0^1 (t - u)_+^0 (s - u)_+^0 du \end{aligned}$$

defined on $[0, 1] \times [0, 1]$ *with* x_+ *being 0 for* $x \leq 0$ *and* x *otherwise. The RKHS* $\mathbb{H}(K)$ *generated by this kernel consists of all functions of the form*

$$f(t) = \int_0^1 (t-u)_+^0 F(u)du = \int_0^t F(u)du$$

with

$$\|f\|^2 = \int_0^1 F^2(u)du.$$

Another way to describe $\mathbb{H}(K)$ *is as the set of all absolutely continuous (with respect to Lebesgue measure) functions that have square integrable derivatives. More general spaces of this variety are the topic of the following section.*

It is also of interest to consider what transpires when we add or difference reproducing kernels. In the case of sums of rks, we state the following result from Aronszajn (1950).

Theorem 2.7.10 *Let* K_1, K_2 *be nonnegative kernels with RKHSs* $\mathbb{H}(K_i), i = 1, 2$ *that have norms* $\|\cdot\|_i, i = 1, 2$*. Then,* $K = K_1 + K_2$ *is the rk of the set of all functions of the form* $f_1 + f_2$ *with* $f_i \in \mathbb{H}(K_i), i = 1, 2$ *under the norm*

$$\|f\|^2 = \min_{f_i \in \mathbb{H}(K_i), i=1,2 : f = f_1 + f_2} \{\|f_1\|_1^2 + \|f_2\|_2^2\}. \tag{2.35}$$

Now let us turn to the case of kernel differences. In that instance, we will write

$$K_1 \ll K_2$$

if $K_2 - K_1$ is nonnegative definite in the sense of (2.29).

Theorem 2.7.11 *If* $K_1 \ll K_2$*,* $\mathbb{H}(K_1) \subset \mathbb{H}(K_2)$ *and the norms* $\|\cdot\|_1$ *and* $\|\cdot\|_2$ *for* $\mathbb{H}(K_1)$ *and* $\mathbb{H}(K_2)$ *satisfy* $\|f_1\|_2 \leq \|f_1\|_1$ *for every* $f_1 \in \mathbb{H}(K_1)$*.*

Proof: As $K_3 = K_2 - K_1$ is nonnegative definite, Theorem 2.7.4 has the consequence that it generates a Hilbert space $\mathbb{H}(K_3)$ for which it is the rk. However, from Theorem 2.7.10, this means that $K_2 = K_3 + K_1$ is the rk for the space of all functions of the form $f_3 + f_1$ with $f_3 \in \mathbb{H}(K_3)$ and $f_1 \in \mathbb{H}(K_1)$. The theorem follows from this fact by taking $f_3 = 0$. □

A much stronger result than Theorem 2.7.11 is the theorem below that was proved by Aronszajn (1950).

Theorem 2.7.12 *Let* K_1, K_2 *be nonnegative kernels. Then,* $\mathbb{H}(K_1) \subset \mathbb{H}(K_2)$ *if there exists a positive constant* B *such that* $K_1 \ll BK_2$*.*

In addition to sums and differences of reproducing kernels, we will also encounter spaces that derive from their products. Thus, suppose that K_1, K_2 are rks for RKHSs $\mathbb{H}_1, \mathbb{H}_2$ consisting of functions on E. Let $\|\cdot\|_i, \langle\cdot,\cdot\rangle_i$ and $\{e_{ij}\}_{j=1}^{\infty}$ be, respectively, the norm and inner product and a CONS for $\mathbb{H}_i, i = 1, 2$. Then, the direct product.

Hilbert space $\mathbb{H} := \mathbb{H}_1 \otimes \mathbb{H}_2$ can be derived in the following manner. For $t = (t_1, t_2) \in E \times E$, we first consider functions of the form

$$g(t) = \sum_{i=1}^{n} g_{1i}(t_1)g_{2i}(t_2) \tag{2.36}$$

with the g_{1i} being functions in $\mathbb{H}_1$ and the g_{2i} deriving from $\mathbb{H}_2$. Then, given $f = \sum_{i=1}^{m} f_{1i}f_{2i}$ for $f_{ij} \in \mathbb{H}_i$, we take the inner product of f and g to be

$$\langle g,f\rangle = \sum_{i=1}^{n}\sum_{j=1}^{m} \langle g_{1i},f_{1j}\rangle_1\langle g_{2i},f_{2j}\rangle_2 \tag{2.37}$$

and, as usual, have $\|g\|^2 = \langle g, g\rangle$. It is not difficult to see that these choices produce a valid norm and inner product for the space of functions of the form (2.36).

To this point, we have succeeded in constructing a pre-Hilbert space. To complete it, we include all functions of the form

$$g = \sum_{i=1}^{\infty}\sum_{j=1}^{\infty} a_{ij}e_{1i}e_{2j} \tag{2.38}$$

such that

$$\|g\|^2 := \sum_{i=1}^{\infty}\sum_{j=1}^{\infty} a_{ij}^2 < \infty.$$

The inner product of g with $f = \sum_{i=1}^{\infty}\sum_{j=1}^{\infty} b_{ij}e_{1i}e_{2j}$ is defined as

$$\langle g,f\rangle = \sum_{i=1}^{\infty}\sum_{j=1}^{\infty} a_{ij}b_{ij}. \tag{2.39}$$

It is clear that sums of finitely many products fall into this new category of functions and that the inner products (2.37) and (2.39) for $\mathbb{H}$ coincide in that instance. With a little additional work, one may verify that every function of type (2.38) admits a representation as a function of type (2.36) and thereby conclude that the formulation provided by (2.38) and (2.39) is, in fact, the completion of our initial pre-Hilbert space construction.

Note that if $g \in \mathbb{H}_1 \otimes \mathbb{H}_2$, for every $s, t \in E$,

$$
\begin{aligned}
|g(s,t)| &\le \sum_{i=1}^{\infty} \sum_{j=1}^{\infty} |a_{ij}||e_{1i}(s)||e_{2j}(t)| \\
&\le \sum_{i=1}^{\infty} |e_{1i}(s)| \left(\sum_{j=1}^{\infty} e_{2j}^2(t) \right)^{1/2} \left(\sum_{j=1}^{\infty} a_{ij}^2 \right)^{1/2}.
\end{aligned}
$$

However,

$$
\begin{aligned}
\sum_{j=1}^{\infty} e_{2j}^2(t) &= \sum_{j=1}^{\infty} \langle e_{2j}, K_2(\cdot, t) \rangle_2 e_{2j}(t) \\
&= K_2(t,t)
\end{aligned}
$$

and, similarly,

$$
\begin{aligned}
\sum_{i=1}^{\infty} |e_{1i}(s)| \left(\sum_{j=1}^{\infty} a_{ij}^2 \right)^{1/2} &\le \left(\sum_{i=1}^{\infty} e_{1i}^2(s) \right)^{1/2} \left(\sum_{i=1}^{\infty} \sum_{j=1}^{\infty} a_{ij}^2 \right)^{1/2} \\
&= \sqrt{K_1(s,s)} \|g\|.
\end{aligned}
$$

Therefore,

$$
|g(s,t)| \le \sqrt{K_1(s,s)} \sqrt{K_2(t,t)} \|g\|.
$$

This means that point evaluation of functions in $\mathbb{H}$ is a bounded linear operation, which, according to Theorem 3.2.3 of Chapter 3, insures that $\mathbb{H}$ is an RKHS. The obvious candidate for the rk is

$$
K(s,t,s',t') := K_1(s,t) K_2(s',t').
$$

For fixed $t, t' \in E$, $K(s,t,s',t')$ is clearly in the space as a function of s, s' because $K_1(\cdot, t) \in \mathbb{H}_1$ and $K_2(\cdot, t') \in \mathbb{H}_2$. As,

$$
\begin{aligned}
g(s,t) &= \sum_{i=1}^{\infty} \sum_{j=1}^{\infty} a_{ij} e_{1i}(s) e_{2j}(t) \\
&= \sum_{i=1}^{\infty} \sum_{j=1}^{\infty} a_{ij} \langle e_{1i}, K_1(\cdot, s) \rangle_1 \langle e_{2j}, K_2(\cdot, t) \rangle_2 \\
&= \sum_{i=1}^{\infty} \sum_{j=1}^{\infty} a_{ij} \langle e_{1i}(\cdot) e_{2j}(\star), K(\cdot, s, \star, t) \rangle
\end{aligned}
$$

we have proved the following theorem.

Theorem 2.7.13 *The direct product of two RKHSs with rks K_1 and K_2 is also an RKHS with K_1K_2 as its rk.*

2.8 Sobolev spaces

Sobolev spaces generally refer to function spaces whose norms involve derivatives. In this section, we look at a particularly simple variety of Sobolev space that consists of univariate functions on the interval $[0, 1]$.

For $q \geq 1$, consider the collection $\mathbb{W}_q[0, 1]$ of functions f on $[0, 1]$ that are $q - 1$ times differentiable, where $f^{(q-1)}$ is absolutely continuous having a derivative $f^{(q)}$ almost everywhere with $f^{(q)} \in \mathbb{L}^2[0, 1]$. Note that $\mathbb{W}_q[0, 1]$ is not complete in the Hilbert space $\mathbb{L}^2[0, 1]$.

Define

$$\phi_i(t) = t^i/i!, \quad i = 0, 1, \ldots,$$

and

$$G_q(t, u) = \frac{(t - u)_+^{q-1}}{(q - 1)!}.$$

For each $f \in \mathbb{W}_q[0, 1]$, we have by Taylor's formula with remainder that

$$f(t) = \sum_{i=0}^{q-1} f^{(i)}(0)\phi_i(t) + \int_0^1 G_q(t, u)f^{(q)}(u)du. \tag{2.40}$$

Thus, a function is in $\mathbb{W}_q[0, 1]$ if and only if it can be expressed as

$$\sum_{i=0}^{q-1} b_i\phi_i(t) + \int_0^1 G_q(t, u)g(u)du \tag{2.41}$$

for some $b_0, \ldots, b_{q-1} \in \mathbb{R}$ and $g \in \mathbb{L}^2[0, 1]$.

There are a number of ways to define an inner product on $\mathbb{W}_q[0, 1]$. The following approach is especially insightful.

Define

$$\mathbb{H}_0 = \text{span}\{\phi_0, \ldots, \phi_{q-1}\}$$

and, for any $f, g \in \mathbb{H}_0$, let

$$\langle f, g\rangle_{\mathbb{H}_0} = \sum_{i=0}^{q-1} f^{(i)}(0)g^{(i)}(0).$$

It is easy to check that $\langle f, g\rangle_{\mathbb{H}_0}$ is an inner product on $\mathbb{H}_0$ and that $\{\phi_0, \ldots, \phi_{q-1}\}$ is an orthonormal basis.

Next consider

$$\mathbb{H}_1 := \left\{ \int_0^1 G_q(t,u)g(u)du : \quad g \in \mathbb{L}^2[0,1] \right\}. \tag{2.42}$$

By (2.40) and (2.41), if $f(t) = \int_0^1 G_q(t,u)g(u)du$ for some $g \in \mathbb{L}^2[0,1]$, then

$$f(0) = f'(0) = \cdots = f^{(q-1)}(0) = 0$$

so that $f^{(q)} = g$. Thus,

$$\langle f, g \rangle_{\mathbb{H}_1} = \int_0^1 f^{(q)}(u) g^{(q)}(u) du$$

provides an inner product on $\mathbb{H}_1$.

Theorem 2.8.1 *The inner-product spaces* $(\mathbb{H}_i, \langle \cdot, \cdot \rangle_{\mathbb{H}_i}), i = 0, 1,$ *are RKHSs with reproducing kernels*

$$K_0(s,t) := \sum_{i=0}^{q-1} \phi_i(s)\phi_i(t),$$

and

$$K_1(s,t) := \int_0^1 G_q(s,u)G_q(t,u)du,$$

respectively.

Proof: The statement about $\mathbb{H}_0$ follows at once from Example 2.7.2 that allows us to focus on $\mathbb{H}_1$. For that case, observe that a sequence $\{f_n\}$ in $\mathbb{H}_1$ is Cauchy if $\{f_n^{(q)}\}$ is Cauchy in $\mathbb{L}^2[0,1]$. As $\mathbb{L}^2[0,1]$ is complete, if $\{f_n\}$ is Cauchy in $\mathbb{H}_1$, then $f_n^{(q)} \to g$ for some g in $\mathbb{L}^2[0,1]$. This, in turn, is equivalent to

$$f(\cdot) := \int_0^1 G_q(\cdot, u)g(u)du$$

in $\mathbb{H}_1$. Thus, $\mathbb{H}_1$ is complete. The form for the rk in this instance is a consequence of Theorem 2.7.7. □

Now construct $\mathbb{W}_q[0,1]$ as the RKHS with kernel

$$K(s,t) := K_0(s,t) + K_1(s,t). \tag{2.43}$$

As $\mathbb{H}_0 \cap \mathbb{H}_1 = \{0\}$, according to Theorem 2.7.10, the inner product of $h_1 = f_1 + g_1$ and $h_2 = f_2 + g_2$ for $f_1, f_2 \in \mathbb{H}_0$ and $g_1, g_2 \in \mathbb{H}_1$ is

$$\begin{aligned}\langle h_1, h_2\rangle_{\mathbb{W}_q[0,1]} &= \langle f_1 + g_1, f_2 + g_2\rangle_{\mathbb{W}_q[0,1]} \\ &= \langle f_1, f_2\rangle_{\mathbb{H}_0} + \langle g_1, g_2\rangle_{\mathbb{H}_1}.\end{aligned}$$

This shows that $\mathbb{H}_0$ and $\mathbb{H}_1$ are orthogonal subspaces of $\mathbb{W}_q[0, 1]$. This space is extremely important in approximation theory due, in part, to its connection to spline functions. See, e.g., Section 6.6.

Now consider another inner product for $\mathbb{W}_q[0, 1]$: namely,

$$\langle f, g\rangle'_{\mathbb{W}_q[0,1]} := \langle f, g\rangle_2 + \langle f^{(q)}, g^{(q)}\rangle_2 \tag{2.44}$$

with $\langle\cdot,\cdot\rangle_2$ and $\|\cdot\|_2$ the $\mathbb{L}^2[0, 1]$ inner product and norm. The norm associated with $\langle\cdot,\cdot\rangle'_{\mathbb{W}_q[0,1]}$ is similarly denoted by $\|\cdot\|'_{\mathbb{W}_q[0,1]}$.

Theorem 2.8.2 *$\|f\|_{\mathbb{W}_q[0,1]}$ and $\|f\|'_{\mathbb{W}_q[0,1]}$ are equivalent norms.*

Proof: Write $f \in \mathbb{W}_q[0, 1]$ as

$$f = \sum_{i=0}^{q-1} b_i\phi_i(t) + \int_0^1 G_q(t, u)f^{(q)}(u)du$$

and observe that

$$\|f\|^2_{\mathbb{W}_q[0,1]} = \sum_{i=0}^{q-1} b_i^2 + \|f^{(q)}\|_2^2.$$

It is easy to see that $\|f\|_2^2 \le C\|f\|^2_{\mathbb{W}_q[0,1]}$ for some finite constant C and, hence, that $\|f\|'^2_{\mathbb{W}_q[0,1]} \le (C + 1)\|f\|^2_{\mathbb{W}_q[0,1]}$.

On the other hand,

$$\begin{aligned}\left\|\sum_{i=0}^{q-1} b_i\phi_i(t)\right\|_2^2 &\le 2\|f\|_2^2 + 2\left\|\int_0^1 G_q(t, u)f^{(q)}(u)du\right\|_2^2 \\ &\le C\|f\|'^2_{\mathbb{W}_q[0,1]}\end{aligned}$$

also for some finite C. Now,

$$\left\|\sum_{i=0}^{q-1} b_i\phi_i(t)\right\|_2^2 = \sum_{i=0}^{q-1}\sum_{j=0}^{q-1} \frac{b_ib_j}{i!j!(i + j + 1)} = b^T\mathscr{A}b,$$

where $b = (b_0, \dots, b_{q-1})^T$ and $\mathscr{A} = \{(i!j!(i+j+1))^{-1}\}$. If there were an $a = (a_0, \dots, a_{q-1})^T$ such that $a^T \mathscr{A} a = 0$ that would mean that $\sum_{i=0}^{q-1} a_i \phi_i(t) \equiv 0$. So, $\mathscr{A}$ is positive-definite and its smallest eigenvalue, λ, must be strictly positive. Consequently,

$$\left\| \sum_{i=0}^{q-1} b_i \phi_i(t) \right\|_2^2 \geq \lambda b^T b = \lambda \sum_{i=1}^{q-1} b_i^2.$$

These derivations show that $\sum_{i=1}^{q-1} b_i^2 \leq C \|f\|'^2_{\mathbb{W}_q[0,1]}$ for some finite C from which we can conclude that $\|f\|^2_{\mathbb{W}_q[0,1]} \leq (C+1)\|f\|'^2_{\mathbb{W}_q[0,1]}$. □

Note that the kernel K_0 is no longer the rk for H_0 and, as a result, K is not the rk for $\mathbb{W}_q[0,1]$ under this alternative norm. This is easily rectified by taking

$$K_0'(s,t) = \sum_{j=1}^{q-1} p_j(s) p_j(t) \tag{2.45}$$

with $p_0, \dots, p_{q-1}$ the Legendre polynomials that one obtains by applying the Gram–Schmidt algorithm to $\phi_0, \dots, \phi_{q-1}$ using the $\mathbb{L}^2[0,1]$ norm. The rk for $\mathbb{W}_q[0,1]$ that results from this is

$$K'(s,t) = K_0'(s,t) + K_1(s,t) \tag{2.46}$$

with K_1 from Theorem 2.8.1 as before and K_0' in (2.45).

To conclude this section, we construct a CONS for $\mathbb{W}_q[0,1]$ under the inner product in (2.44). A first step in that direction is

Theorem 2.8.3 *There exist a complete orthonormal sequence* $\{e_j\}_{j=1}^{\infty}$ *for* $\mathbb{L}^2[0,1]$ *such that for* $i, j \geq 1$

$$\langle e_i^{(q)}, e_j^{(q)} \rangle_2 = \gamma_i \delta_{ij}$$

for values $0 = \gamma_1 = \cdots = \gamma_m < \gamma_{q+1} < \cdots$ *that, for constants* $C_1, C_2 \in (0, \infty)$*, satisfy*

$$C_1 j^{2q} \leq \gamma_{j+q} \leq C_2 j^{2q}, \quad j \geq 1.$$

Proof: We merely sketch some aspects of the proof. A more detailed development is available in Utreras (1988).

Let f be $2q$ times differentiable and satisfy $f^{(j)}(0) = f^{(j)}(1) = 0, q \leq j \leq 2q - 1$. Performing integration by parts q times, we obtain

$$\|f^{(q)}\|_2^2 = \langle f, (-1)^q D^{2q} f \rangle_2.$$

Now consider the solution, with respect to γ and f of the differential equation,

$$(-1)^q D^{2q} f = \gamma f$$

subject to

$$f^{(j)}(0) = f^{(j)}(1) = 0, q \leq j \leq 2q - 1.$$

This gives rise to a sequence $\{(\gamma_j, e_j)\}_{j=1}^{\infty}$ satisfying

$$\langle e_i, e_j \rangle_2 = \delta_{ij},$$
$$\langle e_i^{(q)}, e_j^{(q)} \rangle_2 = \gamma_j \delta_{ij}$$

for $\gamma_j > 0$.

To illustrate the idea consider the case of $q = 1$. Then, the general solution for

$$-D^2 f = \gamma f$$

is

$$f(t) = a \sin(\sqrt{\gamma} t) + b \cos(\sqrt{\gamma} t)$$

for $a, b \in \mathbb{R}$. Now,

$$f'(0) = a\sqrt{\gamma} = 0$$

implies that $a = 0$ and

$$f'(1) = -b \sin(\sqrt{\gamma}) = 0$$

leads to $\gamma = (j\pi)^2$ for $j = 1, \ldots$ If, however, $\gamma = 0$, the general solution is $f(x) = a + ax$ and the boundary conditions imply that $b = 0$. Thus, the solutions are $e_1 = 1$ and

$$e_j(t) = \sqrt{2} \cos((j-1)\pi t)$$

for $j \geq 2$. By Theorem 2.4.18, $\{e_j\}_{j=1}^{\infty}$ is a complete orthonormal basis for $\mathbb{L}^2[0, 1]$. □

Since $\|e_j^{(q)}\|_2 = 0$ for $j = 1, \ldots, q$, $e_1, \ldots, e_q$ must be an orthonormal basis for the polynomials of order q (degree $q - 1$). Thus, $e_1, \ldots, e_q$ is an orthonormal basis for $\mathbb{H}_0$ under the (2.44) norm. However, the functions $e_j, j \geq q + 1$, are not in $\mathbb{H}_1$ as the boundary conditions $e_j^{(k)}(0) = 0, 0 \leq k \leq q - 1$, are not satisfied. Observe that

$$\langle e_i, e_j \rangle'_{\mathbb{W}_q[0,1]} = \langle e_i, e_j \rangle_2 + \langle e_i^{(q)}, e_j^{(q)} \rangle_2 = (1 + \gamma_i)\delta_{ij}. \tag{2.47}$$

Thus, the e_j provide orthogonal functions in $\mathbb{W}_q[0,1]$. As each function in $\mathbb{W}_q[0,1]$ is a function in $\mathbb{L}^2[0,1]$, we can also conclude that $\{e_j\}_{j=1}^{\infty}$ is a basis for $\mathbb{W}_q[0,1]$.

Theorem 2.8.4 *The sequence* $\{e_j/(1+\gamma_j)^{1/2}\}_{j=1}^{\infty}$ *is a CONS for* $\mathbb{W}_q[0,1]$ *under the inner product (2.44) and any* $f \in \mathbb{W}_q[0,1]$ *can be written as*

$$f = \sum_{j=1}^{\infty} (1+\gamma_j)^{-1/2} f_j e_j \tag{2.48}$$

for a square summable coefficient sequence $\{f_j\}_{j=1}^{\infty}$.

The e_j are, of course, all in $C[0,1]$, which means that $\max |e_j(t)|$ is bounded for each j. However, we can say much more than that in this particular instance. An application of results in Salaff (1968) shows that the boundary conditions $f^{(j)}(0) = f^{(j)}(1) = 0, q \le j \le 2q-1$ for the differential operator $(-1)^q D^{2q}$ are regular in the sense of Birkhoff (1908). Thus, we can use results from Stone (1926) to conclude that there is a universal $M \in (0,\infty)$ such that

$$\sup_{j\ge 1} \max_{t\in[0,1]} |e_j(t)| \le M; \tag{2.49}$$

i.e., the $e_j(t)$ are uniformly bounded in both t and j.

3

Linear operator and functionals

In Chapter 2, we provided a review of the basic facts concerning Banach and Hilbert spaces. What is of more interest for our purposes is the properties of functions that operate on such space. This chapter proceeds in that direction.

We are primarily interested in transformations that are linear and these tend to come in two varieties: linear functionals and operators, with the former being a special case of the latter. Both topics fall into the realm of mathematics known as *functional analysis*. This is a very broad area and our exposition cannot hope to do it justice. Rather than attempt to do so, we pick and choose topics that we feel are most relevant to fda and, in particular, those that are needed for subsequent chapters.

More complete treatments of functional analysis are available through many sources. For example, Dunford and Schwarz (1988) is a standard reference for linear operator theory; for an elementary introduction, one can consult the text by Rynne and Youngson (2001).

Before proceeding, it is perhaps worthwhile to comment on our use of the word "functional" as a adjective modifier of both "data" and "analysis" throughout this text. Functional analysis derives its name from its foundation in the study of linear functionals on, typically, Banach spaces. Such spaces need have nothing to do with functions per se. In contrast, functional data is, by definition, a collection of (random) functions that need have no direct connection to linear functionals. Such data would perhaps be better served with titles such as "function data," "curve data," and "sample path data". However, "functional data" is now the accepted moniker and we will adhere to that convention. With this particular caveat in mind, there should be no reason for confusion when the "functional" term is invoked.

Theoretical Foundations of Functional Data Analysis, with an Introduction to Linear Operators, First Edition. Tailen Hsing and Randall Eubank.

3.1 Operators

We now investigate the properties of linear transformations (in the sense of Definition 2.2.6) on normed linear spaces. A quick note on notation may be useful. While, as a rule, we indicate the result of applying a function $\mathscr{T}$ to an element x by $\mathscr{T}(x)$, in the case of linear transformations it is customary to suppress the parentheses. Often, an expression such as $\mathscr{T}(x)$ is written as merely $\mathscr{T}x$.

Let $\mathbb{X}_1, \mathbb{X}_2$ be normed linear spaces. Associated with a linear transformation $\mathscr{T}$ that maps from $\mathbb{X}_1$ into $\mathbb{X}_2$ are the spaces

$$\text{Dom}(\mathscr{T}) = \text{the subset of } \mathbb{X}_1 \text{ on which } \mathscr{T} \text{ is defined,} \tag{3.1}$$

$$\text{Im}(\mathscr{T}) = \{\mathscr{T}x : x \in \text{Dom}(\mathscr{T})\} \tag{3.2}$$

and

$$\text{Ker}(\mathscr{T}) = \{x \in \text{Dom}(\mathscr{T}) : \mathscr{T}x = 0\} \tag{3.3}$$

called, respectively, the *domain*, *range* (or *image*), and *kernel* (or *null space*) of $\mathscr{T}$. We will always assume that $\text{Dom}(\mathscr{T})$ is a linear space, which implies that $\text{Im}(\mathscr{T})$ is also a linear space. Unless otherwise noted, we will take $\text{Dom}(\mathscr{T})$ as the entire $\mathbb{X}_1$ space. The *rank* of an operator $\mathscr{T}$ is defined to be

$$\text{rank}(\mathscr{T}) = \dim(\text{Im}(\mathscr{T}))$$

which may be infinite.

The objective of our study is not just linear transformations but rather those that are bounded in the sense of the following definition.

Definition 3.1.1 *Suppose that $\mathbb{X}_1, \mathbb{X}_2$ are normed vector spaces with norms $\|\cdot\|_i, i = 1, 2$. A linear transformation $\mathscr{T}$ from $\mathbb{X}_1$ to $\mathbb{X}_2$ is bounded if there exists a finite constant $C > 0$ such that*

$$\|\mathscr{T}x\|_2 \le C\|x\|_1$$

for all $x \in \mathbb{X}_1$.

Boundedness has rather profound consequences as a result of the following theorem.

Theorem 3.1.2 *A linear transformation between two normed spaces is uniformly continuous if it is bounded.*

Proof: Let $\mathbb{X}_1$ and $\mathbb{X}_2$ be normed linear spaces with norms $\|\cdot\|_i, i = 1, 2$ and let $\mathscr{T}$ be a linear transformation between the two spaces. If $\mathscr{T}$ is uniformly continuous, then it is continuous at 0 from which it follows that there is a universal $\delta > 0$ such that $\|\mathscr{T}x\|_2 \leq 1$ whenever $\|x\|_1 \leq \delta$. Thus, for example, with any $x \neq 0$, we will have

$$\|\mathscr{T}x\|_2 = \|\mathscr{T}(\delta x/\|x\|_1)\|_2\|x\|_1/\delta \leq \|x\|_1/\delta.$$

For the converse, if $x_n \to x$, the fact that $\|\mathscr{T}(x - x_n)\|_2 \leq C\|x - x_n\|_1$ means that $\mathscr{T}x_n \to \mathscr{T}x$. As C is independent of x, the continuity is uniform. □

Theorem 3.1.2 means that the phrases "continuous linear transformation" and "bounded linear transformation" are synonymous. We will use $\mathfrak{B}(\mathbb{X}_1, \mathbb{X}_2)$ to denote the set of all bounded (and, hence, continuous) linear transformations from $\mathbb{X}_1$ to $\mathbb{X}_2$. This becomes a normed space under the *operator norm*

$$\|\mathscr{T}\| = \sup_{x \in \mathbb{X}_1, \|x\|_1 = 1} \|\mathscr{T}x\|_2. \tag{3.4}$$

Then, for any $x \in \mathbb{X}_1$,

$$\|\mathscr{T}x\|_2 \leq \|\mathscr{T}\|\,\|x\|_1. \tag{3.5}$$

The elements of $\mathfrak{B}(\mathbb{X}_1, \mathbb{X}_2)$ are called bounded linear operators, linear operators, or operators. If $\mathbb{X}_1 = \mathbb{X}_2 = \mathbb{X}$, we use $\mathfrak{B}(\mathbb{X})$ for the set of all bounded operators on $\mathbb{X}$.

Theorem 3.1.3 *Let $\mathbb{X}_1$ and $\mathbb{X}_2$ be normed linear spaces. If $\mathbb{X}_2$ is complete, then $\mathfrak{B}(\mathbb{X}_1, \mathbb{X}_2)$ with norm (3.4) is a Banach space.*

Proof: Let $\{\mathscr{T}_n\}$ be a Cauchy sequence in $\mathfrak{B}(\mathbb{X}_1, \mathbb{X}_2)$. For fixed $x \in \mathbb{X}_1$, consider the sequence $\{\mathscr{T}_n x\}$ in $\mathbb{X}_2$. Applying (3.5) we see that $\{\mathscr{T}_n x\}$ is a Cauchy sequence in $\mathbb{X}_2$. By the completeness of $\mathbb{X}_2$, $\mathscr{T}_n x$ has a limit which we denote by $\mathscr{T}x$. It is obvious that $\mathscr{T}$ is linear. As $\{\mathscr{T}_n\}$ is Cauchy, for any $\epsilon > 0$, there exists an N_ϵ such that $\sup_{n,m \geq N_\epsilon} \|\mathscr{T}_n - \mathscr{T}_m\| < \epsilon/2$. In addition, for any x with $\|x\|_1 = 1$ there exists an $m(x) \geq N_\epsilon$ such that $\|\mathscr{T}_{m(x)}x - \mathscr{T}x\|_2 < \epsilon/2$. Thus, if $n \geq N_\epsilon$,

$$\|\mathscr{T}_n x - \mathscr{T}x\|_2 \leq \|\mathscr{T}_n x - \mathscr{T}_{m(x)}x\|_2 + \|\mathscr{T}_{m(x)}x - \mathscr{T}x\|_2 < \epsilon. \tag{3.6}$$

By the triangle inequality,

$$\|\mathscr{T}x\|_2 \leq \|\mathscr{T}_n x\|_2 + \epsilon,$$

which shows that $\mathscr{T}$ is bounded. Another application of (3.6) shows that $\mathscr{T}_n$ converges to $\mathscr{T}$ in operator norm. □

Example 3.1.4 *Let $\mathbb{X}_1 = \mathbb{R}^p$ and $\mathbb{X}_2 = \mathbb{R}^q$ with $\|\cdot\|$ the Euclidean norm for either space. Then, any linear transformation $\mathscr{T}$ from $\mathbb{R}^p$ to $\mathbb{R}^q$ can be represented as a $q \times p$ matrix/array $\mathscr{T} = \{\tau_{ij}\}_{i=1:q,,j=1:p}$ of real numbers and*

$$\|\mathscr{T}\|^2 = \max_{x^T x=1} x^T \mathscr{T}^T \mathscr{T} x$$

for $x = (x_1, \ldots, x_p)^T$, the transpose of the p-vector x and $\mathscr{T}^T = \{\tau_{ji}\}_{i=1:q,j=1:p}$ the transpose of $\mathscr{T}$. We will see eventually that when $\mathscr{T}$ is symmetric $\|\mathscr{T}\|$ is the absolute value of the eigenvalue of $\mathscr{T}$ with largest magnitude. Somewhat more generally, $\|\mathscr{T}\|$ is the largest singular value of the matrix $\mathscr{T}$.

Example 3.1.5 *Consider the space $C[0, 1]$ equipped with the sup norm in Example 2.1.3. Let t_0 be an arbitrary point in $[0, 1]$ and define the mapping*

$$\mathscr{T} : f \to f(t_0)$$

from $\mathbb{X}_1 := C[0, 1]$ into $\mathbb{X}_2 := \mathbb{R}$. Clearly, $\mathscr{T}$ is linear and

$$\|\mathscr{T}f\|_2 = |f(t_0)| \leq \sup_{t\in[0,1]} |f(t)| = \|f\|_1$$

with equality holding for constant functions. This establishes that $\mathscr{T}$ is also bounded with unit (operator) norm.

Example 3.1.6 *Consider the standard $\mathbb{L}^2[0, 1]$ space and the linear mapping defined by*

$$(\mathscr{T}f)(\cdot) = \int_0^1 K(\cdot, u)f(u)du \tag{3.7}$$

for $f \in \mathbb{L}^2[0, 1]$ and some square-integrable function K on $[0, 1] \times [0, 1]$. Operators of this type are called integral operators (Section 4.6) and the function K is called a kernel. Integral operators are bounded because

$$|(\mathscr{T}f)(t)|^2 \leq \int_0^1 K^2(t, u)du \int_0^1 f^2(u)du = \|f\|_2^2 \int_0^1 K^2(t, u)du$$

with $\|\cdot\|_2$ the $\mathbb{L}^2[0, 1]$ norm. Therefore,

$$\|\mathscr{T}f\|_2^2 \leq \|f\|_2^2 \int_0^1 \int_0^1 K^2(t, u)dudt.$$

The operator norm for $\mathscr{T}$ in this instance may be deduced from Theorem 4.3.4.

In Section 2.6, we defined the Bochner integral that extended the concept of Lebesgue integration to integration over a Banach space. Given a function

f on a measure space $(E, \mathscr{B}, \mu)$ that takes on values in a Banach space $\mathbb{X}$, the Bochner integral of f over E was defined as

$$\int_E f d\mu = \lim_{n\to\infty} \int_E f_n d\mu,$$

where f_n is a sequence of simple functions in the sense of Definition 2.6.1 such that $\lim_{n\to\infty} \int_E \|f_n - f\| d\mu = 0$: i.e., (2.19) is satisfied. We know that Bochner integration is a linear operation from which one might have guessed that the following result would be true.

Theorem 3.1.7 *Let $\mathbb{X}_1, \mathbb{X}_2$ be Banach spaces, f a Bochner integrable function from E to $\mathbb{X}_1$ and $\mathscr{T} \in \mathfrak{B}(\mathbb{X}_1, \mathbb{X}_2)$. Then, $\mathscr{T}f$ is Bochner integrable and*

$$\mathscr{T}\left(\int_E f d\mu\right) = \int_E \mathscr{T} f d\mu.$$

Proof: Let $\{f_n\}$ be a sequence of simple functions that satisfies (2.19). Then, $\mathscr{T}f_n$ is also simple and $\int_E \mathscr{T}f_n d\mu = \mathscr{T}\int_E f_n d\mu$ for all n. Now $\mathscr{T}\int_E f_n d\mu$ (and, hence, $\int_E \mathscr{T}f_n d\mu$) converges to $\mathscr{T}\int_E f d\mu$ by the continuity of $\mathscr{T}$. Approaching matters from the other direction we have

$$\int_E \|\mathscr{T}f - \mathscr{T}f_n\|_2 d\mu \leq \int_E \|\mathscr{T}\| \|f - f_n\|_1 d\mu \to 0.$$

Thus, $\mathscr{T}f$ is Bochner integrable and its Bochner integral is the limit of $\int_E \mathscr{T}f_n d\mu$. □

We say that $\tilde{\mathscr{T}}$ is an extension of a linear transformation $\mathscr{T}$ if $\mathrm{Dom}(\mathscr{T}) \subseteq \mathrm{Dom}(\tilde{\mathscr{T}})$ and $\tilde{\mathscr{T}}x = \mathscr{T}x$ for all $x \in \mathrm{Dom}(\mathscr{T})$. A linear transformation between two normed vector spaces $\mathbb{X}_1$ and $\mathbb{X}_2$ is said to be *densely defined* if $\mathrm{Dom}(\mathscr{T})$ is dense in $\mathbb{X}_1$. When $\mathscr{T}$ is bounded, being densely defined turns out to be enough to define it globally in the sense of being able to produce an extension of $\mathscr{T}$ whose domain is all of $\mathbb{X}_1$. This result is sometimes called the *extension principle*.

Theorem 3.1.8 *Let $\mathbb{X}_1, \mathbb{X}_2$ be Banach spaces and suppose that $\mathscr{T}$ is a bounded linear transformation from* $\mathrm{Dom}(\mathscr{T}) \subseteq \mathbb{X}_1$ *into $\mathbb{X}_2$. Then, there is a unique extension of $\mathscr{T}$ to* $\overline{\mathrm{Dom}(\mathscr{T})}$ *that has the same bound.*

Proof: As any closed subset of a Banach space is also a Banach space, we may as well assume that $\overline{\mathrm{Dom}(\mathscr{T})} = \mathbb{X}_1$. Then, the first step is to realize that any $\tilde{\mathscr{T}} \in \mathfrak{B}(\mathbb{X}_1, \mathbb{X}_2)$ is uniquely determined by its values on a dense subset of $\mathbb{X}_1$; if $\mathbb{D}$ is a dense subset of $\mathbb{X}_1$, for any $x \in \mathbb{X}_1$, there is a sequence $\{x_n\}$ in $\mathbb{D}$ that converges to x and $\tilde{\mathscr{T}}x_n$ converges to $\tilde{\mathscr{T}}x$ by continuity. So, if we can find a bounded extension of $\mathscr{T}$ to $\mathbb{X}_1$, it will necessarily be unique.

To actually create the extension let x be any element of $\mathbb{X}_1$ with $\{x_n\}$ an arbitrary sequence from Dom($\mathscr{T}$) having x as its limit. The sequence $\mathscr{T}x_n$ is Cauchy in $\mathbb{X}_2$ because

$$\|\mathscr{T}x_n - \mathscr{T}x_m\|_2 \le C\|x_n - x_m\|_1$$

with C the assumed finite bound for $\mathscr{T}$. Thus, we know that $\mathscr{T}x_n$ must have a limit in $\mathbb{X}_2$ and we simply define $\tilde{\mathscr{T}}x$ to be that limit. Since $\|\mathscr{T}x_n\|_2 \le C\|x_n\|_1$ for all n, $\|\tilde{\mathscr{T}}\| \le C$. □

3.2 Linear functionals

Given a normed space $\mathbb{X}$, a case of special interest is $\mathfrak{B}(\mathbb{X}, \mathbb{R})$. It is called the *dual space* of $\mathbb{X}$ and its elements are called bounded linear functionals or just *linear functionals*.

In the case of Hilbert spaces, the dual space has a particularly simple form as a result of the Riesz Representation Theorem that is stated in the following. We will have many occasions to draw on the power of this result.

Theorem 3.2.1 *Suppose that $\mathbb{H}$ is a Hilbert space with inner-product and norm $\langle\cdot,\cdot\rangle$, $\|\cdot\|$ and $\mathscr{T} \in \mathfrak{B}(\mathbb{H}, \mathbb{R})$. There is a unique element $e_{\mathscr{T}} \in \mathbb{H}$, called the representer of $\mathscr{T}$, with the property that*

$$\mathscr{T}x = \langle x, e_{\mathscr{T}}\rangle$$

for all $x \in \mathbb{H}$ and $\|\mathscr{T}\| = \|e_{\mathscr{T}}\|$.

Proof: If $\mathscr{T}$ maps all elements to 0, take $e_{\mathscr{T}} = 0$. Otherwise, Theorem 2.5.6 tells us that $\text{Ker}(\mathscr{T})^{\perp}$ is a closed subspace from which we may choose an element y with $\mathscr{T}y = 1$ to obtain

$$\mathscr{T}(x - (\mathscr{T}x)y) = \mathscr{T}x - \mathscr{T}x\mathscr{T}y = 0$$

for every $x \in \mathbb{H}$; i.e., $x - \mathscr{T}(x)y \in \text{Ker}(\mathscr{T})$. As $y \in \text{Ker}(\mathscr{T})^{\perp}$, we have $\langle x - \mathscr{T}(x)y, y\rangle = 0$ and therefore

$$\langle x, y\rangle = \mathscr{T}x\langle y, y\rangle = \mathscr{T}x\|y\|^2.$$

So, $e_{\mathscr{T}} = y/\|y\|^2$ has the requisite properties. If we could now find another $e'_{\mathscr{T}}$ to serve as a representer, their difference would satisfy $\langle x, e_{\mathscr{T}} - e'_{\mathscr{T}}\rangle = 0$ for every $x \in \mathbb{H}$. Thus, $e_{\mathscr{T}}$ is unique. □

The theorem has the consequence that a Hilbert space is self-dual. That is, if $\mathbb{H}$ is a Hilbert space, its dual space is isomorphic to $\mathbb{H}$. However, the relationship is stronger than just that because the last statement of the theorem entails that $\mathbb{H}$ and $\mathfrak{B}(\mathbb{H}, \mathbb{R})$ are congruent or isometrically isomorphic.

Example 3.2.2 *Consider the RKHS setting of Section 2.7. There we had a Hilbert space $\mathbb{H}$ of real valued functions on a set E. If $\mathbb{H}$ is an RKHS with inner product $\langle\cdot,\cdot\rangle$, there is an rk with the reproducing property that for every $f \in \mathbb{H}$,*

$$f(t) = \langle f, K(\cdot,t)\rangle$$

for $t \in E$. In this setting, the evaluation functionals

$$\mathcal{T}_t : f \mapsto f(t)$$

are well defined for any $f \in \mathbb{H}$ and $t \in E$. It is also clear that $\mathcal{T}_t$ is linear and that its representer is $K(\cdot,t)$.

Theorem 3.2.3 *$\mathbb{H}$ is an RKHS if all evaluation functionals are bounded.*

Proof: The necessity follows from Theorem 2.7.6. To go the other direction observe that if $\mathcal{T}_t$ is bounded for each $t \in E$, Theorem 3.2.1 tells us that there is a $g_t \in \mathbb{H}$ such that $f(t) = \mathcal{T}_t f = \langle f, g_t\rangle$. Then, $K(s,t) := g_t(s)$ is the rk. □

Example 3.2.4 *In Theorem 3.1.7, let $\mathbb{X}_1$ be a Hilbert space and take $\mathbb{X}_2 = \mathbb{R}$. With these choices, $\mathcal{T}$ is a linear functional on a Hilbert space and Theorem 3.1.7 translates to*

$$\left\langle \int_E f d\mu, g \right\rangle_1 = \int_E \langle f, g\rangle_1 d\mu \tag{3.8}$$

for all $g \in \mathbb{X}_1$.

Definition 3.2.5 *A function f on a normed space $\mathbb{X}$ is called a sublinear functional if for all x, y in $\mathbb{X}$ and $a \in \mathbb{R}$*

1. *$f(x+y) \le f(x) + f(y)$ and*
2. *$f(ax) = af(x)$.*

For example, any norm on $\mathbb{X}$ is a sublinear functional. The following result is known as the Hahn–Banach Extension Theorem. It plays an important role in optimization theory as detailed, for example, in Luenberger (1969).

Theorem 3.2.6 *Let ℓ be a linear functional defined on a subspace $\mathbb{M}$ of a normed linear space $\mathbb{X}$. Suppose that there is a continuous sublinear functional f defined on $\mathbb{X}$ such that $\ell(x) \le f(x)$ for all $x \in \mathbb{M}$. Then, there is a linear functional $\tilde{\ell}$ defined on $\mathbb{X}$ that agrees with ℓ on $\mathbb{M}$ and satisfies $\tilde{\ell}(x) \le f(x)$ for all $x \in \mathbb{X}$.*

Proof: Let x be any element of $\mathbb{X}$ that is not in $\mathbb{M}$ and let y be an arbitrary element of the subspace obtained by translating $\mathbb{M}$ by x: i.e., y is an element of the subspace in $\mathbb{X}$ consisting of vectors of the form $ax + m$ for some $a \in \mathbb{R}$ and $m \in \mathbb{M}$. To extend ℓ to work on y, we need only define $\tilde{\ell}(y) = \ell(m) + ah(x)$ for some function h satisfying $h(x) \le f(x)$. The trick is in establishing the existence of h.

Let m_1, m_2 be arbitrary elements of $\mathbb{M}$ and observe that by assumption

$$\ell(m_1 + m_2) \le f(m_1 + m_2) \le f(m_1 - x) + f(m_2 + x)$$

because f is sublinear on all of $\mathbb{X}$. As ℓ is linear this translates into the inequality

$$\ell(m_1) - f(m_1 - x) \le f(m_2 + x) - \ell(m_2)$$

and, as m_1, m_2 are arbitrary, it must be that

$$\sup_{m\in\mathbb{M}}[\ell(m) - f(m - x)] \le \inf_{m\in\mathbb{M}}[f(m + x) - \ell(m)].$$

We now define the value of $h(x)$ to be any number between $\sup_{m\in\mathbb{M}}[\ell(m) - f(m - x)]$ and $\inf_{m\in\mathbb{M}}[f(m + x) - \ell(m)]$. To see that this works, suppose that $y = ax + m$ for $m \in \mathbb{M}$ and $a > 0$. Then,

$$\tilde{\ell}(y) = \ell(m) + ah(x) = a\left[\ell\left(\frac{m}{a}\right) + h(x)\right] \le af\left(\frac{m}{a} + x\right) = f(y).$$

The case of $a < 0$ is handled similarly. So, we have succeeded in obtaining an extension of ℓ to the manifold that includes $\mathbb{M}$ and x for an arbitrary $x \in \mathbb{X}$ that is not in $\mathbb{M}$.

Now consider the set Θ of all subspace and linear functional pairs $(\mathbb{A}, \ell_A)$ such that $\mathbb{M} \subseteq \mathbb{A}$ and ℓ_A coincides with ℓ on $\mathbb{M}$. We can provide a partial order on Θ by saying that $(\mathbb{A}_1, \ell_{A_1}), (\mathbb{A}_2, \ell_{A_2}) \in \Theta$ satisfy $(\mathbb{A}_1, \ell_{A_1}) \le (\mathbb{A}_2, \ell_{A_2})$ when $\mathbb{A}_1 \subseteq \mathbb{A}_2$ and ℓ_{A_2} coincides with ℓ_{A_1} on $\mathbb{A}_1$. In particular, when either $(\mathbb{A}_1, \ell_{A_1}) \le (\mathbb{A}_2, \ell_{A_2})$ or $(\mathbb{A}_2, \ell_{A_2}) \le (\mathbb{A}_1, \ell_{A_1})$, $(\mathbb{A}_1, \ell_{A_1})$ and $(\mathbb{A}_2, \ell_{A_2})$ are said to be comparable.

Let $\{(\mathbb{A}_\beta, \ell_{A_\beta})\}_{\beta\in B}$ be any collection of comparable sets in Θ and define

$$\mathbb{A} = \cup_{\beta\in B}\mathbb{A}_\beta$$

with $\ell_A := \ell_{A_\beta}$ for all $\beta \in B$. With this formulation $(\mathbb{A}, \ell_A)$ provides an upper bound on the chain $\{(\mathbb{A}_\beta, \ell_{A_\beta})\}_{\beta\in B}$ in that $(\mathbb{A}_\beta, \ell_{A_\beta}) \le (A, \ell_A)$ for all $\beta \in B$. As a result, we may apply Zorn's lemma to conclude that Θ has a maximal element $(\tilde{\mathbb{X}}, \tilde{\ell})$. The result now follows provided that $\tilde{\mathbb{X}} = \mathbb{X}$. However, if that were not the case, we could apply the previous extension argument to any element in $\mathbb{X}$ that was not in $\tilde{\mathbb{X}}$ and obtain a contradiction of maximality. □

Perhaps the most obvious application of Theorem 3.2.6 is for the case where the linear functional in question is bounded. We state this result formally as follows.

Corollary 3.2.7 *If ℓ is a bounded linear functional on a subspace $\mathbb{M}$ of a normed vector space $\mathbb{X}$, there is an extension $\tilde{\ell}$ of ℓ defined on all of $\mathbb{X}$ with the same norm as ℓ on $\mathbb{M}$.*

Proof: To be precise let us define

$$\|\ell\|_M = \sup_{m \in \mathbb{M}} \frac{|\ell(m)|}{\|m\|}.$$

Then, specify the sublinear functional in Theorem 3.2.6 by $f(x) = \|\ell\|_M \|x\|$ for any $x \in \mathbb{X}$. This means that $|\tilde{\ell}(x)| \le \|\ell\|_M \|x\|$ on $\mathbb{X}$ from which it follows that $\|\tilde{\ell}\| = \|\ell\|_M$. □

Corollary 3.2.8 *Let x be an element of a normed vector space $\mathbb{X}$. Then, there is a linear functional ℓ with unit norm for which $\ell(x) = \|x\|$.*

Proof: Define $\ell(ax) = a\|x\|$ for any $a \in \mathbb{R}$. Then, ℓ is a bounded linear functional on span$\{x\}$ and the result follows from Corollary 3.2.7. □

Example 3.1.5 gives one simple instance of a bounded linear functional for the space $C[0, 1]$. More generally, with the aid of the Hahn–Banach Theorem, it becomes possible to characterize the entire dual of $C[0, 1]$ in terms of the function space $BV[0, 1]$ from Example 2.3.9 that contains all functions of bounded variation on $[0, 1]$.

Theorem 3.2.9 *Let $C[0, 1]$ be the space of continuous functions on $[0, 1]$ equipped with the sup norm. A function ℓ is a bounded linear functional on $C[0, 1]$ if and only if*

$$\ell(f) = \int_0^1 f(t)dw(t) \tag{3.9}$$

for some $w \in BV[0, 1]$.

Proof: Let ℓ be a bounded linear functional on $\mathbb{C}[0, 1]$ and define $\mathbb{X}$ to be the set of bounded functions on the interval $[0, 1]$ with associated norm

$$\|f\| = \sup_{x \in [0,1]} |f(x)|.$$

Clearly, $C[0, 1]$ is a proper subset of $\mathbb{X}$. Thus, from Corollary 3.2.7, ℓ has an extension to a bounded linear functional $\tilde{\ell}$ on $\mathbb{X}$ that has the same norm.

Let f be an arbitrary element of $C[0, 1]$ and consider approximating it by

$$f_n(t) = \sum_{i=1}^{n} f(t_{i-1})[g_{t_i}(t) - g_{t_{i-1}}(t)]$$

for some partition $0 = t_0 < t_2 < \cdots < t_n = 1$ and $g_s(t) = (s - t)_+^0$. Now, $f_n \in \mathbb{X}$ and

$$\|f_n - f\| = \max_{1 \le i \le n} \max_{t_{i-1} \le t \le t_i} |f(t) - f(t_{i-1})| \to 0$$

as $n \to \infty$ because f is uniformly continuous.

As $\tilde{\ell}$ is a bounded linear functional on $\mathbb{X}$, $\tilde{\ell}(f_n)$ is well defined and $\tilde{\ell}(f_n) \to \tilde{\ell}(f)$. However, as $f \in C[0, 1]$, $\tilde{\ell}(f) = \ell(f)$. The fact that $\tilde{\ell}$ is linear means that

$$\tilde{\ell}(f_n) = \sum_{i=1}^{n} f(t_{i-1})[w(t_i) - w(t_{i-1})]$$

with $w(s) = \tilde{\ell}(g_s)$.

Now we claim that w belongs to $BV[0, 1]$. To see that this is so begin by observing that

$$\begin{aligned}\sum_{i=1}^{n} |w(t_i) - w(t_{i-1})| &= \sum_{i=1}^{n} \text{sign}\,(w(t_i) - w(t_{i-1}))\,[w(t_i) - w(t_{i-1})] \\ &= \tilde{\ell}\left(\sum_{i=1}^{n} \text{sign}\,(w(t_i) - w(t_{i-1}))\,[g_{t_i} - g_{t_{i-1}}]\right)\end{aligned}$$

which has the implication that

$$\sum_{i=1}^{n} |w(t_i) - w(t_{i-1})| \le \|\tilde{\ell}\| \left\| \sum_{i=1}^{n} \text{sign}\,(w(t_i) - w(t_{i-1}))\,[g_{t_i} - g_{t_{i-1}}] \right\|.$$

However, $\|\tilde{\ell}\| = \|\ell\|$ and

$$\begin{aligned}&\left\| \sum_{i=1}^{n} \text{sign}\,(w(t_i) - w(t_{i-1}))\,[g_{t_i} - g_{t_{i-1}}] \right\| \\ &\quad = \max_{s \in [0,1]} \left| \sum_{i=1}^{n} \text{sign}\,(w(t_i) - w(t_{i-1}))\,[g_{t_i}(s) - g_{t_{i-1}}(s)] \right| \\ &\quad = 1.\end{aligned}$$

So, $TV(w) \le \|\ell\|$ and our claim has been shown to hold.

As w is of bounded variation and f is continuous

$$\sum_{i=1}^{n} f(t_{i-1})[w(t_i) - w(t_{i-1})] \to \int_0^1 f(t)dw(t).$$

That is, $\lim \tilde{\ell}(f_n) = \ell(f)$ is the Riemann–Stieltjes integral of f with respect to w.

All that remains is to prove that (3.9) defines a linear functional. The linear part is obvious and $|\ell(f)| \leq \|f\| TV(w)$ verifies the boundedness condition.□

The dual space $\mathfrak{B}(\mathbb{X}, \mathbb{R})$ induces a topology different than the norm topology on $\mathbb{X}$ through consideration of the following notion of convergence.

Definition 3.2.10 *A sequence $\{x_n\}_{n=1}^{\infty}$ in a Banach space $\mathbb{X}$ converges* weakly *to x if $\ell(x_n) \to \ell(x)$ for every $\ell \in \mathfrak{B}(\mathbb{X}, \mathbb{R})$.*

Strong or norm convergence implies weak convergence as

$$|\ell(x_n) - \ell(x)| = |\ell(x_n - x)| \leq \|\ell\| \|x_n - x\|.$$

However, the converse is not true. Most of our discussion in this text will concern strong convergence and that is the mode of convergence that should be assumed unless it has been explicitly stated to the contrary.

Using Theorem 3.2.1, weak convergence in a Hilbert space $\mathbb{H}$ can be characterized as having

$$\langle x_n, y\rangle \to \langle x, y\rangle$$

for every $y \in \mathbb{H}$. In this case if we also have $\|x_n\| \to \|x\|$,

$$\|x_n - x\|^2 = \|x_n\|^2 - 2\langle x_n, x\rangle + \|x\|^2 \to 0$$

and x_n converges strongly. The main result we will eventually need about weak convergence in Section 6.2 is the following theorem whose proof is a consequence of the Banach–Alaoglu Theorem (e.g., Rudin, 1991).

Theorem 3.2.11 *Let $\{x_n\}_{n=1}^{\infty}$ be a (norm) bounded sequence in a Hilbert space $\mathbb{H}$. Then, there is subsequence $\{x_{n_k}\}_{k=1}^{\infty}$ such that x_{n_k} converges weakly to some element $x \in \mathbb{H}$.*

3.3 Adjoint operator

The discussion in this section is restricted to operators on Hilbert spaces. In that setting, we seek to develop an abstract extension of the concept of the transpose of a matrix that pertains to linear operators in the most familiar Hilbert space $\mathbb{R}^p$.

Theorem 3.3.1 *Let $\mathbb{H}_1, \mathbb{H}_2$ be Hilbert spaces with inner products $\langle\cdot,\cdot\rangle_i, i = 1, 2$. Corresponding to every $\mathscr{T} \in \mathfrak{B}(\mathbb{H}_1, \mathbb{H}_2)$, there is a unique element $\mathscr{T}^*$ of $\mathfrak{B}(\mathbb{H}_2, \mathbb{H}_1)$ determined by the relation*

$$\langle \mathscr{T}x_1, x_2\rangle_2 = \langle x_1, \mathscr{T}^*x_2\rangle_1 \tag{3.10}$$

for all $x_1 \in \mathbb{H}_1, x_2 \in \mathbb{H}_2$.

Proof: Consider $\langle \mathscr{T} x_1, x_2 \rangle_2$ as a function of x_1 for fixed x_2. It is bounded and linear so that the Riesz representation theorem may be applied to see that there exists a unique $y \in \mathbb{H}_1$ such that $\langle \mathscr{T} x_1, x_2 \rangle_2 = \langle x_1, y \rangle_1$. Thus, we take $\mathscr{T}^* x_2 = y$. This definition gives us a linear mapping. To see that it bounded first note that $\mathscr{T}$ is necessarily the adjoint of $\mathscr{T}^*$. Then,

$$\begin{aligned} \|\mathscr{T}^* x_2\|_1^2 &= |\langle x_2, \mathscr{T}\mathscr{T}^* x_2 \rangle_2| \\ &\le \|\mathscr{T}\| \|\mathscr{T}^* x_2\|_1 \|x_2\|_2. \end{aligned}$$

□

Definition 3.3.2 *For any $\mathscr{T} \in \mathfrak{B}(\mathbb{H}_1, \mathbb{H}_2)$, the unique operator $\mathscr{T}^*$ of $\mathfrak{B}(\mathbb{H}_2, \mathbb{H}_1)$ in (3.10) is said to be the adjoint of $\mathscr{T}$. If $\mathbb{H}_1 = \mathbb{H}_2$, $\mathscr{T}$ is called self-adjoint when $\mathscr{T}^* = \mathscr{T}$.*

Example 3.3.3 *Let $\mathbb{H} = \mathbb{R}^p$. Then any member of $\mathfrak{B}(\mathbb{H})$ can be expressed as a $p \times p$ matrix $\mathscr{T} = \{\tau_{ij}\}_{i,j=1:p}$. Clearly,*

$$\langle \mathscr{T} x, y \rangle = x^T \mathscr{T}^T y = \langle x, \mathscr{T}^T y \rangle.$$

Hence, $\mathscr{T}^* = \mathscr{T}^T$ and any symmetric matrix $\mathscr{T}$ determines a self-adjoint operator.

Example 3.3.4 *Consider the $\mathbb{L}^2[0,1]$ integral operator defined in Example 3.1.6. In that case*

$$\langle \mathscr{T} f, g \rangle = \int_0^1 \int_0^1 K(t,s) f(s) g(t) ds dt = \left\langle f, \int_0^1 K(t,\cdot) g(t) dt \right\rangle$$

and, hence,

$$(\mathscr{T}^* g)(s) = \int_0^1 K(t,s) g(t) dt.$$

If K is symmetric then $\mathscr{T}$ is self-adjoint.

Example 3.3.5 *Let $\mathbb{H}(K)$ be an RKHS of functions on E with $\mathscr{T}_t$ the evaluation functional that takes a function f to $f(t)$ for each $t \in E$. Then,*

$$uf(t) = \langle f, \mathscr{T}_t^* u \rangle,$$

for any $u \in \mathbb{R}$ and $f \in \mathbb{H}(K)$. By the reproducing property, this implies that $\mathscr{T}_t^* u = uK(\cdot, t)$.

Using the adjoint, we can obtain a simple proof of the following result.

Theorem 3.3.6 *If $\mathscr{T} \in \mathfrak{B}(\mathbb{H})$ for some Hilbert space $\mathbb{H}$ and $\{x_n\}$ is a sequence that converges weakly to $x \in \mathbb{H}$, then $\mathscr{T}x_n$ converges weakly to $\mathscr{T}x$.*

Proof: For any $y \in \mathbb{H}$

$$\langle \mathscr{T}x_n, y\rangle = \langle x_n, \mathscr{T}^*y\rangle \to \langle x, \mathscr{T}^*y\rangle = \langle \mathscr{T}x, y\rangle.$$

□

The following result collects various properties we will eventually need involving the adjoint operator.

Theorem 3.3.7 *Let $\mathscr{T} \in \mathfrak{B}(\mathbb{H}_1, \mathbb{H}_2)$ for real Hilbert spaces $\mathbb{H}_1, \mathbb{H}_2$. Then,*

1. $(\mathscr{T}^*)^* = \mathscr{T}$,
2. $\|\mathscr{T}^*\| = \|\mathscr{T}\|$,
3. $\|\mathscr{T}^*\mathscr{T}\| = \|\mathscr{T}\|^2$,
4. $\mathrm{Ker}(\mathscr{T}) = (\mathrm{Im}(\mathscr{T}^*))^\perp$,
5. $\mathrm{Ker}(\mathscr{T}^*\mathscr{T}) = \mathrm{Ker}(\mathscr{T})$ *and* $\overline{\mathrm{Im}(\mathscr{T}^*\mathscr{T})} = \overline{\mathrm{Im}(\mathscr{T}^*)}$,
6. $\mathbb{H}_1 = \mathrm{Ker}(\mathscr{T}) \oplus \overline{\mathrm{Im}(\mathscr{T}^*)} = \mathrm{Ker}(\mathscr{T}^*\mathscr{T}) \oplus \overline{\mathrm{Im}(\mathscr{T}^*\mathscr{T})}$, *and*
7. $\mathrm{rank}(\mathscr{T}^*) = \mathrm{rank}(\mathscr{T})$.

Proof: Let $x_i \in \mathbb{H}_i, i = 1, 2$ and observe that $\langle(\mathscr{T}^*)^*x_1, x_2\rangle_2 = \langle x_1, \mathscr{T}^*x_2\rangle_1 = \langle \mathscr{T}x_1, x_2\rangle_2$ to establish part 1. For part 2, first recall from the proof of Theorem 3.3.1 that $\|\mathscr{T}^*x_2\|_1^2 \le \|\mathscr{T}\|\|\mathscr{T}^*x_2\|_1\|x_2\|_2$. Thus, $\|\mathscr{T}^*x_2\|_1 \le \|\mathscr{T}\|\|x_2\|_2$ provided that $\|\mathscr{T}^*x_2\|_1 \ne 0$. Now exchange the roles of $\mathscr{T}$ and $\mathscr{T}^*$ and apply part 1 to obtain the desired result.

From part 2, we know that $\|\mathscr{T}^*\mathscr{T}\| \le \|\mathscr{T}\|^2$. On the other hand,

$$\|\mathscr{T}x_1\|_2^2 = \langle \mathscr{T}^*\mathscr{T}x_1, x_1\rangle_1 \le \|\mathscr{T}^*\mathscr{T}\|\|x_1\|_1^2,$$

which means that $\|\mathscr{T}\|^2 \le \|\mathscr{T}^*\mathscr{T}\|$ and part 3 has been shown.

For part 4, let $x_1 \in \mathrm{Ker}(\mathscr{T})$ in which case $\langle x_1, \mathscr{T}^*x_2\rangle_1 = \langle \mathscr{T}x_1, x_2\rangle_2 = 0$ for all x_2. This entails that x_1 is orthogonal to any element in $\mathrm{Im}(\mathscr{T}^*)$ and must be in $(\mathrm{Im}(\mathscr{T}^*))^\perp$. Going the other direction let $x_1 \in (\mathrm{Im}(\mathscr{T}^*))^\perp$. Then, since $\mathscr{T}^*\mathscr{T}x_1 \in \mathrm{Im}(\mathscr{T}^*)$, we obtain $\|\mathscr{T}x_1\|_2 = \langle x_1, \mathscr{T}^*\mathscr{T}x_1\rangle_1 = 0$.

Next, if $x_1 \in \mathrm{Ker}(\mathscr{T})$, then $x_1 \in \mathrm{Ker}(\mathscr{T}^*\mathscr{T})$. So, suppose $x_1 \in \mathrm{Ker}(\mathscr{T}^*\mathscr{T})$. However, in that event $0 = \langle \mathscr{T}^*\mathscr{T}x_1, x_1\rangle_1 = \|\mathscr{T}x_1\|_2$ and $x_1 \in \mathrm{Ker}(\mathscr{T})$. This proves the first identity of part 5. The second identity follows from the first, part 4, and the third part of Theorem 2.5.6.

By (2.16), part 4 and part 3 of Theorem 2.5.6,

$$\mathbb{H}_1 = \text{Ker}(\mathscr{T}) \oplus (\text{Ker}(\mathscr{T}))^{\perp} = \text{Ker}(\mathscr{T}) \oplus \overline{\text{Im}(\mathscr{T}^*)},$$

which is the first identity in part 6. Now apply part 5 to obtain the second result.

To show part 7, assume first that $\text{rank}(\mathscr{T}) < \infty$ in which case $\text{Im}(\mathscr{T})$ must be finite-dimensional and hence $\overline{\text{Im}(\mathscr{T})} = \text{Im}(\mathscr{T})$. Applying part 6 for $x \in \mathbb{H}_2$, we have $\mathscr{T}^*x = \mathscr{T}^*x'$ where x' is the projection of x on $\text{Im}(\mathscr{T})$. This shows that

$$\text{Im}(\mathscr{T}^*) \subset \mathscr{T}^*(\text{Im}(\mathscr{T})),$$

which implies that $\text{rank}(\mathscr{T}^*) \leq \text{rank}(\mathscr{T}) < \infty$. One can then reverse the roles of $\mathscr{T}$ and $\mathscr{T}^*$ to conclude that $\text{rank}(\mathscr{T}) \leq \text{rank}(\mathscr{T}^*)$. Hence, $\text{rank}(\mathscr{T}) = \text{rank}(\mathscr{T}^*)$ if either $\text{rank}(\mathscr{T})$ or $\text{rank}(\mathscr{T}^*)$ is finite. If, instead, we start with either $\text{rank}(\mathscr{T})$ or $\text{rank}(\mathscr{T}^*)$ being infinite, the same argument shows that the other rank must also be infinite. □

3.4 Nonnegative, square-root, and projection operators

The role of projection matrices in statistics can be appreciated by examination of any book on linear models where they arise in quadratic forms involving random vectors. Projection matrices are, of course, positive semidefinite. More generally, nonnegative matrices occur naturally as variance–covariance matrices for random vectors. The extension of these ideas for fda occurs when finite dimensional vectors become functions. This leads to operators replacing matrices in covariance and other calculations as will be detailed, for example, in Chapter 7. At present, it suffices to merely set the stage by defining the ideas of nonnegative and projection operators.

Definition 3.4.1 *An operator $\mathscr{T}$ on a Hilbert space $\mathbb{H}$ is said to be nonnegative definite (or just nonnegative) if it is self-adjoint and $\langle \mathscr{T}x, x\rangle \geq 0$ for all $x \in \mathbb{H}$. It is positive definite (or just positive) if $\langle \mathscr{T}x, x\rangle > 0$ for all $x \in \mathbb{H}$. For two operators $\mathscr{T}_1, \mathscr{T}_2$, we write $\mathscr{T}_1 \leq \mathscr{T}_2$ (respectively, $\mathscr{T}_1 < \mathscr{T}_2$) if $\mathscr{T}_2 - \mathscr{T}_1$ is nonnegative (respectively, positive) definite.*

Example 3.4.2 *If $\mathscr{T}$ is any element of $\mathfrak{B}(\mathbb{H})$, $\mathscr{T}^*\mathscr{T}$ is nonnegative definite because it is self-adjoint and $\langle \mathscr{T}^*\mathscr{T}x, x\rangle = \|\mathscr{T}x\|^2$.*

A fundamental result we will need about nonnegative operators is that they admit a square root type decomposition.

Theorem 3.4.3 *Let $\mathcal{T} \in \mathfrak{B}(\mathbb{H})$ for a Hilbert space $\mathbb{H}$. If $\mathcal{T}$ is nonnegative, there is a unique nonnegative operator $\mathcal{S} \in \mathfrak{B}(\mathbb{H})$ that satisfies $\mathcal{S}^2 = \mathcal{T}$ and commutes with any operator that commutes with $\mathcal{T}$.*

Proof: Assume without loss of generality that $\|\mathcal{T}\| \leq 1$ so that we also have $\|I - \mathcal{T}\| \leq 1$. The argument relies on the fact that the Maclaurin series expansion for $\sqrt{1-z} = 1 + \sum_{j=1}^{\infty} c_j z^j$ is absolutely convergent for all $|z| \leq 1$ with all the c_j being negative. This has the consequence that the series $\mathcal{S}_n := I + \sum_{j=1}^{n} c_j (I - \mathcal{T})^j$ is Cauchy and must therefore converge to some operator $\mathcal{S}$ in the Banach space $\mathfrak{B}(\mathbb{H})$. Representing $\mathcal{S}$ as $\mathcal{S} = I + \sum_{j=1}^{\infty} c_j (I - \mathcal{T})^j$, we can rearrange terms by absolute convergence to show that $\mathcal{S}^2 = \mathcal{T}$. Now, $\mathcal{S}$ is nonnegative by the fact

$$\langle x, \mathcal{S}x \rangle = 1 + \sum_{j=1}^{\infty} c_j \left\langle x, (I - \mathcal{T})^j x \right\rangle \geq 1 + \sum_{j=1}^{\infty} c_j = 0$$

as $c_j < 0$ and $0 \leq \langle x, (I - \mathcal{T})^j x \rangle \leq 1$. In addition, $\mathcal{S}_n$ commutes with any operator that commutes with $\mathcal{T}$ and this property passes on to the limit $\mathcal{S}$.

It remains to show that $\mathcal{S}$ is unique. Suppose that there is another operator $\tilde{\mathcal{S}}$ with the prescribed properties. Then

$$\begin{aligned}\left(\tilde{\mathcal{S}} - \mathcal{S}\right) \tilde{\mathcal{S}} \left(\tilde{\mathcal{S}} - \mathcal{S}\right) + \left(\tilde{\mathcal{S}} - \mathcal{S}\right) \mathcal{S} \left(\tilde{\mathcal{S}} - \mathcal{S}\right) &= \left(\tilde{\mathcal{S}}^2 - \mathcal{S}^2\right) \left(\tilde{\mathcal{S}} - \mathcal{S}\right) \\ &= 0.\end{aligned}$$

As both operators on the left-hand side of the last expression are nonnegative, they must each be the zero operator. Thus,

$$\left(\tilde{\mathcal{S}} - \mathcal{S}\right) \tilde{\mathcal{S}} \left(\tilde{\mathcal{S}} - \mathcal{S}\right) - \left(\tilde{\mathcal{S}} - \mathcal{S}\right) \mathcal{S} \left(\tilde{\mathcal{S}} - \mathcal{S}\right) = \left(\tilde{\mathcal{S}} - \mathcal{S}\right)^3 = 0.$$

This shows that $\left(\tilde{\mathcal{S}} - \mathcal{S}\right)^n = 0$ for all $n \geq 3$. So, for all $x \in \mathbb{H}$,

$$\left\langle \left(\tilde{\mathcal{S}} - \mathcal{S}\right)^4 x, x \right\rangle = \left\| \left(\tilde{\mathcal{S}} - \mathcal{S}\right)^2 x \right\|^2 = 0$$

which implies that $\left(\tilde{\mathcal{S}} - \mathcal{S}\right)^2 = 0$. Applying the argument again leads to the conclusion that $\tilde{\mathcal{S}} - \mathcal{S} = 0$. □

From this point forward, for any nonnegative operator $\mathcal{T}$, the notation $\mathcal{T}^{1/2}$ will refer to the *square-root operator* $\mathcal{S}$ described in Theorem 3.4.3.

Now recall the projection theorem (Theorem 2.5.2) for Hilbert spaces. If $\mathbb{M}$ is a closed subspace of a Hilbert space $\mathbb{H}$, it stated that for each $x \in \mathbb{H}$, there was a unique element in $\mathbb{M}$ that minimized the norm difference $\|x - y\|$ over all $y \in \mathbb{M}$. Here we denote this minimizing element by $\mathcal{P}_{\mathbb{M}} x$. When viewed as a function on $\mathbb{H}$, $\mathcal{P}_{\mathbb{M}}$ is referred to as the *projection operator* for the subspace $\mathbb{M}$.

Theorem 3.4.4 *If $\mathbb{M}$ is a closed subspace of a Hilbert space $\mathbb{H}$, $\mathscr{P}_{\mathbb{M}}$ is a self-adjoint element of $\mathfrak{B}(\mathbb{H})$ that satisfies $\mathscr{P}_{\mathbb{M}} = \mathscr{P}^2_{\mathbb{M}}$.*

Proof: To see that $\mathscr{P}_{\mathbb{M}}$ is linear we can use the characterization in (2.14): namely, for $x_1, x_2 \in \mathbb{H}, a_1, a_2 \in \mathbb{R}$, and any $y \in \mathbb{M}$,

$$\begin{aligned}\langle a_1\mathscr{P}_{\mathbb{M}}x_1 + a_2\mathscr{P}_{\mathbb{M}}x_2, y\rangle &= a_1\langle\mathscr{P}_{\mathbb{M}}x_1, y\rangle + a_2\langle\mathscr{P}_{\mathbb{M}}x_2, y\rangle \\ &= a_1\langle x_1, y\rangle + a_2\langle x_2, y\rangle \\ &= \langle a_1x_1 + a_2x_2, y\rangle.\end{aligned}$$

Thus, $a_1\mathscr{P}_{\mathbb{M}}x_1 + a_2\mathscr{P}_{\mathbb{M}}x_2 = \mathscr{P}_{\mathbb{M}}(a_1x_1 + a_2x_2)$. The fact that $\mathscr{P}_{\mathbb{M}}$ is self-adjoint is obtained similarly. Finally, for any $x \in \mathbb{H}$, $\mathscr{P}_{\mathbb{M}}x$ is in $\mathbb{M}$. The minimization feature of projection now has the consequence that $\mathscr{P}^2_{\mathbb{M}} = \mathscr{P}_{\mathbb{M}}\mathscr{P}_{\mathbb{M}} = \mathscr{P}_{\mathbb{M}}$. □

Projection operators are obviously nonnegative. One interesting relationship that derives from Theorem 3.4.4 is that

$$\|\mathscr{P}_{\mathbb{M}}x\| = \|\mathscr{P}^2_{\mathbb{M}}x\| \le \|\mathscr{P}_{\mathbb{M}}\|\|\mathscr{P}_{\mathbb{M}}x\|,$$

which shows that $\|\mathscr{P}_{\mathbb{M}}\| \ge 1$. However, as $\|\mathscr{P}_{\mathbb{M}}x\| \le \|x\|$, we arrive at the following corollary.

Corollary 3.4.5 $\|\mathscr{P}_{\mathbb{M}}\| = 1$.

If the subspace $\mathbb{M}$ of $\mathbb{H}$ has dimension one and is spanned by x with $\|x\| = 1$, then $\mathscr{P}_{\mathbb{M}}$ can be written as $x \otimes x$ where

$$(x \otimes x)y = \langle y, x\rangle x$$

for any $y \in \mathbb{H}$. This can be established with Theorem 2.5.2 upon observing $\langle y, x\rangle = \langle (x \otimes x)y, x\rangle, y \in \mathbb{H}$. This is a special case of more general tensor product notation that we will use frequently throughout the text.

Definition 3.4.6 *Let x_1, x_2 be elements of Hilbert spaces $\mathbb{H}_1$ and $\mathbb{H}_2$, respectively. The tensor product operator $(x_1 \otimes_1 x_2) : \mathbb{H}_1 \mapsto \mathbb{H}_2$ is defined by*

$$(x_1 \otimes_1 x_2)y = \langle x_1, y\rangle_1 x_2$$

for $y \in \mathbb{H}_1$. If $\mathbb{H}_1 = \mathbb{H}_2$ we use $\otimes$ in lieu of $\otimes_1$.

Suppose that $\mathbb{H}_i = \mathbb{R}^{p_i}, i = 1, 2$ for some finite positive integers p_2, p_1. Then,

$$x_1 \otimes_1 x_2 = x_2 x_1^T$$

for $x_i \in \mathbb{R}^{p_i}, i = 1, 2$: namely, $x_1 \otimes_1 x_2$ is the vector outer product of x_2 and x_1.

Theorem 3.4.7 *Let* $x_i \in \mathbb{H}_i, i = 1, 2$. *Then*

$$\|x_1 \otimes_1 x_2\| = \|x_2\|_2 \|x_1\|_1.$$

Proof: For $x_1 \neq 0$,

$$\|x_1 \otimes_1 x_2\| = \sup_{\|u\|=1} \|\langle x_1, u\rangle_1 x_2\|_2 \leq \|x_2\|_2 \|x_1\|_1.$$

The inequality is attained when $u = x_1/\|x_1\|_1$. □

3.5 Operator inverses

A linear mapping $\mathscr{T}$ from a vector space $\mathbb{X}_1$ into a vector space $\mathbb{X}_2$ is *one-to-one* if $\text{Ker}(\mathscr{T}) = \{0\}$ and *onto* if $\text{Im}(\mathscr{T}) = \mathbb{X}_2$. When $\mathscr{T}$ is both one-to-one and onto it is said to be *bijective*. Bijective linear mappings are invertible. That is, there exists a linear mapping $\mathscr{T}^{-1}$ from $\mathbb{X}_2$ to $\mathbb{X}_1$ such that $\mathscr{T}^{-1}\mathscr{T}$ and $\mathscr{T}\mathscr{T}^{-1}$ are the identity transformation on their respective spaces.

Suppose now that $\mathbb{X}_1$ and $\mathbb{X}_2$ are Banach spaces and that $\mathscr{T} \in \mathfrak{B}(\mathbb{X}_1, \mathbb{X}_2)$ is bijective. In this event, we know that $\mathscr{T}^{-1}$ exists. However, we must still ask whether or not it is bounded. The answer is in the affirmative and the specific result that is stated here is sometimes referred to as the Banach Inverse Theorem.

Theorem 3.5.1 *Let* $\mathbb{X}_1$ *and* $\mathbb{X}_2$ *be Banach spaces with* $\mathscr{T} \in \mathfrak{B}(\mathbb{X}_1, \mathbb{X}_2)$. *If* $\mathscr{T}^{-1}$ *exists, i.e.,* $\mathscr{T}$ *is bijective from* $\mathbb{X}_1$ *to* $\mathbb{X}_2$, *it is an element of* $\mathfrak{B}(\mathbb{X}_2, \mathbb{X}_1)$

Proof: The core of the proof is the so-called Open Mapping Theorem which states that if $\mathbb{X}_1$ and $\mathbb{X}_2$ are Banach spaces and $\mathscr{T}$ maps $\mathbb{X}_1$ onto $\mathbb{X}_2$, then there exists some $r \in (0, \infty)$ such that $\{y \in \mathbb{X}_2 : \|y\|_2 \leq r\} \subset \{\mathscr{T}x : \|x\|_1 \leq 1\}$. This result is a consequence of the Baire Category Theorem. A detailed proof can be found in, e.g., Rynne and Youngson (2001).

Let us apply the Open Mapping Theorem here. Take any $y \in \mathbb{X}_2$ with $\|y\|_2 \leq 1$ so that $ry \in \{\mathscr{T}x : \|x\|_1 \leq 1\}$. As $\mathscr{T}$ is one-to-one, there exists an $x \in \mathbb{X}_1$ with $\|x\|_1 \leq 1$ such that $\mathscr{T}x = ry$. Thus,

$$\|\mathscr{T}^{-1}y\|_1 = \frac{1}{r}\|\mathscr{T}^{-1}(ry)\|_1 = \frac{1}{r}\|\mathscr{T}^{-1}\mathscr{T}x\|_1 = \frac{1}{r}\|x\|_1 \leq \frac{1}{r},$$

which shows that $\mathscr{T}^{-1}$ is bounded. □

Example 3.5.2 *Define the integral operator*

$$(\mathscr{T}_q f)(\cdot) = \int_0^1 G_q(\cdot, u) f(u) du$$

for

$$G_q(s,u) = \frac{(s-u)_+^{q-1}}{(q-1)!}$$

and $f \in \mathbb{L}^2[0,1]$. *We saw this operator in Section 2.8 in our discussion of the Sobolev space* $\mathbb{W}_q[0,1]$. *In particular, we showed that any element* $f \in \mathbb{W}_q[0,1]$ *could be represented as*

$$f = \sum_{j=0}^{q-1} f^{(j)}(0)\phi_j + \mathcal{T}_q f^{(q)}$$

for $\phi_j, j = 0, \ldots, q-1$ *a basis for the polynomials of order* q. *This allowed us to write*

$$\mathbb{W}_q[0,1] = \mathbb{H}_0 \oplus \mathbb{H}_1,$$

where $\mathbb{H}_0 = \mathrm{span}\{\phi_0, \ldots, \phi_{q-1}\}$ *and*

$$\mathbb{H}_1 = \{\mathcal{T}_q g : g \in \mathbb{L}^2[0,1]\}.$$

This latter space is a Hilbert space under the inner product

$$\langle f, g\rangle_1 = \langle f^{(q)}, g^{(q)}\rangle \tag{3.11}$$

with $f, g \in \mathbb{H}_1$ *and* $\langle\cdot,\cdot\rangle$ *the* $\mathbb{L}^2[0,1]$ *inner product. The operator* $\mathcal{T}_q$ *can now be seen as a bijective mapping from* $\mathbb{X}_1 = \mathbb{L}^2[0,1]$ *to* $\mathbb{X}_2 = \mathbb{H}_1$.

Define the linear mapping D by

$$(Df) = f'$$

whenever the operation makes sense. If, for example, we consider functions in $\mathbb{L}^2[0,1]$, *this transformation is not bounded. However,* $D^q = \mathcal{T}_q^{-1}$ *and the Banach Inverse Theorem can be invoked to conclude that* $D^q \in \mathfrak{B}(\mathbb{H}_1, \mathbb{L}^2[0,1])$. *In fact, there is a bit more that can be said about the relationship between* $\mathbb{H}_1$ *and* $\mathbb{L}^2[0,1]$.

Theorem 3.5.3 *The Hilbert spaces* $\mathbb{L}^2[0,1]$ *and* $\mathbb{H}_1$ *are congruent under the mapping* $\Psi := \mathcal{T}_q \in \mathfrak{B}(\mathbb{L}^2[0,1], \mathbb{H}_1)$ *and* $\Psi^{-1} = D^q \in \mathfrak{B}(\mathbb{H}_1, \mathbb{L}^2[0,1])$.

Proof: If $f_1, f_2 \in \mathbb{H}_1$, $\langle f_1, f_2\rangle_1 = \langle \mathcal{T}_q f_1^{(q)}, \mathcal{T}_q f_2^{(q)}\rangle_1 = \langle D^q f_1, D^q f_2\rangle$. □

In practice, we need sufficient conditions that can be used to establish that a particular operator is invertible. A useful sufficient condition that insures the existence of an operator's inverse on a Hilbert spaces is

Theorem 3.5.4 *Let $\mathbb{H}$ be a Hilbert space with norm $\|\cdot\|$ and $\mathscr{T} \in \mathfrak{B}(\mathbb{H})$. If $\mathscr{T}$ is self-adjoint and there is some $C > 0$ such that*

$$\|\mathscr{T}f\| \geq C\|f\| \tag{3.12}$$

for all $f \in \mathbb{H}$, then $\mathscr{T}$ is invertible.

Proof: We first show that $\mathrm{Im}(\mathscr{T})$ is closed. Let $y_n \in \mathrm{Im}(\mathscr{T})$ and $y_n \to y \in \mathbb{H}$. Then, there exists $x_n \in \mathbb{H}$ such that $\mathscr{T}x_n = y_n$ with

$$\|y_m - y_n\| = \|\mathscr{T}(x_m - x_n)\| \geq C\|x_m - x_n\|$$

showing that x_n is Cauchy and must converge to some limit $x \in \mathbb{H}$. By continuity, $\mathscr{T}x = y$ and we must have $y \in \mathrm{Im}(\mathscr{T})$. It follows from Theorem 3.3.7 that $\mathbb{H} = \mathrm{Im}(\mathscr{T}) \oplus \mathrm{Ker}(\mathscr{T})$. Assumption (3.12) implies that $\mathrm{Ker}(\mathscr{T}) = \{0\}$ and, hence, $\mathscr{T}$ is both one-to-one and onto. □

If $\mathscr{T} \in \mathfrak{B}(\mathbb{H})$ satisfies $\|\mathscr{T}\| < 1$, then (3.12) holds with $\mathscr{T}$ replaced by $I - \mathscr{T}$, in which case Theorem 3.5.4 can be invoked to establish that $I - \mathscr{T}$ is invertible. The following result provides the details on the form of the inverse operator.

Theorem 3.5.5 *Let $\mathbb{X}$ be a Banach space and $\mathscr{T} \in \mathfrak{B}(\mathbb{X})$. If $\|\mathscr{T}\| < 1$, then $I - \mathscr{T}$ is invertible and*

$$(I - \mathscr{T})^{-1} = I + \sum_{j=1}^{\infty} \mathscr{T}^j. \tag{3.13}$$

Proof: As $\|\mathscr{T}\| < 1$, we have $\sum_{j=0}^{\infty} \|\mathscr{T}\|^j < \infty$. The triangle inequality then insures that the partial sums $\mathscr{S}_k := I + \sum_{j=1}^{k} \mathscr{T}^j$ form a Cauchy sequence in the Banach space $\mathfrak{B}(\mathbb{X})$ and must therefore have some limit which we denote by $\mathscr{S} = I + \sum_{j=1}^{\infty} \mathscr{T}^j$. Observe that

$$\|I - (I - \mathscr{T})\mathscr{S}_k\| = \|I - \mathscr{S}_k(I - \mathscr{T})\| = \|\mathscr{T}^{k+1}\| \leq \|\mathscr{T}\|^{k+1} \to 0.$$

This shows that

$$(I - \mathscr{T})\mathscr{S} = \mathscr{S}(I - \mathscr{T}) = I$$

and completes the proof. □

The Sherman–Morrison–Woodbury matrix identity is useful for solving a variety of problems in linear algebra and matrix theory. The operator version that we give here will prove to be similarly valuable.

Theorem 3.5.6 *For operators $\mathscr{S}, \mathscr{T}, \mathscr{U}$, and $\mathscr{V}$ with $\mathscr{S}, \mathscr{T}$ invertible,*

$$\left(\mathscr{T} + \mathscr{U}\mathscr{S}^{-1}\mathscr{V}\right)^{-1} = \mathscr{T}^{-1} - \mathscr{T}^{-1}\mathscr{U}\left(\mathscr{S} + \mathscr{V}\mathscr{T}^{-1}\mathscr{U}\right)^{-1}\mathscr{V}\mathscr{T}^{-1}. \tag{3.14}$$

Proof: The proof proceeds by showing that the right-hand side of (3.14) is the inverse of $\left(\mathscr{T} + \mathscr{U}\mathscr{S}^{-1}\mathscr{V}\right)$. This latter fact follows from

$$\begin{aligned}
&\left(\mathscr{T} + \mathscr{U}\mathscr{S}^{-1}\mathscr{V}\right)\left(\mathscr{T}^{-1} - \mathscr{T}^{-1}\mathscr{U}\left(\mathscr{S} + \mathscr{V}\mathscr{T}^{-1}\mathscr{U}\right)^{-1}\mathscr{V}\mathscr{T}^{-1}\right) \\
&= I + \mathscr{U}\mathscr{S}^{-1}\mathscr{V}\mathscr{T}^{-1} - \left(\mathscr{U} + \mathscr{U}\mathscr{S}^{-1}\mathscr{V}\mathscr{T}^{-1}\mathscr{U}\right)\left(\mathscr{S} + \mathscr{V}\mathscr{T}^{-1}\mathscr{U}\right)^{-1}\mathscr{V}\mathscr{T}^{-1} \\
&= I + \mathscr{U}\mathscr{S}^{-1}\mathscr{V}\mathscr{T}^{-1} - \mathscr{U}\mathscr{S}^{-1}\left(\mathscr{S} + \mathscr{V}\mathscr{T}^{-1}\mathscr{U}\right)\left(\mathscr{S} + \mathscr{V}\mathscr{T}^{-1}\mathscr{U}\right)^{-1}\mathscr{V}\mathscr{T}^{-1} \\
&= I.
\end{aligned}$$

□

It is fair to say that invertible operators are the exception rather than the rule and this is especially true for the fda applications that are central to this text. The question then arises as to what can be said about the solution of a linear system such as

$$\mathscr{T}x = y \tag{3.15}$$

for a given y when the operator $\mathscr{T}$ is not invertible. In the remainder of this section, we provide an answer for the case where $\mathscr{T}$ is a bounded operator on a Hilbert space $\mathbb{H}$.

Even when $\mathscr{T}$ in (3.15) is neither one-to-one nor onto, there is still an aspect of the operator that can be inverted in some general sense. Specifically, let $\tilde{\mathscr{T}}$ be the restriction of $\mathscr{T}$ to the orthogonal complement of Ker($\mathscr{T}$). Then, this restricted operator is one-to-one and onto its range and, hence, invertible. This line of thought leads us to the concept of a *Moore–Penrose inverse* that also goes by the names Moore–Penrose generalized inverse and pseudo inverse.

Definition 3.5.7 *Let $\mathbb{H}_1, \mathbb{H}_2$ be Hilbert spaces. Suppose that $\mathscr{T} \in \mathfrak{B}(\mathbb{H}_1, \mathbb{H}_2)$ and let $\tilde{\mathscr{T}}$ be the operator $\mathscr{T}$ restricted to* Ker$(\mathscr{T})^{\perp}$. *The Moore–Penrose (generalized) inverse, $\mathscr{T}^{\dagger}$, of $\mathscr{T}$ is a linear transformation with domain*

$$\mathrm{Dom}(\mathscr{T}^{\dagger}) := \mathrm{Im}(\mathscr{T}) + \mathrm{Im}(\mathscr{T})^{\perp}.$$

For $y \in$ Dom$(\mathscr{T}^{\dagger})$,

$$\mathscr{T}^{\dagger}y = \begin{cases} \tilde{\mathscr{T}}^{-1}y, & y \in \mathrm{Im}(\mathscr{T}), \\ 0, & y \in \mathrm{Im}(\mathscr{T})^{\perp}. \end{cases}$$

If Im($\mathscr{T}$) is closed in $\mathbb{H}_2$, then

$$\mathbb{H}_2 = \mathrm{Im}(\mathscr{T}) \oplus \mathrm{Im}(\mathscr{T})^{\perp}$$

by (2.16). In that case, Theorem 3.5.1 implies that $\tilde{\mathscr{T}}^{-1} \in \mathfrak{B}(\text{Im}(\mathscr{T}), \text{Ker}(\mathscr{T})^{\perp})$ and, hence, that $\mathscr{T}^{\dagger} \in \mathfrak{B}(\mathbb{H}_2, \mathbb{H}_1)$. However, the requirement that $\text{Im}(\mathscr{T})$ be closed is a rather restrictive condition; for instance, in the case of the compact operators introduced in Chapter 4 this amounts to the operator being finite-dimensional (cf. Theorem 4.3.7). In general, particularly in fda applications, $\text{Im}(\mathscr{T})$ is not closed but $\mathscr{T}^{\dagger}$ still possesses some of the other properties we might expect from an inverse operator. Specifically, it satisfies the four Moore–Penrose equations (3.16)–(3.19) that are given in the following theorem.

Theorem 3.5.8 *Let $\mathscr{T}^{\dagger}$ be as in Definition 3.5.7. Then,*

$$\mathscr{T}\mathscr{T}^{\dagger}\mathscr{T} = \mathscr{T}, \tag{3.16}$$

$$\mathscr{T}^{\dagger}\mathscr{T}\mathscr{T}^{\dagger} = \mathscr{T}^{\dagger}, \tag{3.17}$$

$$\mathscr{T}^{\dagger}\mathscr{T} = I - \mathscr{P} \tag{3.18}$$

and

$$\mathscr{T}\mathscr{T}^{\dagger} = \mathscr{Q} \tag{3.19}$$

with $\mathscr{P}$ the projection operator for $\text{Ker}(\mathscr{T})$ *and $\mathscr{Q}$ the restriction of the projection operator for* $\overline{\text{Im}(\mathscr{T})}$ *to* $\text{Dom}(\mathscr{T}^{\dagger})$.

Proof: To verify (3.16)–(3.19), we will follow the arguments in Engl et al. (2000). First note that if $y \in \text{Dom}(\mathscr{T}^{\dagger})$ it is necessarily true that

$$\mathscr{T}^{\dagger}y = \mathscr{T}^{\dagger}\mathscr{Q}y = \tilde{\mathscr{T}}^{-1}\mathscr{Q}y. \tag{3.20}$$

This allows us to show (3.19) and (3.17) by observing that

$$\mathscr{T}\mathscr{T}^{\dagger}y = \mathscr{T}\tilde{\mathscr{T}}^{-1}\mathscr{Q}y = \mathscr{Q}y$$

and, hence,

$$\mathscr{T}^{\dagger}\mathscr{T}\mathscr{T}^{\dagger}y = \mathscr{T}^{\dagger}\mathscr{Q}y = \mathscr{T}^{\dagger}y.$$

For (3.18), take any $x \in \mathbb{H}$ and decompose it as $\mathscr{P}x + (I - \mathscr{P})x$ to see that

$$\mathscr{T}^{\dagger}\mathscr{T}x = \tilde{\mathscr{T}}^{-1}\tilde{\mathscr{T}}(I - \mathscr{P})x = (I - \mathscr{P})x.$$

However, from this identity, we obtain (3.16) because

$$\mathscr{T}\mathscr{T}^{\dagger}\mathscr{T}x = \mathscr{T}(I - \mathscr{P})x = \mathscr{T}x.$$

□

Another consequence of (3.20) is

Theorem 3.5.9 $\mathrm{Im}(\mathscr{T}^\dagger) = \mathrm{Ker}(\mathscr{T})^\perp$.

Proof: Expression (3.20) indicates that for any $y \in \mathrm{Dom}(\mathscr{T}^\dagger)$, $\mathscr{T}^\dagger y$ is in the range of $\tilde{\mathscr{T}}^{-1}$, which is synonymous with $\mathrm{Ker}(T)^\perp$. If, on the other hand, $x \in \mathrm{Ker}(T)^\perp$, by definition $\mathscr{T}^\dagger \mathscr{T} x = x$. □

The Moore–Penrose pseudo inverse exhibits various other properties that would be expected for a inverse operator. For example, one may deduce from the definition that

$$\mathscr{T}^\dagger = \mathscr{T}^{-1} \tag{3.21}$$

when $\mathscr{T}$ is invertible and

$$(\mathscr{T}^*)^\dagger = (\mathscr{T}^\dagger)^*. \tag{3.22}$$

If $\mathscr{T}_1, \mathscr{T}_2$ are two bounded operators, it can be shown that

$$(\mathscr{T}_1 \mathscr{T}_2)^\dagger = \mathscr{T}_2^\dagger \mathscr{T}_1^\dagger. \tag{3.23}$$

We now return to the solution of (3.15) and elucidate its connection to the Moore–Penrose inverse. Certainly, we cannot expect there to be an exact or even unique solution to (3.15) when $\mathscr{T}$ is not invertible. As a result, we must be satisfied with some form of approximate solution that has certain specified desirable qualities. In this respect, one possibility is to use a least-squares solution: i.e., an element $\hat{x} \in \mathbb{H}$ that satisfies

$$\|\mathscr{T}\hat{x} - y\| = \inf \{\|\mathscr{T}x - y\| : x \in \mathbb{H}\} \tag{3.24}$$

for a given y. In general, there may be more than one least-squares solution. Indeed, suppose that y in (3.24) is in $\mathrm{Dom}(\mathscr{T}^\dagger)$ and consider the subspace

$$\mathbb{M} = \{x \in \mathbb{H} : \mathscr{T}x = \mathscr{Q}y\}$$

for $\mathscr{Q}$ the projection operator for $\overline{\mathrm{Im}(\mathscr{T})}$. This subspace is not empty because if $y \in \mathrm{Dom}(\mathscr{T}^\dagger)$ we can choose, e.g., $x = \mathscr{T}^\dagger y$ by (3.19). However, for any $x \in \mathbb{M}$

$$\|\mathscr{T}x - y\| = \|\mathscr{Q}y - y\| \leq \|z - y\|$$

for any $z \in \overline{\mathrm{Im}(\mathscr{T})}$, as a result of the projection theorem (Theorem 2.5.2). In particular, this means that

$$\|\mathscr{T}x - y\| \leq \|\mathscr{T}h - y\|$$

for all $h \in \mathbb{H}$; i.e., all the elements of $\mathbb{M}$ give least-squares approximations.

One way to narrow down the possibilities for least-squares approximations is to use the one that has the smallest norm. This "best" approximate solution returns us to the Moore–Penrose inverse.

Theorem 3.5.10 *If $y \in \mathrm{Dom}(\mathcal{T}^\dagger)$, the unique element of minimal norm that satisfies (3.24) is $\hat{x} = \mathcal{T}^\dagger y$.*

Proof: As mentioned earlier, $x = \mathcal{T}^\dagger y$ is in $\mathbb{M}$. Thus, any element in $\mathbb{M}$ can be represented as $\mathcal{T}^\dagger y + z$ for some $z \in \mathrm{Ker}(\mathcal{T})$. The result then follows from the fact that $\mathcal{T}^\dagger y \in \mathrm{Ker}(\mathcal{T})^\perp$ by Theorem 3.5.9 □

Example 3.5.11 *If y in (3.24) is in $\mathrm{Dom}(\mathcal{T}^\dagger)$ and $\hat{x} = \mathcal{T}^\dagger y$, Theorem 2.5.2 has the consequence that $\mathcal{T}\hat{x} - y \in \mathrm{Im}(\mathcal{T})^\perp = \mathrm{Ker}(\mathcal{T}^*)$ due to Theorem 3.3.7. Thus, any least-squares solution must satisfy the* normal equations

$$\mathcal{T}^*\mathcal{T}x = \mathcal{T}^*y. \tag{3.25}$$

Combining Theorem 3.5.10 with (3.18) characterizes the best approximate solution as the minimum norm solution to the normal equations (3.25). Another application of Theorem 3.5.10 with $\mathcal{T}$ and y in (3.15) replaced by $\mathcal{T}^*\mathcal{T}$ and $\mathcal{T}^*y$ leads us to the realization that $\hat{x} = (\mathcal{T}^*\mathcal{T})^\dagger\mathcal{T}^*y$: i.e.,

$$\mathcal{T}^\dagger = (\mathcal{T}^*\mathcal{T})^\dagger\mathcal{T}^*. \tag{3.26}$$

3.6 Fréchet and Gâteaux derivatives

In Section 2.6, we saw that the familiar concept of integrating a functions over an interval of the real line could be extended to various abstract formulations wherein integration takes place in some general Banach space. It should therefore come as no surprise that the operation of differentiating a function of a real variable admits similar types of extensions. In this section, we briefly introduce two such abstract views of differentiation: namely, the Gâteaux and Fréchet derivatives.

For our purposes, it suffices to concentrate on the case of two Banach spaces $\mathbb{X}_1, \mathbb{X}_2$ with respective norms $\|\cdot\|_1$ and $\|\cdot\|_2$. Let f be a function defined on an open subset U of $\mathbb{X}_1$ that takes values in $\mathbb{X}_2$. We say that f is Gâteaux differentiable at $x \in U$ if there is an element $f'(x) \in \mathfrak{B}(\mathbb{X}_1, \mathbb{X}_2)$ such that

$$\lim_{t\to 0} t^{-1}\|f(x+tv) - f(x) - tf'(x)v\|_2 = 0 \tag{3.27}$$

for every $v \in \mathbb{X}_1$. When (3.27) holds, $f'(x)$ is called the Gâteaux derivative of f at x. The derivative, if it exists, is necessarily unique. If there were another operator $\tilde{f}'(x)$ that satisfied (3.27) then, for every v and any $t \neq 0$,

$$\|(\tilde{f}'(x) - f'(x))v\|_2 \leq t^{-1}\|f(x+tv) - f(x) - t\tilde{f}'(x)v\|_2 \\ + t^{-1}\|f(x+tv) - f(x) - tf'(x)v\|_2$$

and the right-hand side tends to 0 as $t \to 0$.

Various analogs of classical results for derivatives of functions of real variables hold for Gâteaux derivatives. For example, we will have use shortly for the following parallel of the mean value theorem.

Theorem 3.6.1 *Given $x, y \in \mathbb{X}_1$, assume that f has a Gâteaux derivative at each point in the set $\{x + t(y - x) : 0 \leq t \leq 1\}$. Then, for every bounded linear functional ℓ on $\mathbb{X}_2$,*

$$\ell\,(f(y) - f(x)) = \ell\left(f'(x + \xi(y - x))(y - x)\right)$$

for some $\xi \in (0, 1)$.

Proof: Set $g(t) = \ell\,(f(x + t(y - x)))$ so that the linearity of ℓ has the consequence that $g'(t) = \ell\left(f'(x + t(y - x))(y - x)\right)$. An application of the ordinary mean value theorem assures the existence of a $\xi \in (0, 1)$ for which $g(1) - g(0) = g'(\xi)$ and proves the result. □

The existence of a Gâteaux derivative is a rather weak assumption as the limit in (3.27) is taken in a fixed direction: namely, in the direction of the vector v. The Fréchet derivative concept arises when we allow the direction to vary thereby producing the condition that

$$\lim_{v \to 0} \frac{\|f(x+v) - f(x) - f'(x)v\|_2}{\|v\|_1} = 0. \tag{3.28}$$

This, in turn, amounts to saying that for any arbitrary sequence $\{v_n\}$ tending to 0

$$\lim_{n \to \infty} \frac{\|f(x+v_n) - f(x) - f'(x)v_n\|_2}{\|v_n\|_1} = 0.$$

If (3.28) is satisfied for some $f'(x) \in \mathfrak{B}(\mathbb{X}_1, \mathbb{X}_2)$, we call $f'(x)$ the Fréchet derivative of f at x.

Clearly, the existence of the Fréchet derivative of f at x implies the existence of the corresponding Gâteaux derivative and the two derivatives must coincide in this instance. The converse is not true in general. For instance, if f is Fréchet differentiable at x then it is continuous at x, as for any $\epsilon > 0$, there exists $\delta > 0$ such that whenever $\|v\|_1 \leq \delta$,

$$\epsilon\|v\|_1 \geq \|f(x+v) - f(x) - f'(x)v\|_2 \\ \geq \|f(x+v) - f(x)\|_2 - \|f'(x)v\|_2$$

or

$$\|f(x+v)-f(x)\|_2 \le (\epsilon + \|f'(x)\|)\|v\|_1.$$

This property is not shared by the Gâteaux derivative. For example, the function

$$f(x,y)=\begin{cases}\frac{x^3y}{x^6+y^2}, & (x,y)\neq(0,0),\\ 0, & (x,y)=(0,0),\end{cases}$$

on $\mathbb{R}^2$ is discontinuous at $(0,0)$ but $f'(0,0)=0$ in the Gâteaux sense. More discussions along this line can be found in Ortega and Rheinboldt (1970).

A sufficient condition for our two derivative notions to agree is the following.

Theorem 3.6.2 *Suppose that f is Gâteaux differentiable in an open subset U of $\mathbb{X}_1$. If f' is continuous at $x\in U$, then $f'(x)$ is the Fréchet derivative of f at x.*

Proof: As f' is continuous at x, for any given $\epsilon>0$, there is $\delta>0$ such that $\|f'(x+v)-f'(x)\|_2<\epsilon$ if $\|v\|_1<\delta$. The fact that U is open means that for sufficiently small δ we will also have $x+tv\in U$ for all $t\in[0,1]$. Theorem 3.6.1 now indicates that

$$\ell\left(f(x+v)-f(x)-f'(x)v\right)=\ell\left(f'(x+\xi v)v-f'(x)v\right)$$

for some $\xi\in(0,1)$. As this is true for all linear functional, we can apply it to the linear functional from Corollary 3.2.8 that returns the norm of $f(x+v)-f(x)-f'(x)v$ with the consequence that

$$\begin{aligned}\|f(x+v)-f(x)-f'(x)v\|_2 &= \|f'(x+\xi v)v-f'(x)v\|_2\\ &\le \|f'(x+\xi v)-f'(x)\|\,\|v\|_1\\ &\le \epsilon\|v\|_1.\end{aligned}$$

□

Higher order derivatives of both the Fréchet and Gâteaux varieties can be defined as the derivative of a derivative of one lower order. For example, if the first Gâteaux derivative of f exists over some open subset of $\mathbb{X}_1$ that contains a point x and there is an element $f''(x)$ of $\mathfrak{B}(\mathbb{X}_1,\mathfrak{B}(\mathbb{X}_1,\mathbb{X}_2))$ that satisfies

$$\lim_{t\to 0} t^{-1}\|f'(x+tv)-f'(x)-tf''(x)v\|=0 \qquad (3.29)$$

for every $v\in\mathbb{X}_1$, we refer to $f''(x)$ as the second Gâteaux derivative of f at x. Note that the norm $\|\cdot\|$ in (3.29) is operator norm. If

$$\lim_{v\to 0}\frac{\|f'(x+v)-f'(x)-f''(x)v\|}{\|v\|_1}=0, \qquad (3.30)$$

$f''(x)$ is the second Fréchet derivative of f at x.

As $f''(x)$ is in $\mathfrak{B}(\mathbb{X}_1, \mathfrak{B}(\mathbb{X}_1, \mathbb{X}_2))$, this means that for every $v_1 \in \mathbb{X}_1, f''(x)v_1$ is in $\mathfrak{B}(\mathbb{X}_1, \mathbb{X}_2)$. Then, for every $v_1, v_2 \in \mathbb{X}_1$, $\left[f''(x)v_1\right] v_2 := f''(x)v_1v_2$ is an element of $\mathbb{X}_2$. That is, the mapping $h(v_1, v_2) : \mathbb{X}_1 \times \mathbb{X}_1 \mapsto \mathbb{X}_2$ defined by $h(v_1, v_2) = f''(s)v_1v_2$ is bilinear which provides a useful view of $f''(x)$ in this instance.

Life becomes somewhat simpler in the case where $\mathbb{X}_2 = \mathbb{R}$. Now $\mathfrak{B}(\mathbb{X}_1, \mathbb{X}_2)$ is the dual space of linear functionals on $\mathbb{X}_1$ and as a result Gâteaux derivatives derive their properties from ordinary calculus in this instance. For example, extrema of linear functionals can be partially characterized by the behavior of their Gâteaux derivatives.

Theorem 3.6.3 *Suppose that f is Gâteaux differentiable over $\mathbb{X}_1$. If f has a local maximum or minimum at $x \in \mathbb{X}_1, f'(x)v = 0$ for every $v \in \mathbb{X}_1$.*

Proof: As $f(x)$ is a local maximum or minimum of f, the function $g(t) = f(x + tv)$ must attain a local maximum of minimum at $t = 0$ for every $v \in \mathbb{X}_1$. Therefore, $g'(0) = 0$. □

The calculation used to prove the previous theorem suggests the following result that proves quite useful for evaluation of Gâteaux and Fréchet derivatives.

Theorem 3.6.4 *Let f be twice Gâteaux differentiable at $x \in \mathbb{X}_1$ and set $g(t) = f(x + tv)$ and $h(t) = f'(x + tv_1)v_2$. Then, $f'(x)v = g'(0)$ and $f''(x)v_1v_2 = h'(0)$.*

Proof: If suffices to observe that

$$\frac{g(t) - g(0)}{t} = \frac{f(x + tv) - f(x)}{t}$$

and that

$$\frac{h(t) - h(0)}{t} = \left[\frac{f'(x + tv_1) - f'(x)}{t}\right] v_2.$$

□

Example 3.6.5 *To illustrate the use Theorem 3.6.4, let $\mathbb{X}_1$ be a Hilbert space with inner product $\langle \cdot, \cdot \rangle$ and define $f(x) = \langle x, x \rangle$. Then, $g(t) = \langle x + tv, x + tv \rangle$ and $g'(0) = 2\langle x, v \rangle = f'(x)v$. So, $f'(x)$ is the linear functional with representer $2x$. It is therefore continuous in x which means that $f'(x)$ is also the Fréchet derivative of f at x as a result of Theorem 3.6.2.*

Now, $h(t) = 2\langle x + tv_1, v_2 \rangle$ and $h'(0) = 2\langle v_1, v_2 \rangle$. Thus, $f''(x)$ is the linear operator that maps $v_1 \in \mathbb{X}_1$ into the linear functional with representer $2v_1$. As this action is constant as a function of x, the derivative is trivially continuous in x and must therefore be the Fréchet derivative as well. In fact, $f''(x)$ is even invertible in this instance with $\left[f''(x)\right]^{-1} \ell = v_\ell / 2$ with $v_\ell \in \mathbb{X}_1$ the representer of the linear functional $\ell \in \mathfrak{B}(\mathbb{X}_1, \mathbb{R})$.

3.7 Generalized Gram–Schmidt decompositions

Projection operators provide the means by which the Gram–Schmidt algorithm from Theorem 2.4.10 can be extended to a much more general context. To explain this idea, let us revisit the original Gram–Schmidt formulation from a projection perspective.

Suppose that $m_1, \dots, m_n$ is a collection of linearly independent vectors in a Hilbert space $\mathbb{H}$. Define

$$\mathbb{M}_j = \text{span}\{m_j\}$$

for $j = 1, \dots, n$. These are all simple, closed subspaces with $x \in \mathbb{M}_j$ meaning that $x = cm_j$ for some $c \in \mathbb{R}$. For each integer $1 \leq k \leq n$, the algebraic direct sum of $\mathbb{M}_1, \dots, \mathbb{M}_k$ described in Definition 2.5.4 is just

$$\begin{aligned} \mathbb{S}_k &= \mathbb{M}_1 + \cdots + \mathbb{M}_k \\ &= \text{span}\{m_1, \dots, m_k\}. \end{aligned}$$

As the m_j are linearly independent, $\mathbb{M}_j \cap \mathbb{M}_k = \{0\}$ and, more generally, this entails that

$$\begin{aligned} \mathbb{M}_i \cap \left(\sum_{j \neq i} \mathbb{M}_j \right) &= \mathbb{M}_i \cap \text{span}\{m_1, \dots, m_{i-1}, m_{i+1}, \dots, m_n\} \\ &= \{0\}. \end{aligned}$$

Now, the Gram–Schmidt algorithm uses the m_j to create a new set of orthonormal vectors $e_1, \dots, e_n$ with $e_1 = m_1 / \|m_1\|$ and

$$e_k = \left(m_k - \sum_{j=1}^{k-1} \langle m_k, e_j \rangle e_j \right) \Big/ \left\| m_k - \sum_{j=1}^{k-1} \langle m_k, e_j \rangle e_j \right\|$$

for $k = 2, \dots, n$. The method of construction ensures that

$$\text{span}\{e_1, \dots, e_k\} = \text{span}\{m_1, \dots, m_k\}.$$

If we let

$$\mathbb{N}_j = \text{span}\{e_j\},$$

we can express this last relation in terms of the orthogonal direct sum notation of Definition 2.5.4 as

$$\mathbb{S}_k = \mathbb{N}_1 \oplus \cdots \oplus \mathbb{N}_k.$$

In particular, we see from this that the $\mathbb{N}_k$ are characterized by

$$\begin{aligned}\mathbb{S}_k \cap \mathbb{S}_{k-1}^{\perp} &= (\mathbb{M}_1 + \cdots + \mathbb{M}_k) \cap (\mathbb{M}_1 + \cdots + \mathbb{M}_{k-1})^{\perp} \\ &= \text{span}\{m_1, \ldots, m_k\} \cap \text{span}\{m_1, \ldots, m_{k-1}\}^{\perp} \\ &= \text{span}\{e_1, \ldots, e_k\} \cap \text{span}\{e_1, \ldots, e_{k-1}\}^{\perp} \\ &= \text{span}\{e_k\} = \mathbb{N}_k.\end{aligned}$$

From Section 3.4, we know that the projection operator for $\mathbb{N}_j$ has the simple form $\mathscr{P}_{\mathbb{N}_j} = e_j \otimes e_j$. Thus, any $x \in \mathbb{M}_k$ can be expressed as

$$x = \sum_{j=1}^{k} \langle x, e_j \rangle e_j = \sum_{j=1}^{k} \mathscr{P}_{\mathbb{N}_j} x. \tag{3.31}$$

One can certainly view this as the culmination of the Gram–Schmidt procedure. However, there is a bit more that can be said if we think about how one might need to use this type of outcome.

In some cases, it may be more convenient to work directly with the orthonormal e_j. This can be true, for example, in norm-based optimization problems where orthogonality may render the calculations more tractable. In such instances, it may be possible to find a desirable coefficient for e_k but, upon doing so, it becomes necessary to trace ones way back to the corresponding element of $\mathbb{M}_k$ that was involved in the original problem formulation. A little adjustment is needed in (3.31) to take care of such eventualities.

Let $\mathscr{P}_{\mathbb{N}_j|\mathbb{M}_k}$ be the projection operator $\mathscr{P}_{\mathbb{N}_j}$ restricted to $\mathbb{M}_k$. Then,

$$\mathscr{P}_{\mathbb{N}_k|\mathbb{M}_k} m_k = \langle m_k, e_k \rangle e_k$$

and, similarly,

$$\mathscr{P}_{\mathbb{M}_k|\mathbb{N}_k} e_k = \langle m_k, e_k \rangle m_k$$

with $\mathscr{P}_{\mathbb{M}_k|\mathbb{N}_k}$ now the projection operator for $\mathbb{M}_k$ restricted to $\mathbb{N}_k$. Therefore,

$$\left(\mathscr{P}_{\mathbb{N}_k|\mathbb{M}_k}\right)^{-1} e_k = \frac{m_k}{\langle m_k, e_k \rangle}$$

and every $x \in \mathbb{M}_k$ can be expressed as

$$\begin{aligned}x &= \sum_{j=1}^{k} \mathscr{P}_{\mathbb{N}_j} \left(\mathscr{P}_{\mathbb{N}_k|\mathbb{M}_k}\right)^{-1} z \\ &= \sum_{j=1}^{k} \mathscr{P}_{\mathbb{N}_j|\mathbb{M}_k} \left(\mathscr{P}_{\mathbb{N}_k|\mathbb{M}_k}\right)^{-1} z\end{aligned} \tag{3.32}$$

for some $z \in \mathbb{N}_k$. Conversely, for any $z \in \mathbb{N}_k$, x in (3.32) is the corresponding element of $\mathbb{M}_k$.

To step beyond the Gram–Schmidt scenario, we follow the path of Sunder (1988) and now consider a Hilbert space $\mathbb{H}$ that can be written as the algebraic direct sum of n closed subspaces $\mathbb{M}_1, \ldots, \mathbb{M}_n$. That is,

$$\mathbb{H} = \sum_{i=1}^{n} \mathbb{M}_i,$$

where $\mathbb{M}_1, \ldots, \mathbb{M}_n$ satisfy

$$\mathbb{M}_i \cap \sum_{j \neq i} \mathbb{M}_j = \{0\}. \tag{3.33}$$

If $x_1 + \cdots + x_n = 0$ for $x_i \in \mathbb{M}_i$ then $x_i = -\sum_{j \neq i} x_j = 0$ for all i by (3.33). Thus, (3.33) defines a notion of linear independence for subspaces. As such, every element in $\mathbb{H}$ can be written as a sum of elements from the $\mathbb{M}_j$ with the components of the sum being determined uniquely.

Now, for $1 \leq k \leq n$, define the partial sums

$$\mathbb{S}_k = \sum_{i=1}^{k} \mathbb{M}_i$$

and set

$$\mathbb{N}_k = \mathbb{S}_k \cap \mathbb{S}_{k-1}^{\perp},$$

where $\mathbb{S}_0 := \{0\}$. Then, $\mathbb{N}_k \perp \mathbb{S}_{k-1}$ for all k and $\mathbb{N}_i \perp \mathbb{N}_j$ for $i \neq j$. One can show by induction that

$$\sum_{i=1}^{k} \mathbb{M}_i = \oplus_{i=1}^{k} \mathbb{N}_i \tag{3.34}$$

for all k, and in particular

$$\mathbb{H} = \oplus_{i=1}^{n} \mathbb{N}_i. \tag{3.35}$$

Let $\mathscr{P}_{\mathbb{N}_k}$ be the orthogonal projection operators onto $\mathbb{N}_k$ for $1 \leq k \leq n$ and for $1 \leq j \leq k \leq n$ define the restriction of $\mathscr{P}_{\mathbb{N}_j}$ to $\mathbb{M}_k$ by

$$\mathscr{P}_{\mathbb{N}_j|\mathbb{M}_k} x = \mathscr{P}_{\mathbb{N}_j} x$$

for $x \in \mathbb{M}_k$.

Theorem 3.7.1 $\mathscr{P}_{\mathbb{N}_k|\mathbb{M}_k}$ *is bijective.*

Proof: Any $x \in \mathbb{N}_k$ can be written as $x = s_{k-1} + x_k$ where $s_{k-1} \in \mathbb{S}_{k-1}$ and $x_k \in \mathbb{M}_k$. As $\mathbb{N}_k \perp \mathbb{S}_{k-1}$,

$$x = \mathscr{P}_{\mathbb{N}_k} x = \mathscr{P}_{\mathbb{N}_k} x_k.$$

So, $\mathscr{P}_{\mathbb{N}_k}$ maps $\mathbb{M}_k$ onto $\mathbb{N}_k$. If $x \in \mathbb{M}_k$ satisfies $\mathscr{P}_{\mathbb{N}_k} x = 0$, it must be that $x \in \mathbb{M}_k \cap \mathbb{N}_k^{\perp}$. By (3.35),

$$\begin{aligned} \mathbb{M}_k \cap \mathbb{N}_k^{\perp} &= \mathbb{M}_k \cap \oplus_{i \neq k} \mathbb{N}_i \\ &= \mathbb{M}_k \cap \oplus_{i \leq k-1} \mathbb{N}_i \\ &= \mathbb{M}_k \cap \mathbb{S}_{k-1} \end{aligned}$$

as $\mathbb{N}_j \perp \mathbb{M}_k$ for $j > k$. The right-hand side of the last relation is $\{0\}$ by the definition of $\mathbb{S}_{k-1}$. This shows that $\mathscr{P}_{\mathbb{N}_k|\mathbb{M}_k}$ is one-to-one. □

In view of (3.34), the inverse of $\mathscr{P}_{\mathbb{N}_k|\mathbb{M}_k}$ can be written as

$$(\mathscr{P}_{\mathbb{N}_k|\mathbb{M}_k})^{-1} = \sum_{j=1}^{k} \mathscr{P}_{\mathbb{N}_j} (\mathscr{P}_{\mathbb{N}_k|\mathbb{M}_k})^{-1}.$$

As a consequence, any $x \in \mathbb{M}_k$ has the representation

$$\begin{aligned} x &= \sum_{j=1}^{k} \mathscr{P}_{\mathbb{N}_j} (\mathscr{P}_{\mathbb{N}_k|\mathbb{M}_k})^{-1} z \\ &= \sum_{j=1}^{k} \mathscr{P}_{\mathbb{N}_j|\mathbb{M}_k} (\mathscr{P}_{\mathbb{N}_k|\mathbb{M}_k})^{-1} z \end{aligned} \tag{3.36}$$

for some $z \in \mathbb{N}_k$ and we have successfully extended (3.32) to our more general setting.

4

Compact operators and singular value decomposition

In this chapter, we continue the discussion of operators that was begun in Chapter 3. However, in doing so, we will narrow our focus to the special case of operators that are *compact* in a sense to be described shortly. When working with Hilbert spaces, this type of operator can be approximated by finite-dimensional operators and, as a result, exhibits similar properties to those we are familiar with from the study of matrices. Not surprisingly, it is compact operators that are pervasive throughout statistics, in general, and fda, in particular.

We begin in Section 4.1 with a general treatment of compact operators. Then, we specialize again; this time to the case of operators between Hilbert spaces in Sections 4.2 and 4.3 and derive both eigenvalue and singular value decompositions (svds) for this setting.

Within the class of compact operators on Hilbert spaces, Hilbert–Schmidt and trace-class operators are of special interest due, in part, to the rapid convergence of their optimal finite-dimensional approximations. Accordingly, we investigate the properties of these operator classes in some depth in Sections 4.4 and 4.5. In functional data, integral operators are especially relevant. A key result for this type of operator is Mercer's Theorem that uses the eigenvalue–eigenvector decomposition of an integral operator to obtain a corresponding series expansion for the operator's kernel. This latter series is very important for functional pca and also provides a simple way to connect integral operators to the Hilbert–Schmidt and trace classes.

Theoretical Foundations of Functional Data Analysis, with an Introduction to Linear Operators, First Edition. Tailen Hsing and Randall Eubank.

4.1 Compact operators

We have mentioned matrices as providing the simplest and most familiar example of linear operators. A natural extension of the matrix idea leads us to the concept of compact operators.

Definition 4.1.1 *A linear transformation $\mathscr{T}$ from a normed space $\mathbb{X}_1$ into another normed space $\mathbb{X}_2$ is compact if for any bounded sequence $\{x_n\} \in \mathbb{X}_1$, $\{\mathscr{T}x_n\}$ contains a convergent subsequence in $\mathbb{X}_2$.*

First note that compact linear transformations are necessarily bounded and are therefore referred to as compact operators. To see this, suppose that instead $\mathscr{T}$ were unbounded. Then, we could find a bounded sequence $\{x_n\}$ in $\mathbb{X}_1$ for which $\|\mathscr{T}x_n\|_2 \geq n$ for each n and, consequently, $\{\mathscr{T}x_n\}$ would not contain a convergent subsequence.

On the other hand, being compact is a special quality that is not shared by every operator as demonstrated by the following result.

Theorem 4.1.2 *The identity operator defined on an infinite-dimensional normed space is not compact.*

Proof: Let us begin with the case of operators on a Hilbert space with CONS $\{e_j\}$ and identity operator I. Then, for $i \neq j$,

$$\|Ie_i - Ie_j\| = \|e_i - e_j\| = \sqrt{2}.$$

So, $\{Ie_j\}$ does not contain a convergent subsequence even though $\{e_j\}$ is a bounded sequence.

As suggested by our Hilbert space argument, verification of this result for a general infinite-dimensional normed space $\mathbb{X}$ relies on the construction of a bounded sequence $\{e_j\}$ for which $\inf_{i,j}\|e_i - e_j\| > 0$. This can be achieved as follows. Let e_1 be any element of $\mathbb{X}$ with unit norm and set $\mathbb{Y} = \text{span}\{e_1\}$. Now choose an arbitrary element x from $\mathbb{Y}^C$ and define

$$d = \inf\{\|x - y\| : y \in \mathbb{Y}\}.$$

As $\mathbb{Y}$ is closed, d must be bigger than zero.

By the definition of infimum, for any $\alpha \in (1, \infty)$, there exists some $z \in \mathbb{Y}$ such that $\|x - z\| < \alpha d$. Thus, we now take $e_2 = (x - z)/\|x - z\|$ which has unit norm and satisfies

$$\|e_2 - y\| = \frac{\|x - (z + \|x - z\|y)\|}{\|x - z\|} > \frac{d}{\alpha d} = \alpha^{-1}$$

for any $y \in \mathbb{Y}$ as $z + \|x - z\|y \in \mathbb{Y}$. In particular, $\|e_1 - e_2\| > \alpha^{-1}$. Next, let $\mathbb{Y} = \text{span}\{e_2, e_1\}$ and repeat the above-mentioned process by defining e_3 with unit norm such that $\|e_3 - y\| > \alpha^{-1}$ for all $y \in \mathbb{Y}$ and $\|e_3 - e_i\| > \alpha^{-1}$, $i = 1, 2$. These basic steps can be repeated ad infinitum to construct the desired sequence. □

Some of the basic properties of compact operators are collected in the following theorem.

Theorem 4.1.3 *The following results apply to compact operators between two normed linear spaces.*

1. *The closure of the range of any compact operator is separable.*
2. *Operators with finite rank are compact.*
3. *The composition of two operators is compact if either operator is compact.*
4. *The set of compact operators that map to any Banach space is closed.*

Proof: Let $\mathscr{T}$ be a compact operator from a normed space $\mathbb{X}_1$ to a normed space $\mathbb{X}_2$. Consider the set $\overline{\mathscr{T}(B(0; r))}$ in $\mathbb{X}_2$ where $B(0; r) = \{x \in \mathbb{X}_1 : \|x\|_1 \leq r\}$ and let $\{y_n\}$ be any sequence in $\overline{\mathscr{T}(B(0; r))}$. Then, for every n, there exists an $x_n \in B(0; r)$ such that

$$\|y_n - \mathscr{T}x_n\|_2 < n^{-1}. \tag{4.1}$$

As $\{x_n\}$ is bounded and $\mathscr{T}$ is compact, $\{\mathscr{T}x_n\}$ contains a convergent subsequence. It then follows from (4.1) that $\{y_n\}$ also contains a convergent subsequence, which shows that $\overline{\mathscr{T}(B(0; r))}$ is sequentially compact. However, the fact that $\mathbb{X}_2$ is a metric space entails that $\overline{\mathscr{T}(B(0; r))}$ is compact and hence separable. In view of the relationship $\text{Im}(\mathscr{T}) \subset \cup_{r=1}^{\infty} \mathscr{T}(B(0; r))$, it follows that $\overline{\text{Im}(\mathscr{T})}$ is also separable and the first result is proved.

To prove the second claim, suppose that $\text{Im}(\mathscr{T})$ is finite-dimensional and $\{x_n\}$ is a bounded sequence in $\mathbb{X}_1$. Then, the Bolzano–Weierstrass Theorem implies that $\{\mathscr{T}x_n\}$ contains a convergent subsequence in $\mathbb{X}_2$ thereby showing compactness.

For part 3, let $\mathscr{T}_1$ and $\mathscr{T}_2$ be operators from $\mathbb{X}_1$ to $\mathbb{X}_2$ and $\mathbb{X}_2$ to $\mathbb{X}_3$, respectively, with $\{x_n\}$ a bounded sequence in $\mathbb{X}_2$. We then need to show that $\{\mathscr{T}_2(\mathscr{T}_1 x_n)\}$ contains a convergent subsequence in $\mathbb{X}_3$ if either $\mathscr{T}_1$ or $\mathscr{T}_2$ is compact.

If $\mathscr{T}_1$ is compact, then $\{\mathscr{T}_1 x_n\}$ contains a convergent subsequence in $\mathbb{X}_2$, and, as $\mathscr{T}_2$ is continuous, the image of this subsequence under $\mathscr{T}_2$ also converges in $\mathbb{X}_3$. If, instead, $\mathscr{T}_2$ is compact then, as $\{\mathscr{T}_1 x_n\}$ is bounded in $\mathbb{X}_2$, $\{\mathscr{T}_2(\mathscr{T}_1 x_n)\}$ contains a convergent subsequence in $\mathbb{X}_3$.

To prove part 4, let $\{\mathscr{T}_n\}$ be a sequence of compact operators from $\mathbb{X}_1$ to $\mathbb{X}_2$, where $\mathbb{X}_2$ is a Banach space. Our goal is to show that if $\|\mathscr{T}_n - \mathscr{T}\| \to 0$ for some $\mathscr{T}$ then $\mathscr{T}$ is also compact. The proof is based on a "diagonalization" argument that we will now explain. Let $\{x_k\}$ be any bounded sequence in $\mathbb{X}_1$. By compactness, $\mathscr{T}_1 x_{1k}$ converges to some y_1 as $k \to \infty$ for some subsequence $\{x_{1k}\} \subset \{x_{0k}\} := \{x_k\}$. Using this same argument, we can conclude $\mathscr{T}_2 x_{2k}$ converges to some y_2 as $k \to \infty$ for some $\{x_{2k}\} \subset \{x_{1k}\}$. Continuing in this manner, we see that for each $n \geq 1$, $\mathscr{T}_n x_{nk}$ converges to some y_n as $k \to \infty$ for a subsequence $\{x_{nk}\} \subset \{x_{(n-1)k}\}$.

We now show that y_n converges. For each n, there is a large enough k_n such that

$$\|\mathscr{T}_n x_{nk} - y_n\|_2 < n^{-1}$$

for all $k \geq k_n$. We can obviously choose the k_n to satisfy $k_{n'} > k_n$ for $n' > n$. So, for $n' > n$ write

$$\begin{aligned} y_n - y_{n'} &= (y_n - \mathscr{T}_n x_{n'k_{n'}}) + (\mathscr{T}_{n'} x_{n'k_{n'}} - y_{n'}) + (\mathscr{T}_n x_{n'k_{n'}} - \mathscr{T} x_{n'k_{n'}}) \\ &\quad + (\mathscr{T} x_{n'k_{n'}} - \mathscr{T}_{n'} x_{n'k_{n'}}), \end{aligned}$$

which gives

$$\|y_n - y_{n'}\|_2 \leq n^{-1} + n'^{-1} + \|\mathscr{T}_n - \mathscr{T}\| + \|\mathscr{T}_{n'} - \mathscr{T}\|.$$

This establishes that $\{y_n\}$ is Cauchy. As $\mathbb{X}_2$ is a Banach space, y_n converges to some y in $\mathbb{X}_2$. The inequality

$$\|\mathscr{T} x_{nk_n} - y\|_2 \leq \|\mathscr{T} - \mathscr{T}_n\| \|x_{nk_n}\|_1 + \|\mathscr{T}_n x_{nk_n} - y_n\|_2 + \|y_n - y\|_2$$

now shows that $\mathscr{T} x_{nk_n}$ converges to y and proves that $\mathscr{T}$ is compact. □

We showed in Theorem 4.1.2 that the identity operator defined on an infinite-dimensional normed space is not compact. The following is a partial extension of that result.

Theorem 4.1.4 *Let $\mathscr{T}$ be a bijective operator in $\mathfrak{B}(\mathbb{X}_1, \mathbb{X}_2)$ for infinite-dimensional Banach spaces $\mathbb{X}_1$ and $\mathbb{X}_2$. Then, $\mathscr{T}$ is not compact.*

Proof: By Theorem 3.5.1, $\mathscr{T}^{-1} \in \mathfrak{B}(\mathbb{X}_2, \mathbb{X}_1)$. If $\mathscr{T}$ is compact, this implies that the identity mapping $I = \mathscr{T}^{-1}\mathscr{T}$ is also compact due to result 3 of Theorem 4.1.3, which contradiction Theorem 4.1.2. □

The fundamental fact that underlies Theorem 4.1.4 is that the range of an infinite-dimensional compact operator is necessarily not closed. We will return to establish this in Theorem 4.3.7 .

Typically, one uses some combination of results 2–4 of Theorem 4.1.3 to establish that a particular operator is compact. In the context of Hilbert spaces, it is possible to strengthen the connection between finite dimensional and general compact operators that was implied by parts 2 and 4 of the theorem.

Theorem 4.1.5 *Let $\mathbb{H}_1, \mathbb{H}_2$ be Hilbert spaces and assume that $\mathcal{T} \in \mathfrak{B}(\mathbb{H}_1, \mathbb{H}_2)$. Then,*

1. *$\mathcal{T}$ is compact if there exists a sequence $\{\mathcal{T}_n\}$ of finite-dimensional operators such that $\|\mathcal{T}_n - \mathcal{T}\| \to 0$ as $n \to \infty$ and*
2. *$\mathcal{T}$ is compact if $\mathcal{T}^*$ is compact.*

Proof: If there are finite-dimensional operators $\mathcal{T}_n$ such that $\|\mathcal{T} - \mathcal{T}_n\| \to 0$, then we conclude that $\mathcal{T}$ is compact by results 2 and 4 of Theorem 4.1.3. For the other direction, using result 1 of Theorem 4.1.3, we can find an CONS $\{e_j\}$ for $\overline{\text{Im}(\mathcal{T})}$ and define the finite-dimensional operator

$$\mathcal{T}_n x := \sum_{j=1}^{n} \langle \mathcal{T}x, e_j\rangle_2 e_j = \mathcal{P}_n \mathcal{T}x,$$

for $x \in \mathbb{H}_1$ with $\mathcal{P}_n$ the projection operator onto $\text{span}\{e_1, \ldots, e_n\}$. Suppose that $\|\mathcal{T} - \mathcal{T}_n\| \nrightarrow 0$. Then, there exist an $\epsilon > 0$ and a unit-norm sequence $\{x_{n'}\}$ such that

$$\|(\mathcal{T} - \mathcal{T}_{n'})x_{n'}\|_2 > \epsilon \tag{4.2}$$

for all n'. Now, the compactness of $\mathcal{T}$ implies that $\mathcal{T}x_{n''} \to y$ for some subsequence $\{x_{n''}\} \subset \{x_{n'}\}$ and some y. Thus, write

$$(\mathcal{T} - \mathcal{T}_{n''})x_{n''} = (I - \mathcal{P}_{n''})\mathcal{T}x_{n''} = (I - \mathcal{P}_{n''})y + (I - \mathcal{P}_{n''})(\mathcal{T}x_{n''} - y),$$

where I is the identity operator on $\overline{\text{Im}(\mathcal{T})}$. As $I - \mathcal{P}_n$ is bounded and $(I - \mathcal{P}_n)y \to 0$ (although $\|I - \mathcal{P}_n\| \nrightarrow 0$), the right-hand side converges to 0 and contradicts (4.2). This shows that $\|\mathcal{T} - \mathcal{T}_n\| \to 0$ and completes the proof of part 1.

By parts 2 and 7 of Theorem 3.3.7, $\mathcal{T}_n^*$ in the above-mentioned construction is also finite-dimensional and $\|\mathcal{T}_n^* - \mathcal{T}^*\| = \|\mathcal{T}_n - \mathcal{T}\|$. Thus, part 2 follows from part 1. □

Part 2 of Theorem 4.1.5 and part 1 of Theorem 4.1.3 have the implication that if $\mathcal{T}$ is a compact operator between Hilbert spaces $\mathbb{H}_1$ and $\mathbb{H}_2$ then $\overline{\text{Im}(\mathcal{T}^*)}$ is separable. Recall from Theorem 3.3.7 that $\mathbb{H}_1 = \text{Ker}(\mathcal{T}) \oplus \overline{\text{Im}(\mathcal{T}^*)}$. As a consequence, when considering the properties of a compact operator in $\mathfrak{B}(\mathbb{H}_1, \mathbb{H}_2)$, there is no loss of generality in assuming that both $\mathbb{H}_1$ and $\mathbb{H}_2$

are separable. We will therefore make that assumption throughout the rest of this chapter.

4.2 Eigenvalues of compact operators

The use of eigenvalues and eigenvectors for real positive definite matrices is central to the development of the mva pca concept from Section 1.1. If $\mathscr{T}$ is a symmetric $p \times p$ positive semidefinite matrix, we know that it can be expressed as

$$\mathscr{T} = \sum_{j=1}^{p} \lambda_j e_j \otimes e_j \tag{4.3}$$

with $\lambda_1 \geq \lambda_2 \geq \cdots \geq \lambda_p \geq 0$, e_j an eigenvector corresponding to λ_j and $e_j \otimes e_j = e_j e_j^T$ the outer product of e_j with itself (see Definition 3.4.6). In this section, we aim to extend the eigenvalue–eigenvector decomposition to work with compact operators on Hilbert spaces. As one might suspect, the tools we develop here will eventually be employed in functional data parallels of the mva pca concept.

Let us focus on operators in $\mathfrak{B}(\mathbb{H})$ for some Hilbert space $\mathbb{H}$. The first step is to give a definition of an eigenvalue and eigenvector that is appropriate for that setting.

Definition 4.2.1 *Let $\mathscr{T} \in \mathfrak{B}(\mathbb{H})$ and suppose that there exists $\lambda \in \mathbb{R}$ and a nonzero $e \in \mathbb{H}$ such that*

$$\mathscr{T} e = \lambda e. \tag{4.4}$$

Then, λ is an eigenvalue and e is a corresponding eigenvector (or eigenfunction when $\mathbb{H}$ is a function space) of $\mathscr{T}$.

It is easy to see that $\mathrm{Ker}(\mathscr{T} - \lambda I)$ is a closed linear subspace of $\mathbb{H}$ and is therefore also a Hilbert space. Clearly, $\mathrm{Ker}(\mathscr{T} - \lambda I)$ is nontrivial if and only if λ is an eigenvalue, in which case we call $\mathrm{Ker}(\mathscr{T} - \lambda I)$ the eigenspace of λ.

Theorem 4.2.2 *Let $\mathscr{T} \in \mathfrak{B}(\mathbb{H})$ and suppose that $e_j \in \mathrm{Ker}(\mathscr{T} - \lambda_j I)$ for $j = 1, 2, \ldots$ where the λ_j are distinct and nonzero. Then,*

1. *the e_j are linearly independent and*
2. *if $\mathscr{T}$ is self-adjoint, the e_j are mutually orthogonal.*

Proof: For part 1, we need to show that if for some $c_1, \ldots, c_n$

$$\sum_{j=1}^{n} c_j e_j = 0 \quad \text{and} \quad \sum_{j=1}^{n} \lambda_j c_j e_j = 0 \tag{4.5}$$

then $c_1 = \cdots = c_n = 0$. We will establish this for all n by induction. We start with $n = 2$. Then, (4.5) implies

$$\lambda_1(c_1e_1 + c_2e_2) - (\lambda_1c_1e_1 + \lambda_2c_2e_2) = (\lambda_1 - \lambda_2)c_2e_2 = 0,$$

from which we conclude that $c_2 = 0$ as $\lambda_1 \neq \lambda_2$. This entails that $c_1 = 0$ as well since $c_1e_1 = -c_2e_2 = 0$.

Next, suppose we have shown the claim for some $n = k$ but we have found coefficients $c_1, \ldots, c_{k+1}$ such (4.5) holds for $n = k + 1$. In that case,

$$\lambda_{k+1}\sum_{j=1}^{k+1} c_je_j - \sum_{j=1}^{k+1} \lambda_jc_je_j = \sum_{j=1}^{k}(\lambda_{k+1} - \lambda_j)c_je_j = 0,$$

which, by the induction assumption, implies that $c_j = 0$ for all $k = 1, \ldots, k + 1$.

For part 2, write

$$\langle e_i, e_j\rangle = \langle e_i, \lambda_j^{-1}\mathscr{T}e_j\rangle = \lambda_j^{-1}\langle \mathscr{T}e_i, e_j\rangle = \lambda_j^{-1}\lambda_i\langle e_i, e_j\rangle.$$

As $\lambda_i \neq \lambda_j$, this can only be true if $\langle e_i, e_j\rangle = 0$. □

Theorem 4.2.3 *Let $\mathscr{T} \in \mathfrak{B}(\mathbb{H})$ be a compact operator. Then,*

1. $\mathrm{Ker}(\mathscr{T} - \lambda I)$ *is finite-dimensional for any* $\lambda \neq 0$,
2. *the number of distinct eigenvalues of $\mathscr{T}$ with absolute values bigger than any positive number is finite, and*
3. *the set of nonzero eigenvalues of $\mathscr{T}$ is countable.*

Proof: We need only prove parts 1 and 2 as part 3 is a direct consequence of these two results. To verify part 1, suppose that $\mathrm{Ker}(\mathscr{T} - \lambda I)$ is infinite-dimensional for some $\lambda \neq 0$. Then, the same construction that was employed in the proof of Theorem 4.1.2 can be used here to find a sequence of unit-norm elements $\{e_j\}$ in $\mathrm{Ker}(\mathscr{T} - \lambda I)$ such that $\inf_{i,j}\|e_i - e_j\| > 0$ and, hence,

$$\inf_{i,j} \|\mathscr{T}e_i - \mathscr{T}e_j\| = \lambda \inf_{i,j} \|e_i - e_j\| > 0.$$

This shows that $\{\mathscr{T}e_j\}$ does not contain a convergent subsequence and contradicts the assumption that $\mathscr{T}$ is compact.

For part 2, let $\lambda_1, \lambda_2, \ldots$ be a sequence of distinct eigenvalues with $|\lambda_j| \geq \lambda > 0$ for all j. Define $\mathbb{M}_n = \mathrm{span}\{e_1, \ldots, e_n\}$ for $e_j \in \mathrm{Ker}(\mathscr{T} - \lambda_jI)$ and take $\mathbb{M}_0 := \{0\}$. The e_j are linearly independent by part 1 of Theorem 4.2.2 which means that $\mathbb{M}_n \cap \mathbb{M}_{n-1}^{\perp}$ is a one-dimensional linear space for each n. We can now use the Gram–Schmidt recursion to create an orthonormal sequence $\{x_n\}$ such that $x_n \in \text{Ker}(\mathscr{T} - \lambda_n I) \cap \mathbb{M}_{n-1}^{\perp}$ for all n and, for $n > m$,

$$\|\mathscr{T}x_n - \mathscr{T}x_m\|^2 = \|\mathscr{T}x_n\|^2 + \|\mathscr{T}x_m\|^2 \geq 2\lambda^2.$$

Again, this contradicts the assumption that $\mathscr{T}$ is compact. □

With these preliminaries, we are now ready to present the eigenvalue–eigenvector decomposition for a self-adjoint compact operator.

Theorem 4.2.4 *Let $\mathscr{T}$ be a compact, self-adjoint operator on $\mathbb{H}$. The set of nonzero eigenvalues for $\mathscr{T}$ is either finite or consists of a sequence which tends to zero. Each nonzero eigenvalue has finite multiplicity and eigenvectors corresponding to different eigenvalues are orthogonal. Let $\lambda_1, \lambda_2, \ldots$ be the eigenvalues ordered so that $|\lambda_1| \geq |\lambda_2| \geq \cdots$ and let $e_1, e_2, \ldots$ be the corresponding orthonormal eigenvectors obtained using the Gram–Schmidt orthogonalization process as necessary for repeated eigenvalues. Then, $\{e_j\}$ is a CONS for $\overline{\text{Im}(\mathscr{T})}$ and*

$$\mathscr{T} = \sum_{j\geq 1} \lambda_j e_j \otimes e_j; \tag{4.6}$$

i.e, for every $x \in \mathbb{H}$,

$$\mathscr{T}x = \sum_{j\geq 1} \lambda_j \langle x, e_j \rangle e_j. \tag{4.7}$$

Proof: In view of Theorems 4.2.2 and 4.2.3, we only need to show (4.7). By part 6 of Theorems 3.3.7 and the fact that $\mathscr{T}$ is self-adjoint

$$\mathbb{H} = \text{Ker}(\mathscr{T}) \oplus \overline{\text{Im}(\mathscr{T})}.$$

As (4.7) holds for all $x \in \overline{\text{span}\{e_j : j \geq 1\}}$ or $\text{Ker}(\mathscr{T})$, it suffices to show that $\{e_j\}_{j=1}^{\infty}$ is a CONS for $\overline{\text{Im}(\mathscr{T})}$: namely, that

$$\overline{\text{Im}(\mathscr{T})} = \overline{\text{span}\{e_j : j \geq 1\}}. \tag{4.8}$$

For any finite n and nonzero $c_1, \ldots, c_n \in \mathbb{R}$, we have $\sum_{j=1}^{n} c_j e_j \in \overline{\text{Im}(\mathscr{T})}$ as

$$\sum_{j=1}^{n} c_j e_j = \mathscr{T}\left(\sum_{j=1}^{n} \lambda_j^{-1} c_j e_j\right).$$

This shows that $\text{span}\{e_j : j \geq 1\} \subset \overline{\text{Im}(\mathscr{T})}$ so that $\overline{\text{span}\{e_j : j \geq 1\}} \subset \overline{\text{Im}(\mathscr{T})}$.

From (2.16), we can write

$$\overline{\mathrm{Im}(\mathscr{T})} = \overline{\mathrm{span}\{e_j : j \geq 1\}} \oplus \mathbb{N},$$

where $\mathbb{N}$ contains those elements in $\overline{\mathrm{Im}(\mathscr{T})}$ that are orthogonal to $\overline{\mathrm{span}\{e_j : j \geq 1\}}$. Clearly, $\mathscr{T}$ maps $\mathrm{span}\{e_j : j \geq 1\}$ into its closure. It also maps $\mathbb{N}$ into $\mathbb{N}$. To see that this latter claim is valid let $x \in \mathbb{N}$ and $y \in \overline{\mathrm{span}\{e_j : j \geq 1\}}$. Then,

$$\langle \mathscr{T}x, y\rangle = \langle x, \mathscr{T}y\rangle = 0$$

and $\mathscr{T}x \in \overline{\mathrm{span}\{e_j : j \geq 1\}}^{\perp} = \mathbb{N}$.

Let $\mathscr{T}_{\mathbb{N}}$ be the restriction of $\mathscr{T}$ to $\mathbb{N}$ and observe that any nonzero eigenvalue of $\mathscr{T}_{\mathbb{N}}$ must also be an eigenvalue for the original operator $\mathscr{T}$. If $\mathscr{T}_{\mathbb{N}}$ is not the zero operator, it is easy to see that either $\|\mathscr{T}_{\mathbb{N}}\|$ or $-\|\mathscr{T}_{\mathbb{N}}\|$ must be one of its eigenvalues and, hence, an eigenvalue for $\mathscr{T}$ as well. However, all the nonzero eigenvalues for $\mathscr{T}$ were already captured in the collection $\{\lambda_j, j \geq 1\}$ which leaves $\mathscr{T}_{\mathbb{N}}$ being the zero operator as the only option. Thus, $\mathbb{N} \subset \overline{\mathrm{Im}(\mathscr{T})} \cap \mathrm{Ker}(\mathscr{T}) = \{0\}$ and (4.8) is proved. □

The sum in (4.6) is taken over all positive values for the index. This allows for the possibility that the operator has either finitely or infinitely many nonzero eigenvalues. Our interest is primarily in the latter instance and, as a result, we will usually write our sums as having an infinite upper limit with the understanding that all but finitely many eigenvalues will be zero for a finite dimensional operator.

It is clear that all eigenvalues of a nonnegative definite operator are nonnegative, for if (λ, e) is an eigenvalue/eigenvector pair of $\mathscr{T}$ then

$$\langle \mathscr{T}e, e\rangle = \lambda\|e\|^2.$$

If $\mathscr{T}$ is compact and self-adjoint, then Theorem 4.2.4 entails that $\mathscr{T}$ is nonnegative definite if and only if all eigenvalues of $\mathscr{T}$ are nonnegative.

The eigenvalues of compact operators provide solutions to various variational problems. For example, in the case of nonnegative operators, we have the following rather immediate result.

Theorem 4.2.5 *If $\mathscr{T}$ is compact and nonnegative definite with associated eigenvalue/eigenvector sequence $\{(\lambda_j, e_j)\}_{j=1}^{\infty}$, then*

$$\lambda_k = \max_{e \in \mathrm{span}\{e_1,\ldots,e_{k-1}\}^{\perp}} \frac{\langle \mathscr{T}e, e\rangle}{\|e\|^2}, \tag{4.9}$$

for all k with $\mathrm{span}\{e_1, \ldots, e_{k-1}\}^{\perp}$ *being the entirety of $\mathbb{H}$ when $k = 1$.*

This result holds somewhat more generally. From Theorem 4.2.4, a self-adjoint compact operator with the representation (4.6) satisfies

$$\mathscr{T}^n = \sum_{j\geq 1} \lambda_j^n e_j \otimes e_j \tag{4.10}$$

for any positive integer n: namely, $(\lambda_j^n, e_j), j \geq 1$, are the eigenvalue–eigenvector pairs of $\mathscr{T}^n$. Thus, the following is a straightforward consequence of Theorem 4.2.5.

Theorem 4.2.6 *Let $\mathscr{T} \in \mathfrak{B}(\mathbb{H})$ be a compact, self-adjoint operator with eigenvalue-eigenvector pairs $\{(\lambda_j, e_j)\}_{j=1}^{\infty}$ arranged so that $\lambda_1^2 \geq \lambda_2^2 \geq \cdots$. Then,*

$$|\lambda_k| = \max_{e \in \text{span}\{e_1,\ldots,e_{k-1}\}^{\perp}} \frac{\|\mathscr{T}e\|}{\|e\|}. \tag{4.11}$$

Note that Theorem 4.2.6 characterizes the operator norm of a self-adjoint compact operator as being the absolute value of its eigenvalue with the largest magnitude.

If the compact, self-adjoint operator $\mathscr{T}$ is also nonnegative-definite, then all the eigenvalues are nonnegative and the representation in (4.10) can be extended to noninteger powers. For example, this allows us to define the self-adjoint operator

$$\mathscr{T}^{1/2} = \sum_{j=1}^{\infty} \sqrt{\lambda_j} e_j \otimes e_j,$$

which satisfies $\mathscr{T}^{12}\mathscr{T}^{1/2} = \mathscr{T}$ (cf. Theorem 3.4.3). Note that $\mathscr{T}^{1/2}$ is necessarily a compact element of $\mathfrak{B}(\mathbb{H})$.

A more profound characterization of eigenvalues than the one in Theorem 4.2.5 is the Courant–Fischer minimax principle that can be described in the following manner.

Theorem 4.2.7 *Let $\mathscr{T}$ be a nonnegative, compact operator on a Hilbert space $\mathbb{H}$ with eigenvalues $\lambda_1 \geq \lambda_2 \geq \cdots \geq 0$. Then,*

$$\lambda_k = \max_{v_1,\ldots,v_k \in \mathbb{H}} \min_{v \in \text{span}\{v_1,\ldots,v_k\}} \frac{\langle \mathscr{T}v, v\rangle}{\|v\|^2} \tag{4.12}$$

and

$$\lambda_k = \min_{v_1,\ldots,v_{k-1} \in \mathbb{H}} \max_{v \in \text{span}\{v_1,\ldots,v_{k-1}\}^{\perp}} \frac{\langle \mathscr{T}v, v\rangle}{\|v\|^2}, \tag{4.13}$$

where the max and min in (4.12) and (4.13) are attained when v_k is the eigenvector e_k that corresponds to λ_k.

This is a truly remarkable result and, accordingly, a bit of interpretation is in order before we proceed to the proof. For example, the minimax relation (4.13) states that one can pick any $k-1$ dimensional subspace $\mathbb{V}_{k-1}$ and the maximum of $\mathscr{T}$'s Rayleigh quotient

$$R_{\mathscr{T}}(v) := \frac{\langle \mathscr{T} v, v\rangle}{\|v\|^2} \tag{4.14}$$

across all vectors that are in the orthogonal complement of this subspace can be no smaller than λ_k: i.e.,

$$\lambda_k \leq \max_{v \in \mathbb{V}_{k-1}^{\perp}} \frac{\langle \mathscr{T} v, v\rangle}{\|v\|^2}$$

for all $\mathbb{V}_{k-1}$ with equality for $\mathbb{V}_{k-1} = \text{span}\{e_1, \ldots, e_{k-1}\}$ and $v = e_k$.

Proof: Define $\mathbb{M}_{k-1} = \text{span}\{e_1, \ldots, e_{k-1}\}$. Now suppose that $\mathbb{V}_k$ is any k dimensional subspace of $\overline{\text{Im}(\mathscr{T})}$ and note that $\mathbb{V}_k \cap \mathbb{M}_{k-1}^{\perp}$ is nonempty. If v is any element in this intersection, we can write it as $v = \sum_{j=k}^{\infty} a_j e_j$ to see that

$$\frac{\langle \mathscr{T} v, v\rangle}{\|v\|^2} = \frac{\sum_{j=k}^{\infty} \lambda_j a_j^2}{\sum_{j=k}^{\infty} a_j^2} \leq \lambda_k.$$

Thus,

$$\min_{v \in \mathbb{V}_k} \frac{\langle \mathscr{T} v, v\rangle}{\|v\|^2} \leq \min_{v \in \mathbb{V}_k \cap \mathbb{M}_{k-1}^{\perp}} \frac{\langle \mathscr{T} v, v\rangle}{\|v\|^2} \leq \lambda_k,$$

where the continuity of the Rayleigh quotient allows us to conclude that these minima all exist. However, the choice of $\mathbb{V}_k$ was arbitrary which means that (4.12) holds because equality is obtained with the choice $\mathbb{V}_k = \text{span}\{e_1, \ldots, e_k\}$ by Theorem 4.2.5.

Things work similarly in proving (4.13). The difference is that $\mathbb{V}_{k-1}^{\perp} \cap \mathbb{M}_k$ is now the nonempty set of interest. In this case, an element in the intersection looks like $v = \sum_{j=1}^{k} a_j e_j$, which produces the inequality

$$\frac{\langle \mathscr{T} v, v\rangle}{\|v\|^2} = \frac{\sum_{j=1}^{k} \lambda_j a_j^2}{\sum_{j=1}^{k} a_j^2} \geq \lambda_k$$

that is needed to complete the proof. □

A consequence of Theorem 4.2.7 that will eventually prove useful is given in the following theorem.

Theorem 4.2.8 *Let $\mathscr{T}, \tilde{\mathscr{T}}$ be nonnegative definite, compact operators with eigenvalue sequences $\{\lambda_j\}$ and $\{\tilde{\lambda}_j\}$, respectively. Then*

$$\sup_{k\geq 0} |\lambda_{k+1} - \tilde{\lambda}_{k+1}| \leq \|\mathscr{T} - \tilde{\mathscr{T}}\|.$$

Proof: By Theorem 4.2.7,

$$\lambda_{k+1} = \min_{v_1,\ldots,v_k \in \mathbb{H}} \max_{v \in \text{span}\{v_1,\ldots,v_k\}^{\perp}} \frac{\langle \mathscr{T} v, v\rangle}{\|v\|^2}.$$

Clearly, for any $v_1, \ldots, v_k$,

$$\begin{aligned}
\max_{v \in \text{span}\{v_1,\ldots,v_k\}^{\perp}} \frac{\langle \mathscr{T} v, v\rangle}{\|v\|^2} &= \max_{v \in \text{span}\{v_1,\ldots,v_k\}^{\perp}} \frac{\langle (\tilde{\mathscr{T}} + \mathscr{T} - \tilde{\mathscr{T}})v, v\rangle}{\|v\|^2} \\
&\leq \max_{v \in \text{span}\{v_1,\ldots,v_k\}^{\perp}} \frac{\langle \tilde{\mathscr{T}} v, v\rangle}{\|v\|^2} \\
&\quad + \max_{v \in \mathbb{H}} \frac{\langle (\mathscr{T} - \tilde{\mathscr{T}})v, v\rangle}{\|v\|^2},
\end{aligned}$$

where the second term of the right-hand side of the inequality is clearly bounded by $\|\mathscr{T} - \tilde{\mathscr{T}}\|$. Taking the minimum over $v_1, \ldots, v_k \in \mathbb{H}$ on both sides gives

$$\lambda_{k+1} \leq \tilde{\lambda}_{k+1} + \|\mathscr{T} - \tilde{\mathscr{T}}\|.$$

Interchanging $\mathscr{T}$ and $\tilde{\mathscr{T}}$ gives the result. □

The previous two results, as stated, would only seem to be applicable to nonnegative compact operators. However, this is a bit deceiving and they can actually be used with any self-adjoint, compact operator once one separates its eigenvalues into those that are all positive or all negative. Using Theorem 4.2.4, we can write any self-adjoint compact operator as

$$\mathscr{T} = \mathscr{T}_+ - \mathscr{T}_-,$$

where

$$\mathscr{T}_+ = \sum_{\lambda_j > 0} \lambda_j e_j \otimes e_j$$

and

$$\mathscr{T}_- = \sum_{\lambda_j < 0} (-\lambda_j) e_j \otimes e_j.$$

Results for nonnegative definite compact operators can now be applied to $\mathscr{T}_-$ as well as $\mathscr{T}_+$ to, e.g., produce variational results for the negative eigenvalues.

4.3 The singular value decomposition

The spectral decomposition for self-adjoint compact operators on a Hilbert space can be extended to include situations involving compact operators between two Hilbert spaces. The result is generally referred to as the svd or singular value expansion (sve) of an operator.

Let $\mathbb{H}_1$ and $\mathbb{H}_2$ be Hilbert spaces with inner products $\langle\cdot,\cdot\rangle_i$ and norms $\|\cdot\|_i$, $i = 1, 2$. Suppose that $\mathscr{T}$ in $\mathfrak{B}(\mathbb{H}_1, \mathbb{H}_2)$ is compact. We know by Theorem 4.1.3 and Theorem 4.1.5 that $\mathscr{T}^*\mathscr{T}$ is also compact. Invoking Theorem 4.2.4, we conclude that $\mathscr{T}^*\mathscr{T}$ has eigenvalue–eigenvector pairs $(\lambda_j^2, f_{1j}), j = 1, 2, \ldots$, where $\lambda_1^2 \geq \lambda_2^2 \geq \cdots \geq 0$ and the f_{1j}'s are orthonormal. By the same argument, $\mathscr{T}\mathscr{T}^*$ is also compact and nonnegative definite with

$$(\mathscr{T}\mathscr{T}^*)\mathscr{T}f_{1j} = \mathscr{T}(\mathscr{T}^*\mathscr{T}f_{1j}) = \lambda_j^2\mathscr{T}f_{1j}, j = 1, 2, \ldots.$$

Thus, λ_j^2 is an eigenvalue for $\mathscr{T}\mathscr{T}^*$ with the corresponding unit-norm eigenvector $f_{2j} := \mathscr{T}f_{1j}/\lambda_j$, where $\lambda_j = \sqrt{\lambda_j^2}$. One can reverse this argument and conclude that every eigenvalue of $\mathscr{T}\mathscr{T}^*$ is also an eigenvalue of $\mathscr{T}^*\mathscr{T}$. Consequently, $\mathscr{T}^*\mathscr{T}$ and $\mathscr{T}\mathscr{T}^*$ have the same eigenvalues. This provides the intuitive basis for our statement of the following result.

Theorem 4.3.1 *Let $\mathbb{H}_1$ and $\mathbb{H}_2$ be Hilbert spaces and let $\mathscr{T}$ be a compact operator from $\mathbb{H}_1$ into $\mathbb{H}_2$. Then,*

$$\mathscr{T} = \sum_{j=1}^{\infty} \lambda_j (f_{1j} \otimes_1 f_{2j}), \tag{4.15}$$

where (4.15) is interpreted as meaning that

$$\mathscr{T}x = \sum_{j=1}^{\infty} \lambda_j \langle x, f_{1j}\rangle_1 f_{2j} \tag{4.16}$$

for any $x \in \mathbb{H}_1$ with

(i) $\{\lambda_j^2\}$ the nonascending eigenvalues of $\mathscr{T}^*\mathscr{T}$ and $\mathscr{T}\mathscr{T}^*$,

(ii) $\{f_{1j}\}$ the orthonormal eigenvectors of $\mathscr{T}^*\mathscr{T}$, and

(iii) $\{f_{2j}\}$ the orthonormal eigenvectors of $\mathscr{T}\mathscr{T}^*$ satisfying $\mathscr{T}^* f_{2j} = \lambda_j f_{1j}$ for $\lambda_j = \sqrt{\lambda_j^2}$.

Proof: In view of the discussion preceding the theorem and Theorem 3.3.7, we only need to verify (4.16) for $x \in (\mathrm{Ker}(\mathscr{T}))^{\perp}$. Applying parts 4 and 5 of Theorem 3.3.7 and part 3 of Theorem 2.5.6 gives

$$(\mathrm{Ker}(\mathscr{T}))^{\perp} = (\mathrm{Ker}(\mathscr{T}^*\mathscr{T}))^{\perp} = (\mathrm{Im}(\mathscr{T}^*\mathscr{T}))^{\perp\perp} = \overline{\mathrm{Im}(\mathscr{T}^*\mathscr{T})}.$$

However, in the proof of Theorem 4.2.4, we established relation (4.8) that translates here to saying that

$$(\mathrm{Ker}(\mathscr{T}))^{\perp} = \overline{\mathrm{span}\{f_{1j} : j \geq 1\}}.$$

Thus,

$$x = \sum_{j=1}^{\infty} \langle x, f_{1j} \rangle_1 f_{1j}$$

for any $x \in (\mathrm{Ker}(\mathscr{T}))^{\perp}$ and (4.16) follows immediately for any such x. □

The triples $(\lambda_j, f_{1j}, f_{2j})$, $j = 1, 2, \ldots$ in Theorem 4.3.1 are sometimes called a *singular system* for the operator $\mathscr{T}$. The $\{\lambda_j\}$ are termed the *singular values* of $\mathscr{T}$ while the $\{f_{1j}\}$ and $\{f_{2j}\}$ are the right and left *singular vectors* or *singular functions* (as appropriate), respectively.

The matrix svd is a special case of the operator svd. We state this formally as

Corollary 4.3.2 *If $\mathscr{T}$ is a $p \times q$ matrix of rank $k \leq \min(p, q)$, there exist orthonormal matrices*

$$\mathscr{U} = [u_1, \ldots, u_p] \in \mathbb{R}^{p\times p}$$

and

$$\mathscr{V} = [v_1, \ldots, v_q] \in \mathbb{R}^{q\times q}$$

such that

$$\mathscr{T} = \mathscr{U}\Lambda\mathscr{V}^T = \sum_{j=1}^{k} \lambda_j u_j v_j^T,$$

where $\Lambda = \mathrm{diag}(\lambda_1, \ldots, \lambda_k, 0, \ldots, 0)$ *with* $\lambda_1 \geq \lambda_2 \geq \cdots \geq \lambda_k > 0$.

Proof: As $\mathscr{T}$ is of finite rank, it is necessarily compact. Thus, by Theorem 4.3.1, $\mathscr{T} = \sum_{j=1}^{k} \lambda_j (v_j \otimes_1 u_j)$ with

(i) $\lambda_1^2 \geq \cdots \lambda_k^2 > 0$ the nonzero eigenvalues of $\mathscr{T}^T\mathscr{T}$, and $\mathscr{T}\mathscr{T}^T$,

(ii) $v_1, \ldots, v_q$ the eigenvectors of $\mathscr{T}^T\mathscr{T}$ and

(iii) $u_1, \ldots, u_p$ the eigenvectors of $\mathscr{T}\mathscr{T}^T$.

The tensor-product operator $\otimes_1$ in this case is just the vector outer product: i.e., $v_j \otimes_1 u_j = u_j v_j^T$. □

The singular values for a compact operator inherit certain optimality properties from the fact that they are eigenvalues. For instance, an svd version of Theorem 4.2.7 takes the following form.

Theorem 4.3.3 *Let $\mathscr{T} \in \mathfrak{B}(\mathbb{H}_1, \mathbb{H}_2)$ be a compact with singular values $\lambda_1 \geq \lambda_2 \geq \cdots \geq 0$. Then*

$$\lambda_k = \max_{v_1,\ldots,v_k \in \mathbb{H}_1} \min_{f \in \text{span}\{v_1,\ldots,v_k\}} \frac{\|\mathscr{T}f\|_2}{\|f\|_1} \tag{4.17}$$

and

$$\lambda_k = \min_{v_1,\ldots,v_{k-1} \in \mathbb{H}_1} \max_{f \in \text{span}\{v_1,\ldots,v_{k-1}\}^\perp} \frac{\|\mathscr{T}f\|_2}{\|f\|_1}, \tag{4.18}$$

where the max and min in (4.12) and (4.13) are attained with v_k equal to the right singular vector f_{1k} corresponding to λ_k.

Proof: The proof is immediate from Theorem 4.2.7 as the λ_i^2 are the nonascending eigenvalues of $\mathscr{T}^*\mathscr{T}$. □

A simple but important case of the previous result occurs when $k = 1$. In that instance, we obtain

Theorem 4.3.4 *Let $\mathbb{H}_1$ and $\mathbb{H}_2$ be Hilbert spaces and let $\mathscr{T}$ be a compact operator from $\mathbb{H}_1$ into $\mathbb{H}_2$ whose largest singular value is λ_1. Then,*

$$\|\mathscr{T}\| = \lambda_1.$$

The following result provides a fundamental characterization of compact operators.

Theorem 4.3.5 *An operator $\mathscr{T} \in \mathfrak{B}(\mathbb{H}_1, \mathbb{H}_2)$ is compact if and only if the sve (4.15) holds.*

Proof: Theorem 4.3.1 establishes the sve for any compact operator. To proceed in the other direction, assume that (4.15) holds and define

$$\mathscr{T}_n = \sum_{j=1}^{n} \lambda_j (f_{1j} \otimes_1 f_{2j}).$$

Then, $\mathscr{T}_n$ is finite dimensional and hence compact. By Theorem 4.3.4, $\|\mathscr{T} - \mathscr{T}_n\| = \lambda_{n+1} \to 0$. The compactness of $\mathscr{T}$ is now a consequence of part 1 of Theorem 4.1.5. □

An alternative version of (4.17) that directly connects the left and right singular vectors takes the following form.

Theorem 4.3.6 *Let $\mathscr{T} \in \mathfrak{B}(\mathbb{H}_1, \mathbb{H}_2)$ be a compact with associated singular system $\{(\lambda_j, f_{1j}, f_{2j})\}_{j=1}^{\infty}$. Then,*

$$\lambda_k = \max_{\substack{v_{11},\ldots,v_{1k}\in\mathbb{H}_1 \\ v_{21},\ldots,v_{2k}\in\mathbb{H}_2}} \min_{\substack{f_1\in\text{span}\{v_{11},\ldots,v_{1k}\} \\ f_2\in\text{span}\{v_{21},\ldots,v_{2k}\}}} \frac{|\langle \mathscr{T}f_1, f_2\rangle_2|}{\|f_1\|_1\|f_2\|_2}, \tag{4.19}$$

where the maximum is attained with $v_{1k} = f_{1k}$ and $v_{2k} = f_{2k}$.

Proof: Analogous to the arguments for Theorem 4.2.7, we define

$$\mathbb{M}_{i(k-1)} = \text{span}\{f_{i1}, \ldots, f_{i(k-1)}\}$$

for $i = 1, 2$ and choose k dimensional subspaces $\mathbb{V}_{ik}$ with $\mathbb{V}_{ik} \subset \mathbb{H}_i, i = 1, 2$. Then, the subspaces $\mathbb{M}_{i(k-1)}^{\perp} \cap \mathbb{V}_{ik}, i = 1, 2$, are not empty and we can express any element in $\mathbb{M}_{i(k-1)}^{\perp} \cap \mathbb{V}_{ik}$ as $f_i = \sum_{j=k}^{\infty} a_{ij} f_{ij}$ with $\|f_i\|_i^2 = \sum_{j=k}^{\infty} a_{ij}^2$. Thus,

$$\frac{\langle \mathscr{T}f_1, f_2\rangle_2^2}{\|f_1\|_1^2\|f_2\|_2^2} = \frac{\left(\sum_{j=k}^{\infty} \lambda_j a_{1j} a_{2j}\right)^2}{\left(\sum_{j=k}^{\infty} a_{1j}^2\right)\left(\sum_{j=k}^{\infty} a_{2j}^2\right)} \leq \lambda_k^2$$

from the Cauchy–Schwarz inequality. This means that

$$\min_{f_1\in\mathbb{V}_{1k}, f_2\in\mathbb{V}_{2k}} \frac{\langle \mathscr{T}f_1, f_2\rangle_2^2}{\|f_1\|_1^2\|f_2\|_2^2} \leq \lambda_k^2$$

for arbitrary choices of $\mathbb{V}_{ik}$ with equality when $\mathbb{V}_{ik} = \mathbb{M}_{ik}, i = 1, 2$, as a result of Theorem 4.3.1. □

For compact operators, the Moore–Penrose inverse from Section 3.5 can be characterized in terms of the operator's singular system. Assume that $\{e_{ij}\}_{j=1}^{\infty}, i = 1, 2$ are CONSs for $\mathbb{H}_1$ and $\mathbb{H}_2$ so that, for example, any $x \in \mathbb{H}_1$ can be expressed as $x = \sum_{j=1}^{\infty} \langle x, e_{1j}\rangle_1 e_{1j}$ with $\sum_{j=1}^{\infty} \langle x, e_{1j}\rangle_1^2 < \infty$. The sve then gives

$$\mathscr{T}x = \sum_{j=1}^{\infty} \lambda_j \langle x, e_{1j}\rangle_1 e_{2j}.$$

A condition that characterizes $y = \sum_{j=1}^{\infty} \langle y, e_{2j} \rangle_2 e_{2j} \in \text{Im}(\mathscr{T})$ is now seen to be that

$$\sum_{j=1}^{\infty} \frac{\langle y, e_{2j} \rangle_2^2}{\lambda_j^2} < \infty. \tag{4.20}$$

This is known as the Picard condition and, when it holds, we can write

$$\mathscr{T}^{\dagger} y = \sum_{j=1}^{\infty} \frac{\langle y, e_{2j} \rangle_2}{\lambda_j} e_{1j}. \tag{4.21}$$

The following result builds on the previous discussion and uses Theorem 4.1.4 to provide some additional insight into the properties of compact operators.

Theorem 4.3.7 *Suppose that $\mathscr{T}$ is an infinite-dimensional compact operator between two Hilbert spaces. Then,* $\text{Im}(\mathscr{T})$ *is not closed.*

Proof: Let $\mathscr{T}$ have the singular system $\{(\lambda_j, f_{1j}, f_{2j})\}_{j=1}^{\infty}$. As $\lambda_j \downarrow 0$ as $j \to \infty$, we can choose a subsequence $\{j_k\}$ such that $\lambda_{j_k} \leq k^{-2}$ for all k. Define

$$y = \sum_{k=1}^{\infty} \lambda_{j_k} f_{2j_k}.$$

Clearly, $y \in \overline{\text{Im}(\mathscr{T})}$ as $f_{2j_k} = \mathscr{T}(f_{1j_k}/\lambda_{j_k})$. However, the Picard condition does not hold for y. Thus, $y \notin \text{Im}(\mathscr{T})$. □

4.4 Hilbert–Schmidt operators

There are various subclasses of the set of compact operators that arise in our work ahead. One of these is the collection of Hilbert–Schmidt operators that are the object of interest for this section. We begin by stating a result that will prove useful in several places below in this section.

Theorem 4.4.1 *Let $\mathscr{T}_1$ and $\mathscr{T}_2$ be operators in $\mathfrak{B}(\mathbb{H}_1, \mathbb{H}_2)$. Suppose that there exist CONSs $\{e_{1i}\}$ and $\{e_{2j}\}$ for $\mathbb{H}_1$ and $\mathbb{H}_2$, respectively, such that either*

$$\sum_{i=1}^{\infty} (\|\mathscr{T}_1 e_{1i}\|_2^2 + \|\mathscr{T}_2 e_{1i}\|_2^2) < \infty, \tag{4.22}$$

or

$$\sum_{j=1}^{\infty}(\|\mathscr{T}_1^* e_{2j}\|_1^2 + \|\mathscr{T}_2^* e_{2j}\|_1^2) < \infty. \tag{4.23}$$

Then,

$$\begin{aligned}\sum_{i=1}^{\infty} \langle \mathscr{T}_1 e_{1i}, \mathscr{T}_2 e_{1i}\rangle_2 &= \sum_{i=1}^{\infty}\sum_{j=1}^{\infty} \langle \mathscr{T}_1 e_{1i}, e_{2j}\rangle_2 \langle \mathscr{T}_2 e_{1i}, e_{2j}\rangle_2 \\ &= \sum_{i=1}^{\infty}\sum_{j=1}^{\infty} \langle e_{1i}, \mathscr{T}_1^* e_{2j}\rangle_1 \langle e_{1i}, \mathscr{T}_2^* e_{2j}\rangle_1 \\ &= \sum_{j=1}^{\infty} \langle \mathscr{T}_1^* e_{2j}, \mathscr{T}_2^* e_{2j}\rangle_1, \end{aligned} \tag{4.24}$$

where this identity holds for all choices of the CONSs $\{e_{1i}\}$ and $\{e_{2j}\}$.

Proof: First we verify (4.24) for the special case $\mathscr{T}_1 = \mathscr{T}_2 = \mathscr{T}$. The argument is straightforward using the expansions

$$\mathscr{T} e_{1i} = \sum_{j=1}^{\infty} \langle \mathscr{T} e_{1i}, e_{2j}\rangle_2 e_{2j}$$

and

$$\mathscr{T}^* e_{2j} = \sum_{i=1}^{\infty} \langle \mathscr{T}^* e_{2j}, e_{1i}\rangle_1 e_{1i}.$$

Thus, we have

$$\begin{aligned}\sum_{i=1}^{\infty} \|\mathscr{T} e_{1i}\|_2^2 &= \sum_{i=1}^{\infty}\sum_{j=1}^{\infty} \langle \mathscr{T} e_{1i}, e_{2j}\rangle_2^2 \\ &= \sum_{i=1}^{\infty}\sum_{j=1}^{\infty} \langle e_{1i}, \mathscr{T}^* e_{2j}\rangle_1^2 \\ &= \sum_{j=1}^{\infty} \|\mathscr{T}^* e_{2j}\|_1^2. \end{aligned} \tag{4.25}$$

As the last expression does not depend on $\{e_{1i}\}$, we can conclude that all our previous formulae are also independent of this choice for the CONS; the same argument then leads to the conclusion that the expressions are independent of the choice for $\{e_{2j}\}$ as well.

Next, consider the three identities obtained by replacing $\mathscr{T}$ in (4.25) with $\mathscr{T}_1 + \mathscr{T}_2$, $\mathscr{T}_1$ and $\mathscr{T}_2$. Using (4.22) or (4.23), we can obtain (4.24) by subtracting the second and third of these identities from the first. □

Definition 4.4.2 *Let $\{e_i\}$ be a CONS for $\mathbb{H}_1$ and $\mathscr{T} \in \mathfrak{B}(\mathbb{H}_1, \mathbb{H}_2)$. If $\mathscr{T}$ satisfies*

$$\sum_{i=1}^{\infty} \|\mathscr{T} e_i\|_2^2 < \infty,$$

then $\mathscr{T}$ is called a Hilbert–Schmidt (HS) operator. The collection of HS operators in $\mathfrak{B}(\mathbb{H}_1, \mathbb{H}_2)$ is denoted by $\mathfrak{B}_{HS}(\mathbb{H}_1, \mathbb{H}_2)$.

Theorem 4.4.1 shows that Definition 4.4.2 does not depend on the choice of CONS. It also reveals that $\mathscr{T}$ is HS if $\mathscr{T}^*$ is HS. Our next result has the consequence that Hilbert–Schmidt operators constitute a subclass of the collection of all compact operators.

Theorem 4.4.3 *A Hilbert–Schmidt operator is compact.*

Proof: Let $\mathscr{T}$ be a HS operator and for $x \in \mathbb{H}_1$ define

$$\mathscr{T}_n x = \sum_{i=1}^{n} \langle \mathscr{T} x, e_{2i}\rangle_2 e_{2i},$$

where $\{e_{2i}\}$ is a CONS for $\mathbb{H}_2$. Clearly, the range of $\mathscr{T}_n$ is finite-dimensional and so $\mathscr{T}_n$ is compact. It suffices to show that $\|\mathscr{T} - \mathscr{T}_n\| \to 0$. However, $(\mathscr{T}_n - \mathscr{T})x = \sum_{i=n+1}^{\infty} \langle \mathscr{T} x, e_{2i}\rangle_2 e_{2i}$ and, hence, if $\|x\|_1 \leq 1$ an application of the Cauchy–Schwarz inequality produces

$$\|(\mathscr{T}_n - \mathscr{T})x\|_2^2 = \sum_{i=n+1}^{\infty} \langle \mathscr{T} x, e_{2i}\rangle_2^2 = \sum_{i=n+1}^{\infty} \langle x, \mathscr{T}^* e_{2i}\rangle_1^2 \leq \sum_{i=n+1}^{\infty} \|\mathscr{T}^* e_{2i}\|_1^2,$$

which tends to 0 as n tends to ∞. □

Clearly, $\mathfrak{B}_{HS}(\mathbb{H}_1, \mathbb{H}_2)$ is a linear space. We can also construct an associated inner product in the following manner.

Definition 4.4.4 *The inner product of $\mathscr{T}_1, \mathscr{T}_2 \in \mathfrak{B}_{HS}(\mathbb{H}_1, \mathbb{H}_2)$ is*

$$\langle \mathscr{T}_1, \mathscr{T}_2\rangle_{HS} = \sum_{j=1}^{\infty} \langle \mathscr{T}_1 e_j, \mathscr{T}_2 e_j\rangle_2, \tag{4.26}$$

where $\{e_j\}$ is any CONS for $\mathbb{H}_1$. The HS norm of $\mathscr{T}$, $\|\mathscr{T}\|_{HS}$, is the norm determined by the HS inner product: namely,

$$\|\mathscr{T}\|_{HS} = \left\{ \sum_{j=1}^{\infty} \|\mathscr{T} e_j\|_2^2 \right\}^{1/2}.$$

Theorem 4.4.1 again shows that the definition for the HS norm and inner product is independent of the choice of CONS. It also entails that

$$\|\mathscr{T}\|_{HS}^2 = \sum_{i=1}^{\infty} \sum_{j=1}^{\infty} \langle \mathscr{T} e_{1i}, e_{2j} \rangle_2^2 \tag{4.27}$$

for arbitrary CONSs $\{e_{1i}\}$ and $\{e_{2j}\}$ of $\mathbb{H}_1$ and $\mathbb{H}_2$, respectively. In particular by choosing the CONSs to be the singular vectors of $\mathscr{T}$ we obtain

$$\|\mathscr{T}\|_{HS}^2 = \sum_{i=1}^{\infty} \lambda_j^2, \tag{4.28}$$

where the λ_j are the singular values of $\mathscr{T}$.

In the finite-dimensional case, the HS norm is often referred to as the Frobenius norm. Suppose that $\mathscr{T} = \{\tau_{ij}\}_{i=1:q,j=1:p}$ is a $q \times p$ real matrix: i.e., it is a linear operator that maps $\mathbb{H}_1 = \mathbb{R}^p$ into $\mathbb{H}_2 = \mathbb{R}^q$. Then, its squared HS norm is

$$\|\mathscr{T}\|_{HS}^2 = \sum_{i=1}^{q} \sum_{j=1}^{p} \tau_{ij}^2 = \text{trace}(\mathscr{T}^T \mathscr{T}) \tag{4.29}$$

with trace the matrix trace. To see this, choose $e_1, \dots, e_p$ so that e_j is a vector of all zeros except for a one as its jth element. Then, $\mathscr{T} e_j$ is the jth column of $\mathscr{T}$ with $\|\mathscr{T} e_j\|^2 = e_j^T \mathscr{T}^T \mathscr{T} e_j = \sum_{i=1}^{q} \tau_{ij}^2$.

The following theorem validates our choice for the inner product in Definition 4.4.4.

Theorem 4.4.5 *The linear space $\mathfrak{B}_{HS}(\mathbb{H}_1, \mathbb{H}_2)$ is a separable Hilbert space when equipped with the HS inner product. For any choice of CONS $\{e_{1i}\}$ and $\{e_{2j}\}$ for $\mathbb{H}_1$ and $\mathbb{H}_2$, respectively, $\{e_{1i} \otimes_1 e_{2j}\}$ is a CONS of $\mathfrak{B}_{HS}(\mathbb{H}_1, \mathbb{H}_2)$.*

Proof: As $\|\mathscr{T}\|_{HS}^2 = \sum_{i=1}^{\infty} \sum_{j=1}^{\infty} a_{ij}^2$ for $a_{ij} = \langle \mathscr{T} e_{1i}, e_{2j} \rangle_2$, completeness under the HS norm is a consequence of the completeness of the ℓ^2 space (Theorem 2.3.5). It only remains to show that $\{e_{1i} \otimes_1 e_{2j}\}$ is a CONS. Orthonormality is obvious in this instance. Suppose then that $\mathscr{T}$ satisfies $\langle \mathscr{T}, e_{1i} \otimes_1 e_{2j} \rangle_{HS} = 0$ for all i,j. This is equivalent to $\langle \mathscr{T} e_{1i}, e_{2j} \rangle_2 = 0$ for all i,j, which, in turn, implies that $\mathscr{T} = 0$ due to (4.27). Theorem 2.4.12 can now be invoked to complete the proof. □

Example 4.4.6 *Let $\mathbb{H}_1, \mathbb{H}_2$ be separable Hilbert spaces. Given a measure space $(E, \mathscr{B}, \mu)$, Theorem 2.6.5 has the consequence that a measurable mapping $\mathscr{G}$ on E taking values in $\mathfrak{B}_{HS}(\mathbb{H}_1, \mathbb{H}_2)$ is Bochner integrable if $\|\mathscr{G}\|_{HS}$ is integrable: i.e., if*

$$\int_E \|\mathscr{G}\|_{HS} d\mu < \infty.$$

Our aim is to show that for any such function

$$\int_E (\mathscr{G}f) d\mu = \left(\int_E \mathscr{G} d\mu \right) f \tag{4.30}$$

for all $f \in \mathbb{H}_1$. To see why this holds, for any fixed $f \in \mathbb{H}_1$ define a mapping $\mathscr{H}$ that maps $\mathscr{T} \in \mathfrak{B}_{HS}(\mathbb{H}_1, \mathbb{H}_2)$ to $\mathscr{T}f \in \mathbb{H}_2$. Then, we can rewrite (4.30) as

$$\int_E \mathscr{H}(\mathscr{G}) d\mu = \mathscr{H}\left(\int_E \mathscr{G} d\mu \right). \tag{4.31}$$

Observe that the operator norm of $\mathscr{G}$ is bounded by $\|f\|_1$ and hence $\mathscr{H}$ is in $\mathfrak{B}(\mathfrak{B}_{HS}(\mathbb{H}_1, \mathbb{H}_2), \mathbb{H}_2)$. Thus, (4.31) follows immediately from Theorem 3.1.7.

A truncated sve provides a best approximation to HS operators in the sense of the following theorem.

Theorem 4.4.7 *Let $\mathscr{T}$ be a Hilbert–Schmidt operator between two Hilbert spaces $\mathbb{H}_1, \mathbb{H}_2$ with singular system $\{(\lambda_j, f_{1j}, f_{2j})\}_{j=1}^{\infty}$. Then, for any finite integer k,*

$$\left\| \mathscr{T} - \sum_{j=1}^{k} x_j \otimes_1 y_j \right\|_{HS} \geq \left\| \mathscr{T} - \sum_{j=1}^{k} \lambda_j f_{1j} \otimes_1 f_{2j} \right\|_{HS} \tag{4.32}$$

for any set of functions $x_j \in \mathbb{H}_1$, $y_j \in \mathbb{H}_2$, $j = 1, \ldots, k$.

This result has various names such as the Schmidt–Mirsky Theorem or the Eckart–Young Theorem in the case of finite dimensions. A review of its early history can be found in Stewart (1993). In the special case where $\mathbb{H}_1$ and $\mathbb{H}_2$ coincide, Theorem 4.4.7 translates into saying that a truncated eigenvalue–eigenvector decomposition provides a best approximation to a nonnegative self-adjoint Hilbert–Schmidt operator. One can, in fact, view Theorem 4.6.8 as representing a refinement of this type of result.

Proof: It suffices to show that

$$\left\| \mathscr{T} - \sum_{j=1}^{k} x_j \otimes_1 y_j \right\|_{HS}^2 \geq \|\mathscr{T}\|_{HS}^2 - \sum_{j=1}^{k} \lambda_j^2.$$

Without loss, we can assume that the (y_j, x_j) are orthonormal. In that case, if $\{e_j\}$ is any CONS for $\mathbb{H}_1$

$$
\begin{aligned}
&\|\mathcal{T} - \sum_{i=1}^{k} x_i \otimes_1 y_i\|_{HS}^2 \\
&\quad = \sum_{j=1}^{\infty} \left\langle \left(\mathcal{T}^* \mathcal{T} + \sum_{i=1}^{k} (x_i - \mathcal{T}^* y_i) \otimes_1 (x_i - \mathcal{T}^* y_i)^* \right) e_j, e_j \right\rangle_1 \\
&\quad\quad - \sum_{i=1}^{k} \|\mathcal{T}^* y_i\|_1^2.
\end{aligned}
$$

As $(x_i - \mathcal{T}^* y_i) \otimes_1 (x_i - \mathcal{T}^* y_i)^*$ is nonnegative, the result will follow once we establish that $\sum_{i=1}^{k} \|\mathcal{T}^* y_i\|_1^2 \leq \sum_{i=1}^{k} \lambda_i^2$.

Now the sve of $\mathcal{T}^*$ gives $\mathcal{T}^* y_i = \sum_{j=1}^{\infty} \lambda_j \langle f_{2j}, y_i \rangle_2 f_{1j}$. This leads to the identity

$$
\begin{aligned}
\|\mathcal{T}^* y_i\|_1^2 = \lambda_k^2 &+ \left(\sum_{j=1}^{k} \lambda_j^2 \langle f_{2j}, y_i \rangle_2^2 - \lambda_k^2 \sum_{j=1}^{k} \langle f_{2j}, y_i \rangle_2^2 \right) \\
&- \left(\lambda_k^2 \sum_{j=k+1}^{\infty} \langle f_{2j}, y_i \rangle_2^2 - \sum_{j=k+1}^{\infty} \lambda_j^2 \langle f_{2j}, y_i \rangle_2^2 \right) \\
&- \lambda_k^2 \left(1 - \sum_{j=1}^{\infty} \langle f_{2j}, y_i \rangle_2^2 \right).
\end{aligned}
$$

The last two terms are nonpositive meaning that

$$
\begin{aligned}
\sum_{i=1}^{k} \|\mathcal{T}^* y_i\|_1^2 &\leq k \lambda_k^2 + \sum_{i=1}^{k} \sum_{j=1}^{k} (\lambda_j^2 - \lambda_k^2) \langle f_{2j}, y_i \rangle_2^2 \\
&= \sum_{j=1}^{k} \left[\lambda_k^2 + (\lambda_j^2 - \lambda_k^2) \sum_{i=1}^{k} \langle f_{2j}, y_i \rangle_2^2 \right] \\
&\leq \sum_{j=1}^{k} \lambda_j^2
\end{aligned}
$$

as a result of the orthonormality of the y_i and Parseval 's relation. □

4.5 Trace class operators

In the previous section, we introduced HS operators that provide an important subset of the compact operators. Another such subset is the trace class or nuclear operators that we examine in this section.

Let $\mathscr{T}$ be any bounded linear operator on some Hilbert space $\mathbb{H}$. By Theorem 3.4.3, we can define the square root of $\mathscr{T}^*\mathscr{T}$ as a bounded, nonnegative, self-adjoint operator $(\mathscr{T}^*\mathscr{T})^{1/2}$ such that $(\mathscr{T}^*\mathscr{T})^{1/2}(\mathscr{T}^*\mathscr{T})^{1/2} = \mathscr{T}^*\mathscr{T}$.

Definition 4.5.1 *Let $\mathscr{T} \in \mathfrak{B}(\mathbb{H}_1, \mathbb{H}_2)$ for separable Hilbert spaces $\mathbb{H}_1$ and $\mathbb{H}_2$. Then, $\mathscr{T}$ is trace class if for some CONS $\{e_j\}_{j=1}^{\infty}$ of $\mathbb{H}_1$, the quantity*

$$\|\mathscr{T}\|_{TR} := \sum_{j=1}^{\infty} \langle (\mathscr{T}^*\mathscr{T})^{1/2} e_j, e_j \rangle_1 \tag{4.33}$$

is finite. In this case, $\|\mathscr{T}\|_{TR}$ is said to be the trace norm of $\mathscr{T}$.

An argument similar to that used for Theorem 4.4.1 shows that this definition does not depend on the choice of CONS.

For any trace class operator $\mathscr{T}$, we have

$$\|\mathscr{T}\|_{TR} = \|(\mathscr{T}^*\mathscr{T})^{1/4}\|_{HS}^2,$$

where $(\mathscr{T}^*\mathscr{T})^{1/4} = ((\mathscr{T}^*\mathscr{T})^{1/2})^{1/2}$. This establishes that if $\mathscr{T}$ is trace class then $(\mathscr{T}^*\mathscr{T})^{1/4}$ is HS and hence compact. As

$$\mathscr{T}^*\mathscr{T} = ((\mathscr{T}^*\mathscr{T})^{1/4})^4,$$

and compositions of compact operators are compact (Theorem 4.1.3), we see that $\mathscr{T}^*\mathscr{T}$ is compact. The development that led to the sve and Theorem 4.3.5 now allows us to conclude that trace class operators are compact and that

$$\|\mathscr{T}\|_{TR} = \sum_{i=1}^{\infty} \lambda_i, \tag{4.34}$$

where the λ_i are the singular values of $\mathscr{T}$. Thus, from (4.28)

$$\|\mathscr{T}\|_{HS} \le \left(\lambda_1 \sum_{i=1}^{\infty} \lambda_i\right)^{1/2} = \sqrt{\lambda_1 \|\mathscr{T}\|_{TR}}$$

and it follows that

Theorem 4.5.2 *Trace class operators are HS.*

Now let us focus on operators in $\mathfrak{B}(\mathbb{H})$. If $\mathscr{T}$ is self-adjoint and has the associated eigenvalue sequence $\{\lambda_j\}$,

$$\|\mathscr{T}\|_{TR} = \sum_{j=1}^{\infty} |\lambda_j|.$$

In this case, $\sum_{j=1}^{\infty} \lambda_j$ is well defined and finite. It is called the *trace* of $\mathscr{T}$ and denoted by trace$(\mathscr{T})$. One can also readily verify that

$$\text{trace}(\mathscr{T}) = \sum_{j=1}^{\infty} \langle \mathscr{T} e_j, e_j \rangle$$

for any CONS $\{e_j\}$. If, moreover, $\mathscr{T}$ is nonnegative, then

$$\text{trace}(\mathscr{T}) = \|\mathscr{T}\|_{TR} = \sum_{j=1}^{\infty} \lambda_j. \tag{4.35}$$

In the finite dimensional case, trace$(\mathscr{T})$ is the same as the matrix trace. Unsurprisingly, the operator trace shares the basic features of the matrix trace in that, for example, given two trace class operators $\mathscr{T}_1, \mathscr{T}_2$ and $a_1, a_2 \in \mathbb{R}$, $\text{trace}(a_1\mathscr{T}_1 + a_2\mathscr{T}_2) = a_1\text{trace}(\mathscr{T}_1) + a_2\text{trace}(\mathscr{T}_2)$, and $\text{trace}(\mathscr{T}_2\mathscr{T}_1) = \text{trace}(\mathscr{T}_2\mathscr{T}_1)$. Proofs of these relations are given in, e.g., Reed and Simon (1980).

The connections between the eigenvalues of an operator and its HS and trace norms (when they exist) can be used to obtain various bounds on the size of its eigenvalues. When this feature is applied to the difference between operators, one result that emerges is von Neumann's trace inequality that has implications for the perturbation of operator eigenvalues. Results of this nature are explored in more detail in Chapter 5.

Theorem 4.5.3 *Suppose that $\mathscr{T}$ and $\tilde{\mathscr{T}}$ are HS operators from $\mathbb{H}_1$ to $\mathbb{H}_2$ with nonascending singular values $\lambda_j, \tilde{\lambda}_j, j = 1, 2, \ldots$ Then,*

$$\sum_{j=1}^{\infty} (\lambda_j - \tilde{\lambda}_j)^2 \leq \text{trace}\{(\mathscr{T} - \tilde{\mathscr{T}})^*(\mathscr{T} - \tilde{\mathscr{T}})\}. \tag{4.36}$$

Proof: The proof is based on developments in Grigorieff (1991). Let f_{1j} and f_{2j} be the right and left singular vectors of $\mathscr{T}$ corresponding to λ_j and, similarly, take $\tilde{f}_{1j}$ and $\tilde{f}_{2j}$ to be the right and left singular vectors of $\tilde{\mathscr{T}}$ corresponding to $\tilde{\lambda}_j$. Now write

$$\begin{aligned}
&\text{trace}\{(\mathscr{T} - \tilde{\mathscr{T}})^*(\mathscr{T} - \tilde{\mathscr{T}})\} \\
&\quad = \sum_{j=1}^{\infty} \lambda_j^2 + \sum_{j=1}^{\infty} \tilde{\lambda}_j^2 - 2\sum_{j=1}^{\infty} \langle \mathscr{T} f_{1j}, \tilde{\mathscr{T}} f_{1j}\rangle_2 \\
&\quad = \sum_{j=1}^{\infty} \lambda_j^2 + \sum_{j=1}^{\infty} \tilde{\lambda}_j^2 - 2\sum_{j=1}^{\infty} \lambda_j \langle f_{2j}, \tilde{\mathscr{T}} f_{1j}\rangle_2 \\
&\quad = \sum_{j=1}^{\infty} \lambda_j^2 + \sum_{j=1}^{\infty} \tilde{\lambda}_j^2 - 2\sum_{j=1}^{\infty}\sum_{k=1}^{\infty} \lambda_j \tilde{\lambda}_k \langle f_{1j}, \tilde{f}_{1k}\rangle_1 \langle f_{2j}, \tilde{f}_{2k}\rangle_2,
\end{aligned}$$

where we applied the expansion

$$\tilde{\mathscr{T}} f_{1j} = \sum_{k=1}^{\infty} \langle f_{1j}, \tilde{f}_{1k}\rangle_1 \tilde{\mathscr{T}} \tilde{f}_{1k} = \sum_{k=1}^{\infty} \tilde{\lambda}_k \langle f_{1j}, \tilde{f}_{1k}\rangle_1 \tilde{f}_{2k}$$

in the last step. Observe that

$$\left|\sum_{k=1}^{\infty} \langle f_{1j}, \tilde{f}_{1k}\rangle_1 \langle f_{2j}, \tilde{f}_{2k}\rangle_2\right| = \left|\left\langle \left(\sum_{k=1}^{\infty} \tilde{f}_{1k} \otimes_1 \tilde{f}_{2k}\right) f_{1j}, f_{2j}\right\rangle_2\right| \leq 1 \quad (4.37)$$

for all j and

$$\left|\sum_{j=1}^{\infty} \langle f_{1j}, \tilde{f}_{1k}\rangle_1 \langle f_{2j}, \tilde{f}_{2k}\rangle_2\right| = \left|\left\langle \left(\sum_{j=1}^{\infty} f_{1j} \otimes_1 f_{2j}\right) \tilde{f}_{1k}, \tilde{f}_{2k}\right\rangle_2\right| \leq 1 \quad (4.38)$$

for all k.

We will now prove (4.36) while assuming that rank($\mathscr{T}$) and rank($\tilde{\mathscr{T}}$) are both bounded by some finite integer n. If this is not the case, we can take limits on the inequalities proved for truncated operators using the singular value decompositions.

Using the above-mentioned derivations, in the finite rank case (4.36) becomes

$$\lambda^T \mathscr{A} \tilde{\lambda} \leq \lambda^T \tilde{\lambda}, \quad (4.39)$$

where

$$\begin{aligned}
\lambda &= (\lambda_1, \ldots, \lambda_n)^T, \\
\lambda &= (\tilde{\lambda}_1, \ldots, \tilde{\lambda}_n)^T
\end{aligned}$$

and

$$\mathscr{A} = \{\langle f_{1j}, \tilde{f}_{1k}\rangle_1 \langle f_{2j}, \tilde{f}_{2k}\rangle_2\}_{j,k=1:n}.$$

For $i = 1, \ldots, n$, let e_i be an n-dimensional vector whose first i elements are equal to 1 and the rest equal to 0. As λ_i and $\tilde{\lambda}_i$ are nonascending, both λ and $\tilde{\lambda}$ can be written as linear combinations of $e_1, \ldots, e_n$ with nonnegative coefficients. Without loss of generality, assume that λ_1 and $\tilde{\lambda}_1$ are bounded by 1. It then suffices to establish (4.39) for $\lambda = e_k$ and $\tilde{\lambda} = e_\ell$ for arbitrary k and ℓ. However, this is immediate as the row and column sums of $\mathscr{A}$ are all bounded by 1 as a result of (4.37) and (4.38). □

4.6 Integral operators and Mercer's Theorem

We briefly discussed integral operators in Example 3.1.6. In this section, we use the results developed in this chapter to expand on our previous treatment of the topic.

Let $(E, \mathscr{B}, \mu)$ be a measure space for some finite measure μ. Suppose that K is a measurable function on $E \times E$ such that $\int\int_{E\times E} K^2(s,t)d\mu(s)d\mu(t)$ is finite and define the integral operator $\mathscr{K}$ by

$$(\mathscr{K}f)(\cdot) := \int_E K(s,\cdot)f(s)d\mu(s) \tag{4.40}$$

for $f \in \mathbb{L}^2(E, \mathscr{B}, \mu)$. The function K is referred to as the *kernel* of $\mathscr{K}$. By the Cauchy–Schwarz inequality, for $f \in \mathbb{L}^2(E, \mathscr{B}, \mu)$, the function $(\mathscr{K}f)(\cdot)$ is measurable and satisfies

$$\int_E (\mathscr{K}f)^2(t)d\mu(t) \le \int\int_{E\times E} K^2(s,t)d\mu(s)d\mu(t) \int_E f^2(s)d\mu(s).$$

Thus, $\mathscr{K} \in \mathfrak{B}(\mathbb{L}^2(E, \mathscr{B}, \mu))$ with

$$\|\mathscr{K}\| \le \left(\int\int_{E\times E} K^2(s,t)d\mu(s)d\mu(t)\right)^{1/2}.$$

We will generally take $E = [0, 1]$ for our applications. However, the results that follow are applicable to any compact metric space and, accordingly, we will tacitly assume that E has that type of structure and that $\mathscr{B} = \mathscr{B}(E)$ is the Borel σ-field of E: namely, the smallest σ-field containing all the open sets in E. Without loss of generality, we will also assume that the support of μ is the entire space E; if this is not the case then our results remain true with E replaced by the the support of μ, which is also compact.

The integral operators that we focus on are those whose operator kernels K are continuous on $E \times E$. Unless otherwise stated, this will also be included as a standard assumption in the rest of the section. The compactness of E then has the consequence that K is uniformly continuous. This continuity is translated to the image of $\mathscr{K}$ as indicated in the following result.

Lemma 4.6.1 *For each* $f \in \mathbb{L}^2(E, \mathscr{B}, \mu)$, $(\mathscr{K}f)(\cdot)$ *is uniformly continuous.*

Proof: As K is uniformly continuous, for any given $\epsilon > 0$, there exist $\delta > 0$ such that $|K(s, s_2) - K(s, s_1)| < \epsilon$ for all $s, s_2, s_1 \in E$ with $|s_2 - s_1| < \delta$. Thus,

$$\left| \int_E K(s, s_2) f(s) d\mu(s) - \int_E K(s, s_1) f(s) d\mu(s) \right| \leq \epsilon \|f\|. \qquad \square$$

The following result provides our first characterization of integral operators.

Theorem 4.6.2 $\mathscr{K}$ *is compact.*

Proof: The proof employs Theorem 4.1.5. To do so, we need to construct an approximating sequence of finite-dimensional operators. One such sequence can be obtained by use of the Stone–Weierstrass Theorem (see, e.g., Royden and Fitzpatrick 2010), which tells us that for any $\epsilon > 0$ there is some finite n_ϵ and continuous functions $g_i, h_i, i = 1, \ldots, n_\epsilon$ with the property that

$$\sup_{s,t \in E} |K(s, t) - K_{n_\epsilon}(s, t)| < \epsilon$$

for $K_n(s, t) = \sum_{i=1}^n g_i(s) h_i(t)$.

Define $\mathscr{K}_n$ to be the integral operator with kernel K_n. Observe that $\text{Im}(\mathscr{K}_n) \subset \text{span}\{h_1, \ldots, h_n\}$. Thus, $\mathscr{K}_n$ is finite-dimensional and compact. By the Cauchy–Schwarz inequality,

$$\begin{aligned} \|(\mathscr{K}_{n_\epsilon} - \mathscr{K}) f\|^2 &= \int_E \left(\int_E [K_{n_\epsilon}(s, t) - K(s, t)] f(s) d\mu(s) \right)^2 d\mu(t) \\ &\leq \epsilon^2 \|f\|^2 \mu^2(E). \end{aligned} \qquad \square$$

Assume that K is symmetric in which case we know from Example 3.3.4 that $\mathscr{K}$ is self-adjoint. Theorem 4.2.4 gives us the eigenvalue–eigenvector decomposition $\mathscr{K} = \sum_{j=1}^\infty \lambda_j e_j \otimes e_j$, where λ_j, e_j satisfy the descriptions in Theorem 4.2.4. Lemma 4.6.1 ensures that the version of e_j determined by

$$e_j(t) = \lambda_j^{-1} \int_E K(s, t) e_j(s) d\mu(s)$$

is continuous in t. In the following, we will always assume that this is the choice that has been made for e_j.

Example 4.6.3 *An operator that arises in various settings (e.g., in the construction of the* $\mathbb{W}_1[0,1]$ *space of Section 2.8) is the* $\mathbb{L}^2[0,1]$ *integral operator that has the kernel*

$$K(s,t) = \min(s,t)$$

for $s, t \in [0,1]$. *Note that* K *is the covariance function of the standard Brownian motion process on* $[0,1]$.

The operator that corresponds to K *is given explicitly by*

$$(\mathscr{K}f)(t) = \int_0^1 K(t,s)f(s)ds = \int_0^t f(s)ds.$$

Let us now find its eigenvalues and eigenfunctions by solving

$$\int_0^1 \min(s,t)e(s)ds = \lambda e(t) \tag{4.41}$$

for λ *and* e. *In this regard, first note that (4.41) is the same as*

$$\int_0^t se(s)ds + t\int_t^1 e(s)ds = \lambda e(t).$$

Differentiating both sides of this relation produces

$$te(t) + \int_t^1 e(s)ds - te(t) = \lambda e'(t);$$

i.e.,

$$\int_t^1 e(s)ds = \lambda e'(t). \tag{4.42}$$

Differentiating (4.42) again reveals that $e(t) = -\lambda e''(t)$ *for which the general solution is*

$$e(s) = a\sin\left(s/\sqrt{\lambda}\right) + b\cos\left(s/\sqrt{\lambda}\right).$$

From (4.41), $e(0) = 0$ *and hence* $b = 0$ *in the above. Similarly, (4.42) implies that* $e'(1) = 0$ *so that*

$$a\cos\left(1/\sqrt{\lambda}\right) = 0,$$

which leads to

$$1/\sqrt{\lambda} = (2j-1)\pi/2, j = 1, 2, \ldots$$

Thus, the eigenvalues are

$$\lambda_j = \frac{4}{((2j-1)\pi)^2}$$

and the orthonormal eigenfunctions are

$$e_j(t) = \sqrt{2}\sin\left(\frac{(2j-1)\pi}{2}t\right).$$

Recall that a kernel $K(s,t)$ is nonnegative definite if it satisfies (2.29). The following result relates this notion to that of nonnegative-definite integral operators.

Theorem 4.6.4 *An integral operator is nonnegative definite if and only if its kernel is nonnegative definite.*

Proof: Let K be the operator kernel of the integral operator $\mathscr{K}$. Given $n > 0$ let δ_n be chosen so that $|K(s_2, t_2) - K(s_1, t_1)| < n^{-1}$ whenever $d((s_1, t_1), (s_2, t_2)) < \delta_n$, where d is some metric for the product space $E \times E$. As E is a compact metric space, the Heine–Borel Theorem (Theorem 2.1.17) has the implication that there exists a finite partition $\{E_{ni}\}$ of E such that each E_{ni} has diameter less than δ_n. Let v_i be an arbitrary point of E_{ni} and, for all $(s,t) \in E_{ni} \times E_{nj}$, define $K_n(s,t)$ to be $K(v_i, v_j)$. The uniform continuity of K now has the consequence that

$$\max_{(s,t)\in E\times E} |K(s,t) - K_n(s,t)| < n^{-1}.$$

Now let $\mathscr{K}_n$ be the integral operator with kernel K_n. With this choice, we find that, for any $f \in \mathbb{L}^2(E, \mathscr{B}, \mu)$,

$$|\langle \mathscr{K}f, f\rangle - \langle \mathscr{K}_n f, f\rangle| \leq n^{-1}\|f\|^2 \tag{4.43}$$

and

$$\langle \mathscr{K}_n f, f\rangle = \sum_{i=1}^{n}\sum_{j=1}^{n} K(v_i, v_j)\int_{E_{ni}} f(t)d\mu(t)\int_{E_{nj}} f(t)d\mu(t). \tag{4.44}$$

If K is nonnegative definite, then (4.43) and (4.44) entail that $\langle \mathscr{K}f, f\rangle \geq 0$.

Conversely, suppose that

$$\sum_{i=1}^{m}\sum_{j=1}^{m} a_i a_j K(v_i, v_j) < 0$$

for some a_i, v_i. As K is uniformly continuous, there exist disjoint sets $E_1, \ldots, E_m \in \mathscr{B}$ with $\mu(E_i) > 0, v_i \in E_i$ for all i, and

$$\max_{u_i, v_i \in E_i, i=1,\ldots,m} \sum_{i=1}^{m}\sum_{j=1}^{m} a_i a_j K(u_i, v_j) < 0.$$

This implies that

$$\sum_{i=1}^{m}\sum_{j=1}^{m} a_i a_j (\mu(E_i)\mu(E_j))^{-1} \int_{E_i}\int_{E_j} K(u, v) d\mu(u) d\mu(v) < 0$$

due to the mean-value theorem. Upon observing that the last expression is simply $\langle \mathscr{K} f, f\rangle$ for $f = \sum_{i=1}^{m} a_i(\mu(E_i))^{-1} I_{E_i}$, we conclude that $\mathscr{K}$ is not nonnegative definite. □

The following result is the celebrated Mercer's Theorem, which says essentially that for an integral operator $\mathscr{K}$ with a symmetric and nonnegative-definite kernel, equivalent series expansions can be obtained for both the operator and its kernel.

Theorem 4.6.5 *Let the continuous kernel K be symmetric and nonnegative definite and $\mathscr{K}$ the corresponding integral operator. If (λ_j, e_j) are the eigenvalue and eigenfunction pairs of $\mathscr{K}$, then K has the representation*

$$K(s, t) = \sum_{j=1}^{\infty} \lambda_j e_j(s) e_j(t),$$

for all s, t, with the sum converging absolutely and uniformly.

The following lemma contains most of the technical details that are needed to prove Theorem 4.6.5.

Lemma 4.6.6 *Under the conditions of Theorem 4.6.5,*

1. $\sum_{j=1}^{\infty} \lambda_j e_j^2(t) \le K(t, t)$ *for all t,*
2. $\sum_{j=1}^{\infty} |\lambda_j e_j(s) e_j(t)| \le \{K(t, t)K(s, s)\}^{1/2}$ *for all s, t,*
3. $\lim_{n\to\infty} \sup_{s,t} \sum_{j=n+1}^{\infty} |\lambda_j e_j(s) e_j(t)| = 0$, *and*

4. *the function* $\sum_{j=1}^{\infty} \lambda_j e_j(s) e_j(t)$ *is well defined and uniformly continuous in* (s,t)*, with the sum converging absolutely and uniformly.*

Proof: Let

$$K_n(s,t) = K(s,t) - \sum_{j=1}^{n} \lambda_j e_j(s) e_j(t)$$

and take $\mathscr{K}_n$ to be the integral operator with kernel K_n. Note that K_n is continuous. Then, for any f,

$$\langle \mathscr{K}_n f, f \rangle = \langle \mathscr{K} f, f \rangle - \sum_{j=1}^{n} \lambda_j \langle f, e_j \rangle^2 = \sum_{j=n+1}^{\infty} \lambda_j \langle f, e_j \rangle^2 \geq 0$$

and $\mathscr{K}_n$ must be nonnegative definite. This implies by Theorem 4.6.4 that K_n is nonnegative definite and hence $K_n(t,t) \geq 0$ thereby proving part 1.

Part 2 is now a consequence of part 1 as, for any set J of positive integers,

$$\sum_{j \in J} |\lambda_j e_j(s) e_j(t)| \leq \left(\sum_{j \in J} \lambda_j e_j^2(s) \right)^{1/2} \left(\sum_{j \in J} \lambda_j e_j^2(t) \right)^{1/2} \tag{4.45}$$

from the Cauchy–Schwarz inequality.

To show part 3 first observe that (4.45) ensures that for all s, t

$$\sum_{j=n+1}^{\infty} |\lambda_j e_j(s) e_j(t)| \leq \left(\sum_{j=n+1}^{\infty} \lambda_j e_j^2(s) \right)^{1/2} \left(\sum_{j=n+1}^{\infty} \lambda_j e_j^2(t) \right)^{1/2}, \tag{4.46}$$

which tends to 0 monotonically due to Lebesgue's dominated convergence theorem. As E compact, uniform convergence of the left-hand side of (4.46) follows from Dini's Theorem.

Fix any $\epsilon > 0$. Using part 3, we can conclude that there exists n_ϵ such that

$$\sup_{s,t} \sum_{j=n_\epsilon+1}^{\infty} |\lambda_j e_j(s) e_j(t)| < \epsilon. \tag{4.47}$$

In addition, the uniform continuity of $e_1, \ldots, e_{n_\epsilon}$ guarantees the existence of $\delta > 0$ such that

$$\left| \sum_{j=1}^{n_\epsilon} \lambda_j e_j(s) e_j(t) - \sum_{j=1}^{n_\epsilon} \lambda_j e_j(s') e_j(t') \right| < \epsilon \tag{4.48}$$

whenever $d((s,t),(s',t')) < \delta$. Part 4 is now a straightforward consequence of (4.47) and (4.48). □

Proof of Theorem 4.6.5. For different continuous kernels $K_1(s,t)$ and $K_2(s,t)$, it is straightforward to construct a function f such that $\int_E K_2(s,t)f(s)d\mu(s)$ and $\int_E K_1(s,t)f(s)d\mu(s)$ differ. Thus, K is the unique operator kernel that defines $\mathscr{K}$. Now, the integral operator with the continuous kernel $\sum_{j=1}^{\infty} \lambda_i e_i(s)e_i(t)$ has the same eigen decomposition as $\mathscr{K}$ and is therefore the same operator. Thus, $K(s,t) = \sum_{j=1}^{\infty} \lambda_i e_i(s)e_i(t)$ for all s,t with the right-hand side converging absolutely and uniformly as a consequence of Lemma 4.6.6. □

With the aid of Mercer's Theorem and the developments in Section 4.5, we can now see that the integral operator $\mathscr{K}$ in (4.40) is trace class. Simple formulae for the trace and HS norms of $\mathscr{K}$ are given in the subsequent theorem.

Theorem 4.6.7 *Under the conditions of Theorem 4.6.5,*

$$\text{trace}(\mathscr{K}) = \int_E K(s,s)d\mu(s) \tag{4.49}$$

and

$$\|\mathscr{K}\|_{HS}^2 = \int\int_{E\times E} K^2(s,t)d\mu(s)d\mu(t). \tag{4.50}$$

Proof: By (4.35) and Theorem 4.6.5, we see that

$$\begin{aligned}\text{trace}(\mathscr{K}) &= \sum_{j=1}^{\infty} \lambda_i = \sum_{j=1}^{\infty} \lambda_i \int_E e_i^2(t)d\mu(t)\\ &= \int_E \left(\sum_{j=1}^{\infty} \lambda_i e_i^2(t) \right) d\mu(t)\\ &= \int_E K(t,t)d\mu(t).\end{aligned}$$

Exchange of the order of summation and integration here is allowed by Fubini's Theorem.

Now, let $K_n(s,t) = \sum_{i=1}^{n} \lambda_i e_i(s)e_i(t)$ and take $\mathscr{K}_n$ to be the corresponding integral operator. First,

$$\text{trace}(\mathscr{K}_n^2) = \sum_{i=1}^{n} \lambda_i^2 \uparrow \sum_{i=1}^{\infty} \lambda_i^2 = \text{trace}(\mathscr{K}^2),$$

where the right-hand side is $\|\mathscr{K}\|_{HS}^2$ by (4.28). In addition,

$$\text{trace}(\mathscr{K}_n^2) = \int\int_{E\times E} K_n^2(s,t)d\mu(s)d\mu(t).$$

The right side of this expression tends to $\int\int_{E\times E} K^2(s,t)d\mu(s)d\mu(t)$ as a result of Theorem 4.6.5 and Lebesgue's dominated convergence theorem. □

The following result describes an optimality feature of the decomposition of K under Mercer's Theorem. We say that a kernel $K(s,t)$ has rank r if the corresponding integral operator has rank r.

Theorem 4.6.8 *Let K be a symmetric and nonnegative-definite kernel with the eigen decomposition*

$$K(s,t) = \sum_{j=1}^{\infty} \lambda_j e_j(s) e_j(t).$$

Then, for any positive integer r for which $\lambda_r > 0$,

$$\min_{\mathrm{rank}(W)=r} \int\int_{E\times E} \{K(s,t) - W(s,t)\}^2 d\mu(s)d\mu(t) = \sum_{j=r+1}^{\infty} \lambda_j^2,$$

where the minimum is achieved by $W(s,t) = \sum_{j=1}^{r} \lambda_j e_j(s) e_j(t)$.

Proof: Let $\mathscr{K}$ be the integral operator with kernel K. Theorem 4.4.7 implies that the operator $\mathscr{W}$ that minimizes $\|\mathscr{K} - \mathscr{W}\|_{HS}^2$ is $\mathscr{W} = \sum_{j=1}^{r} \lambda_j e_j \otimes e_j$ for which $\|\mathscr{K} - \mathscr{W}\|_{HS}^2 = \sum_{j=r+1}^{\infty} \lambda_j^2$. Note that $\mathscr{W}$ is an integral operator with kernel

$$W(s,t) = \sum_{j=1}^{r} \lambda_j e_j(s) e_j(t).$$

Part 2 of Theorem 4.6.7 then gives

$$\|\mathscr{K} - \mathscr{W}\|_{HS}^2 = \int\int_{E\times E} \{K(s,t) - W(s,t)\}^2 d\mu(s)d\mu(t)$$

from which the result follows readily. □

4.7 Operators on an RKHS

As one might expect, the bounded operators defined on an RKHS can be characterized by their actions on the rk. To describe this property, let $\mathscr{T}$ be an operator on $\mathbb{H}(K)$ for some rk K. Then, let $\mathscr{T}K(\cdot,s)$ be the function produced by applying $\mathscr{T}$ to the first argument of K for a fixed value of the second argument. Now define the operator's kernel to be

$$R(\cdot,s) = \mathscr{T}^* K(\cdot,s). \tag{4.51}$$

With this definition, we obtain

$$\langle f, R(\cdot, s)\rangle = \langle \mathscr{T} f, K(\cdot, s)\rangle = (\mathscr{T} f)(s)$$

for $f \in \mathbb{H}(K)$. Thus, $R(\cdot, s)$ is the representer of the linear functional $(\mathscr{T} f)(s)$. An application of the reproducing property then reveals that the kernel corresponding to the adjoint of $\mathscr{T}$ is $R^*(\cdot, s) = R(s, \cdot)$. An operator on an RKHS is therefore self-adjoint if its kernel is symmetric.

Formula (4.51) provides a recipe for assigning a kernel to every operator on an RKHS. Such kernels persist through linear operations; i.e., the kernel for the linear combination $a_1\mathscr{T}_1 + a_2\mathscr{T}_2, a_1, a_2 \in \mathbb{R}$ is $a_1R_1 + a_2R_2$ with R_i the kernel for $\mathscr{T}_i, i = 1, 2$. Taking $\mathscr{T} = \mathscr{T}_2\mathscr{T}_1$, we see that

$$\begin{aligned}\mathscr{T}^* K(\cdot, s) &= \mathscr{T}_1^* R_2(\cdot, s) = \langle \mathscr{T}_1^* R_2(\star, s), K(\star, \cdot)\rangle \\ &= \langle R_2(\star, s), \mathscr{T}_1 K(\star, \cdot)\rangle = \langle R_2(\star, s), R_1(\cdot, \star)\rangle.\end{aligned}$$

This provides a characterization of the composition of two operators on an RKHS in terms of their kernels.

One might now wonder when an operator's kernel generates an RKHS. We know that this is equivalent to the kernel being nonnegative definite. However, more can be said on this issue.

Theorem 4.7.1 *An operator on an RKHS with rk K is nonnegative if its kernel R is nonnegative in which case there is a nonnegative constant B such that $BK - R$ is nonnegative.*

Proof: An operator $\mathscr{T}$ being nonnegative means that $\langle \mathscr{T} f, f\rangle \geq 0$ for every $f \in \mathbb{H}(K)$. Take $f = f_n$ for $f_n(\cdot) = \sum_{i=1}^n a_i K(\cdot, t_i)$ and use the reproducing property to see that this implies the kernel for $\mathscr{T}$ is nonnegative definite. The converse follows from the fact that functions of the form f_n are dense in $\mathbb{H}(K)$.

As $\mathscr{T}$ is nonnegative $0 \leq \langle \mathscr{T} f, f\rangle \leq B\|f\|^2$ with, e.g., $B = \|\mathscr{T}\|^2$. Therefore, $BI - \mathscr{T}$ is a nonnegative operator and its kernel

$$(BI - \mathscr{T})^* K(\cdot, s) = BK(\cdot, s) - R(\cdot, s)$$

must be a nonnegative function. □

There is no reason to restrict attention to bounded operators on a single RKHS and the idea of an operator's kernel works equally well when there are two RKHSs. To see this, let $\mathbb{H}(K_1)$ and $\mathbb{H}(K_2)$ be RKHSs corresponding to two rks K_1, K_2 on sets E_1, E_2 and suppose that $\mathscr{T} \in \mathfrak{B}(\mathbb{H}(K_1), \mathbb{H}(K_2))$. Then, the operator's kernel is

$$R(\cdot, s) = \mathscr{T}^* K_2(\cdot, s)$$

because

$$(\mathscr{T}f)(s) = \langle \mathscr{T}f, K_2(\cdot, s)\rangle_2 = \langle f, R(\cdot, s)\rangle_1$$

with $\langle \cdot, \cdot \rangle_i$ the inner product for $\mathbb{H}(K_i), i = 1, 2$. Similarly,

$$(\mathscr{T}^*f)(t) = \langle \mathscr{T}^*f, K_1(\cdot, t)\rangle_1 = \langle f, \mathscr{T}K_1(\cdot, t)\rangle_2$$

and the kernel for the adjoint of $\mathscr{T}$ is $R^*(\cdot, t) := \mathscr{T}K_1(\cdot, t)$.

Now let $\{e_{ij}\}_{j=1}^{\infty}$ be CONSs for $\mathbb{H}(K_i), i = 1, 2$.

Theorem 4.7.2 *An operator $\mathscr{T} : \mathbb{H}(K_1) \mapsto \mathbb{H}(K_2)$ is HS if for each $t \in E_1$ and $s \in E_2$*

$$R(t, s) = \sum_{i=1}^{\infty} \sum_{j=1}^{\infty} a_{ij} e_{1i}(t) e_{2j}(s) \tag{4.52}$$

with $\sum_{i=1}^{\infty} \sum_{j=1}^{\infty} a_{ij}^2 < \infty$.

Proof: Assume first that (4.52) holds. Then,

$$(\mathscr{T}e_{1i})(s) = \langle e_{1i}, R(\cdot, s)\rangle_1 = \sum_{j=1}^{\infty} a_{ij} e_{2j}(s),$$

which means that

$$\|\mathscr{T}\|_{HS}^2 = \sum_{i=1}^{\infty} \|\mathscr{T}e_{1i}\|_2^2 = \sum_{i=1}^{\infty} \sum_{j=1}^{\infty} a_{ij}^2 < \infty$$

with $\|\cdot\|_2$ the $\mathbb{H}(K_2)$ norm. So, $\mathscr{T}$ is HS.

Conversely, suppose that $\mathscr{T}$ is HS. As $R(\cdot, s)$ is an element of $\mathbb{H}(K_1)$

$$\begin{aligned}
R(t, s) &= \sum_{i=1}^{\infty} \langle R(\cdot, s), e_{1i}\rangle_1 e_{1i}(t) \\
&= \sum_{i=1}^{\infty} \langle K_2(\cdot, s), \mathscr{T}e_{1i}\rangle_2 e_{1i}(t) \\
&= \sum_{i=1}^{\infty} (\mathscr{T}e_{1i})(s) e_{1i}(t) \\
&= \sum_{i=1}^{\infty} \sum_{j=1}^{\infty} \langle e_{2j}, \mathscr{T}e_{1i}\rangle_2 e_{1i}(t) e_{2j}(s)
\end{aligned}$$

and

$$\sum_{i=1}^{\infty}\sum_{j=1}^{\infty}\langle e_{2j}, \mathscr{T} e_{1i}\rangle_2^2 = \sum_{i=1}^{\infty} \|\mathscr{T} e_{1i}\|_2^2 = \|\mathscr{T}\|_{HS}^2.$$

□

Arguing similarly to the proof of the previous theorem we see that an operator $\mathscr{T} \in \mathfrak{B}(\mathbb{H}(K_1), \mathbb{H}(K_2))$ is trace class if (4.52) holds with

$$\sum_{i=1}^{\infty}\sum_{j=1}^{\infty} |a_{ij}| < \infty.$$

In this latter instance, the convergence of the series is uniform in s, t when K_1 and K_2 are bounded because

$$\begin{aligned}
&|R(t,s) - \sum_{i=1}^{n}\sum_{j=1}^{n} a_{ij}e_{1i}(t)e_{2j}(s)| \\
&\le \sum_{i=n+1}^{\infty}\sum_{j=n+1}^{\infty} |a_{ij}e_{1i}(t)e_{2j}(s)| \\
&= \sum_{i=n+1}^{\infty}\sum_{j=n+1}^{\infty} |a_{ij}\langle e_{1i}, K_1(\cdot,t)\rangle_1\langle e_{2j}, K_2(\cdot,s)\rangle_2| \\
&\le |K_1(t,t)|^{1/2}|K_2(s,s)|^{1/2} \sum_{i=n+1}^{\infty}\sum_{j=n+1}^{\infty} |a_{ij}|.
\end{aligned}$$

4.8 Simultaneous diagonalization of two nonnegative definite operators

A classic result from matrix theory that arises in various statistical venues such as linear models concerns the simultaneous diagonalization of two nonnegative definite matrices. In this section, we briefly discuss a situation where it is possible to extend this idea to a more general context involving a compact operator.

The result we will prove can be stated as follows.

Theorem 4.8.1 *Let $\mathscr{C}$ and $\mathscr{W}$ be self-adjoint operators on a separable Hilbert space $\mathbb{H}$. Suppose that $\mathscr{C}$ is compact and nonnegative definite, $\mathscr{W}$ is positive definite, and $\mathscr{G} := \mathscr{C} + \mathscr{W}$ is invertible. Let $\{(\eta_j, v_j)\}_{j=1}^{\infty}$ be the eigenvalue–eigenvector pairs of $\mathscr{G}^{-1/2}\mathscr{C}\mathscr{G}^{-1/2}$, where the η_j are necessarily* *in* $[0, 1)$, *and define*

$$\begin{aligned}\gamma_j &= \eta_j/(1-\eta_j),\\ u_j &= (1-\eta_j)^{-1/2}\mathscr{G}^{-1/2}v_j.\end{aligned}$$

Then,

$$\begin{aligned}\mathscr{C}\,u_j &= \eta_j\mathscr{G}\,u_j\\ &= \gamma_j\mathscr{W}\,u_j\\ \langle u_i, \mathscr{W}\,u_j\rangle &= \delta_{ij},\end{aligned}$$

and

$$x = \sum_{j=1}^{\infty}\langle x, \mathscr{W}\,u_j\rangle u_j$$

for all x.

Proof: As

$$\mathscr{G}^{-1/2}\mathscr{C}\,\mathscr{G}^{-1/2} = I - \mathscr{G}^{-1/2}\mathscr{W}\,\mathscr{G}^{-1/2},$$

all the eigenvalues η_j must be in $[0, 1)$. Now

$$\mathscr{G}^{-1/2}\mathscr{W}\,\mathscr{G}^{-1/2}v_j = \mathscr{G}^{-1/2}(\mathscr{G}-\mathscr{C}\,)\mathscr{G}^{-1/2}v_j = (1-\eta_j)v_j.$$

Thus,

$$\begin{aligned}\langle u_i, \mathscr{W}\,u_j\rangle &= (1-\eta_i)^{-1/2}(1-\eta_j)^{-1/2}\langle v_i, \mathscr{G}^{-1/2}\mathscr{W}\,\mathscr{G}^{-1/2}v_j\rangle\\ &= \delta_{ij}\end{aligned}$$

and

$$\begin{aligned}\mathscr{C}\,u_j &= (1-\eta_j)^{-1/2}\mathscr{C}\,\mathscr{G}^{-1/2}v_j\\ &= (1-\eta_j)^{-1/2}\mathscr{G}^{1/2}\mathscr{G}^{-1/2}\mathscr{C}\,\mathscr{G}^{-1/2}v_j\\ &= \eta_j\mathscr{G}\,u_j = \eta_j(\mathscr{C} + \mathscr{W}\,)u_j.\end{aligned}$$

The last identity also gives

$$(1-\eta_j)\mathscr{C}\,u_j = \eta_j\mathscr{W}\,u_j.$$

Finally, as $\{v_j\}$ is a CONS for $\mathbb{H}$, any $x \in \mathbb{H}$ satisfies

$$\begin{aligned}
x &= \mathscr{G}^{-1/2}\mathscr{G}^{1/2}x \\
&= \mathscr{G}^{-1/2}\left(\sum_{j=1}^{\infty}\langle \mathscr{G}^{1/2}x, v_j\rangle v_j\right) \\
&= \sum_{j=1}^{\infty}\langle \mathscr{G}^{1/2}x, v_j\rangle \mathscr{G}^{-1/2}v_j \\
&= \sum_{j=1}^{\infty}(1-\eta_j)^{1/2}\langle \mathscr{G}^{1/2}x, \mathscr{G}^{1/2}u_j\rangle \mathscr{G}^{-1/2}v_j \\
&= \sum_{j=1}^{\infty}(1-\eta_j)\langle x, \mathscr{G}u_j\rangle u_j,
\end{aligned}$$

and $\mathscr{G}u_j = (1-\eta_j)^{-1}\mathscr{W}u_j$. □

As a corollary to Theorem 4.8.1, we can state the following.

Corollary 4.8.2 *Let $\mathscr{C}$, $\mathscr{W}$ be symmetric $n \times n$ matrices with $\mathscr{W}$ positive definite. Then, there exists a matrix $\mathscr{U}$ such that $\mathscr{U}^T\mathscr{C}\mathscr{U}$ is diagonal and $\mathscr{U}^T\mathscr{W}\mathscr{U}$ is the identity.*

Proof: If $\mathscr{C}$ is nonnegative definite, then the conclusions follow immediately from Theorem 4.8.1. If $\mathscr{C}$ is not nonnegative definite, then first consider $\tilde{\mathscr{C}} := \mathscr{C} + B\mathscr{W}$, where B is a large enough constant that makes $\tilde{\mathscr{C}}$ positive definite. □

5

Perturbation theory

This chapter delves into perturbation theory for compact operators. The material collected here will subsequently furnish some of the tools that will be needed for establishing large sample properties associated with methods for principle components estimation in Chapter 9.

The definitive treatise on operator perturbation theory is that of Kato (1995). Our particular treatment of this topic focuses on two scenarios that parallel the developments in Chapter 4 and is partly motivated by the results in Dauxious, Pousse, and Romain (1982), Hall and Hosseini-Nasab (2005, 2009), and Riesz and Sz.-Nagy (1990).

First, in Section 5.1, we consider the more standard case of self-adjoint, compact operators on a Hilbert space. In that setting, we obtain bounds and expansions that allow us to measure the effect that perturbing an operator will have on its eigenvalues and eigenvectors. Section 5.2 then explores the more complicated case of operators that are not self-adjoint and presents similar results to those of Section 5.1 for singular values and vectors.

5.1 Perturbation of self-adjoint compact operators

Theorems 4.2.8 and 4.5.3 represent anomalies of sorts when viewed in terms of the overall theme of Chapter 4. They both dealt with the eigenvalues of two operators rather than just one and then provided bounds for the difference in the operators' eigenvalues in terms of some measure of the size of the overall difference between the two operators. Results of this nature can be particularly useful when the eigenvalues are the object of interest and it is possible to directly measure or bound the size of the operator difference.

Somewhat more generally, we might consider two operators $\mathscr{T}$ and $\tilde{\mathscr{T}}$ and define the perturbation operator $\Delta = \tilde{\mathscr{T}} - \mathscr{T}$ so that $\tilde{\mathscr{T}} = \mathscr{T} + \Delta$. This type of

Theoretical Foundations of Functional Data Analysis, with an Introduction to Linear Operators, First Edition. Tailen Hsing and Randall Eubank.

formulation gives the impression that $\tilde{\mathscr{T}}$ is some type of approximation to $\mathscr{T}$ with Δ representing an error or residual term that stems from the approximation. Thus, $\mathscr{T}$ and its associated eigenvalues and eigenvectors represent the targets and we wish to ascertain how well the eigenvalues and eigenvectors of $\tilde{\mathscr{T}}$ can serve if they are used in a surrogate capacity where they are employed in place of those for $\mathscr{T}$. This is the direction we will follow in this section.

The problem setting is one where we have a compact, self-adjoint operator $\mathscr{T}$ on a Hilbert space $\mathbb{H}$. From Section 4.2, we know that $\mathscr{T}$ admits a representation in terms of the eigenvalue–eigenvector expansion (4.6). We will sometimes need to list the eigenvalues in this representation as repeated and other times as nonrepeated. Repeated eigenvalues $\lambda_1 \geq \lambda_2 \geq \cdots$ are those that are listed along with information about their multiplicity while nonrepeated eigenvalues $\lambda_1 > \lambda_2 > \cdots$ are given without this information. Unless otherwise stated, λ_j will only represent a nonrepeated eigenvalue. Using this notational paradigm, we can alternatively express the eigen decomposition of $\mathscr{T}$ as

$$\mathscr{T} = \sum_{j=1}^{\infty} \lambda_j \mathscr{P}_j, \tag{5.1}$$

where $\mathscr{P}_j$ is the projection operator for the eigenspace of λ_j. Note that if λ_j has multiplicity one then Theorem 3.4.7 tells us that $\mathscr{P}_j = e_j \otimes e_j$ for e_j the eigenvector corresponding to λ_j. More generally, if the multiplicity of λ_j is more than one, from Theorem 4.2.4, we know that it must be some finite number, n_j. In that case, we can use the Gram–Schmidt method of Theorem 2.4.10 to create orthonormal vectors $e_{j1}, \ldots, e_{jn_j}$ that span the eigenspace. Then,

$$\mathscr{P}_j = \sum_{k=1}^{n_j} e_{jk} \otimes e_{jk}.$$

We initially need to quantify the effect of perturbations on an operator's projection operators. This will eventually provide tools that can be used to make similar assessments about eigenvalues and eigenvectors.

To deal with projection operators effectively, it becomes expedient to bring in the concept of the resolvent operator. This requires us to expand the Hilbert space formulation we have used up to this point. Specifically, we now need the scalar field for $\mathbb{H}$ to be the set of complex numbers $\mathbb{C}$. Definition 2.4.1 must be adjusted to reflect this change with its condition 4 being restated as

$$\langle x, y \rangle = \overline{\langle y, x \rangle},$$

where $\overline{a}$ is the complex conjugate of $a \in \mathbb{C}$.

The set of $z \in \mathbb{C}$ for which $\mathscr{T} - zI$ is not invertible is called the spectrum of $\mathscr{T}$ and denoted by $\sigma(\mathscr{T})$. For compact operators on an infinite-dimensional space $\mathbb{H}$, $\sigma(\mathscr{T})$ contains all the eigenvalues plus 0 whether 0 is an eigenvalue or not. The complement of $\sigma(\mathscr{T})$ is called the resolvent set of $\mathscr{T}$ and denoted by $\rho(\mathscr{T})$. For $z \in \rho(\mathscr{T})$, direct multiplication using (5.1) establishes that

$$\begin{aligned} \mathscr{R}(z) &:= (\mathscr{T} - zI)^{-1} \\ &= \sum_{j=1}^{\infty} \frac{1}{\lambda_j - z} \mathscr{P}_j \end{aligned} \tag{5.2}$$

is a bounded operator that we will refer to as the *resolvent* of $\mathscr{T}$. In fact, as

$$\begin{aligned} \|\mathscr{R}(z) f\|^2 &\le \max_j \frac{1}{|z - \lambda_j|^2} \sum_{j=1}^{\infty} \|\mathscr{P}_j f\|^2 \\ &= \max_j \frac{1}{|z - \lambda_j|^2} \|f\|^2, \end{aligned}$$

we have the exact expression

$$\|\mathscr{R}(z)\| = 1 / \min_j |z - \lambda_j|. \tag{5.3}$$

The resolvent plays a fundamental role in perturbation theory as we will shortly see. However, to access this utility, we need to be able to compute contour integrals of (5.2) as a function of its z argument. Development of the mathematical framework that will allow us to do so is our following task. We have already seen abstract integration from another perspective in Section 2.6.

In complex analysis, contour integrals are integrals of functions over paths or curves in the complex plane. Thus, we now take Γ to be a simple closed curve, also known as a Jordan curve, in $\mathbb{C}$ of length l_Γ and let $\{\mathscr{T}(z) : z \in \Gamma\}$ be an indexed collection of operators on $\mathbb{H}$; that is, $\mathscr{T}(z) \in \mathfrak{B}(\mathbb{H})$ for every $z \in \Gamma$ and we need look no further than (5.2) to see a specific example of such a collection. The length of the arc on Γ connecting points z to z' on the curve counterclockwise will be denoted as $d_\Gamma(z, z')$.

As a first step, let us choose points $z_0, \ldots, z_n$ on Γ and order them counterclockwise with $z_0 = z_n$. If ξ_j is any point on the arc going from z_j to z_{j+1}, we define the operator

$$\mathscr{I}(\{z_j, \xi_j\}_{j=0}^n) = \sum_{j=0}^{n-1} \mathscr{T}(\xi_j)(z_{j+1} - z_j). \tag{5.4}$$

This bears a formal similarity to a Riemann sum and one might anticipate that something of this nature could be extended to a limit in an analogous manner to the development of the Riemann integral of a real function. Conditions that can be employed to make such a notion rigorous stem from the following simple result.

Lemma 5.1.1 *Let* $\{z_j, \xi_j\}_{j=0}^n$ *and* $\{\tilde{z}_j, \tilde{\xi}_j\}_{j=0}^n$ *be two ordered sets of points on* Γ *and define*

$$\delta = \max_j d_\Gamma(z_j, z_{j+1}) + \max_j d_\Gamma(\tilde{z}_j, \tilde{z}_{j+1}).$$

Then,

$$\|\mathscr{I}(\{z_j, \xi_j\}_{j=0}^n) - \mathscr{I}(\{\tilde{z}_j, \tilde{\xi}_j\}_{j=0}^n)\| \le \sup_{d_\Gamma(z,z')\le\delta} \|\mathscr{T}(z) - \mathscr{T}(z')\| l_\Gamma. \tag{5.5}$$

Proof: Merge the two sets into a single ordered set $\{\check{z}_j, \check{\xi}_j\}$ so that

$$\mathscr{I}(\{z_j, \xi_j\}_{j=0}^n) - \mathscr{I}(\{\tilde{z}_j, \tilde{\xi}_j\}_{j=0}^n) = \sum_j \left(\mathscr{T}(\check{\xi}_j) - \mathscr{T}(\check{\xi}'_j)\right)(\check{z}_j - \check{z}_{j-1}).$$

The result then follows from the triangle inequality. □

This leads us to the integral definition that we seek.

Definition 5.1.2 *Let* $\{z_{jn}, \xi_{jn}\}_{j=0}^n, n = 1, \ldots$ *be a sequence of ordered sets with*

$$\lim_{n\to\infty} \max_j d_\Gamma(z_{jn}, z_{(j+1)n}) = 0 \tag{5.6}$$

and assume that

$$\lim_{\delta\downarrow 0} \sup_{d_\Gamma(z,z')\le\delta} \|\mathscr{T}(z) - \mathscr{T}(z')\| = 0. \tag{5.7}$$

The contour integral of $\mathscr{T}(z)$ *over* Γ *is*

$$\oint_\Gamma \mathscr{T}(z)dz = \lim_{n\to\infty} \mathscr{I}(\{z_{jn}, \xi_{jn}\}_{j=0}^n). \tag{5.8}$$

Lemma 5.1.1 has the implication that $\mathscr{I}(\{z_{nj}, \xi_{nj}\}_{j=0}^n)$ is a Cauchy sequence under conditions (5.6) and (5.7) and must therefore have a limit in $\mathfrak{B}(\mathbb{H})$. To see that this limit is unique, take any two sequences that satisfy (5.6) and merge them. Condition (5.6) is still satisfied by this new sequence and the corresponding Riemann type sum of operators must therefore converge to a limit that agrees with the one obtained from either of the sequences that were used in its construction.

The fact that $\oint_\Gamma \mathscr{T}(z)dz$ arises from (5.4) makes it easy to see that contour integration of operators behaves much like ordinary Riemann integration.

In particular, it is a linear operation: if $\{\mathscr{T}_1(z) : z \in \Gamma\}, \{\mathscr{T}_2(z) : z \in \Gamma\}$ are two collections of functions $\oint_\Gamma (\alpha_1 \mathscr{T}_1(z) + \alpha_2 \mathscr{T}_2(z))dz = \alpha_1 \oint_\Gamma \mathscr{T}_1(z)dz + \alpha_2 \oint_\Gamma \mathscr{T}_2(z)dz$ for $\alpha_1, \alpha_2 \in \mathbb{R}$.

When contour integration is applied to the resolvent, the outcome is a formula that gives a useful characterization of projection operators.

Theorem 5.1.3 *Let $\mathscr{T}$ be a compact, self-adjoint operator and let Γ represent the boundary of a disk D that contains $\{\lambda_k : p \le k \le q\}$ and no other eigenvalues with $\{\lambda_k : p \le k \le q\}$ strictly in the interior of D. Then, the projection operator, $\mathscr{P}$, for the union of the eigenspaces corresponding to these eigenvalues has the representation*

$$\mathscr{P} = -\frac{1}{2\pi i} \oint_\Gamma \mathscr{R}(z)dz, \tag{5.9}$$

where $\mathscr{R}$ is the resolvent of $\mathscr{T}$ in (5.2).

Proof: We first show the contour integral in (5.9) is well defined by verifying (5.7). It follows that

$$\|\mathscr{R}(z) - \mathscr{R}(z')\| = \max_j \left| \frac{1}{z - \lambda_j} - \frac{1}{z' - \lambda_j} \right| = \max_j \left| \frac{z - z'}{(z - \lambda_j)(z' - \lambda_j)} \right|.$$

As $\epsilon := \inf_{z \in \Gamma, j \ge 1} |z - \lambda_j| > 0$,

$$\sup_{d_\Gamma(z,z') \le \delta} \|\mathscr{R}(z) - \mathscr{R}(z')\| \le \frac{\delta}{\epsilon^2}.$$

Hence, (5.7) holds and the contour integral in (5.9) is well defined.

By (5.2), we obtain

$$-\frac{1}{2\pi i} \oint_\Gamma \mathscr{R}(z)dz = -\frac{1}{2\pi i} \sum_{k=1}^{\infty} \mathscr{P}_k \oint_\Gamma \frac{dz}{\lambda_k - z}.$$

Cauchy's integral formula gives

$$\oint_\Gamma \frac{dz}{\lambda_k - z} = (-2\pi i) I(p \le k \le q)$$

and completes the proof. □

We now wish to compare the projection operators for $\mathscr{T}$ and $\tilde{\mathscr{T}}$. However, there is a bit of technical detail that must be dispensed with before such comparisons can be made mathematically tractable.

First, if all the eigenvalues are distinct for both $\tilde{\mathscr{T}}$ and $\mathscr{T}$, then $\tilde{\mathscr{P}}_j$ clearly should be the projection operator corresponding to the jth largest eigenvalues of $\tilde{\mathscr{T}}$. However, things become more complicated if one or both of $\mathscr{T}$ and $\tilde{\mathscr{T}}$ have eigenvalues with multiplicities greater than one and the two sets of eigenvalues must be paired in some manner. To do this, we will use the eigenvalues of the "target" operator $\mathscr{T}$ and their multiplicities as the basis for the pairing.

Let $\lambda_1 > \lambda_2 > \cdots$ be the nonrepeated eigenvalues of $\mathscr{T}$ and, as before, let n_j be the multiplicity of λ_j. Now let $\tilde{\lambda}_1 \geq \tilde{\lambda}_2 \geq \cdots$ be the repeated eigenvalues of $\tilde{\mathscr{T}}$. Then, $\tilde{\mathscr{P}}_1$ is defined as the projection operator corresponding to the union of the eigenspaces for $\tilde{\lambda}_k, 1 \leq k \leq n_1$, and $\tilde{\mathscr{P}}_2$ is the projection operator for the union of the eigenspaces for $\tilde{\lambda}_k, n_1 + 1 \leq k \leq n_2$, etc. In general, define

$$N_j = \sum_{k=1}^{j} n_k$$

and take

$$\tilde{\mathscr{P}}_j = \sum_{k=N_{j-1}+1}^{N_j} \tilde{e}_k \otimes \tilde{e}_k \tag{5.10}$$

with $\tilde{e}_k$ the eigenvector associated with $\tilde{\lambda}_k$. Using this approach, the $\tilde{\mathscr{P}}_j$ are obtained by matching the repeated eigenvalues of $\tilde{\mathscr{T}}$ with those from $\mathscr{T}$, with the latter as the basis for matching. This ensures that the dimension of $\tilde{\mathscr{P}}_j$ is the same as that of $\mathscr{P}_j$ which would not necessarily occur if we simply matched the eigenspaces corresponding to the nonrepeated eigenvalues of $\mathscr{T}$ with those from $\tilde{\mathscr{T}}$.

Theorem 5.1.4 *Let the nonrepeated eigenvalues of the compact, self-adjoint operators $\mathscr{T}$ be $\lambda_1 > \lambda_2 > \cdots$ and take Γ as the circle in the complex plane centered at λ_j with radius $\eta_j := (1/2)\, min_{k \neq j} |\lambda_k - \lambda_j|$. Let $\tilde{\mathscr{T}}$ be another compact, self-adjoint operator and assume that*

$$\|\Delta\| < \eta_j \tag{5.11}$$

for $\Delta = \tilde{\mathscr{T}} - \mathscr{T}$. Then, for $\tilde{\mathscr{P}}_j$ in (5.10)

$$\tilde{\mathscr{P}}_j - \mathscr{P}_j = \mathscr{S}_j \Delta \mathscr{P}_j + \mathscr{P}_j \Delta \mathscr{S}_j + \frac{1}{2\pi i} \oint_\Gamma \mathscr{M}(z) dz, \tag{5.12}$$

where

$$\mathscr{S}_j = \sum_{k \neq j} \frac{1}{\lambda_k - \lambda_j} \mathscr{P}_k \quad \text{and} \quad \mathscr{M}(z) = \mathscr{R}(z) \sum_{k=2}^{\infty} \{-\Delta \mathscr{R}(z)\}^k. \tag{5.13}$$

Proof: Denote the resolvent for $\tilde{\mathscr{T}}$ by $\tilde{\mathscr{R}}$. Theorem 4.2.8 ensures that Γ encircles λ_j and all the eigenvalues of $\tilde{\mathscr{T}}$ corresponding to λ_j but no other eigenvalues. So, Theorem 5.1.3 can be applied to produce

$$\tilde{\mathscr{P}}_j - \mathscr{P}_j = -\frac{1}{2\pi i}\oint_\Gamma \left(\tilde{\mathscr{R}}(z) - \mathscr{R}(z)\right) dz. \tag{5.14}$$

Now

$$\tilde{\mathscr{R}}(z) = (\Delta + \mathscr{T} - zI)^{-1} = \mathscr{R}(z)(\Delta\mathscr{R}(z) + I)^{-1}$$

and (5.3) entails that

$$\sup_{z\in\Gamma} \|\Delta\mathscr{R}(z)\| \le \frac{\|\Delta\|}{\eta_j} < 1.$$

Thus, (3.13) allows us to write

$$\tilde{\mathscr{R}}(z) - \mathscr{R}(z) = \mathscr{R}(z)\sum_{k=1}^{\infty}\{-\Delta\mathscr{R}(z)\}^k.$$

Using this in (5.14) leads to

$$\tilde{\mathscr{P}}_j - \mathscr{P}_j = -\frac{1}{2\pi i}\oint_\Gamma \mathscr{R}(z)\Delta\mathscr{R}(z)dz + \frac{1}{2\pi i}\oint_\Gamma \mathscr{M}(z)dz.$$

As

$$\frac{1}{2\pi i}\oint_\Gamma \frac{1}{\lambda_k - z}\frac{1}{\lambda_l - z}dz = \begin{cases} -\dfrac{1}{\lambda_k - \lambda_j}, & \text{if } k \ne j \text{ and } l = j, \\ -\dfrac{1}{\lambda_l - \lambda_j}, & \text{if } k = j \text{ and } l \ne j, \\ 0, & \text{otherwise,} \end{cases}$$

the theorem follows from

$$\begin{aligned} -\frac{1}{2\pi i}\oint_\Gamma \mathscr{R}(z)\Delta\mathscr{R}(z)dz &= -\frac{1}{2\pi i}\sum_{k=1}^{\infty}\sum_{l=1}^{\infty}\oint_\Gamma \frac{1}{\lambda_k - z}\frac{1}{\lambda_l - z}dz\mathscr{P}_k\Delta\mathscr{P}_l \\ &= \sum_{k\ne j}\frac{1}{\lambda_k - \lambda_j}(\mathscr{P}_k\Delta\mathscr{P}_j + \mathscr{P}_j\Delta\mathscr{P}_k). \end{aligned}$$

□

Corollary 5.1.5 *Under the conditions of Theorem 5.1.4,*

$$\|\tilde{\mathscr{P}}_j - \mathscr{P}_j - (\mathscr{S}_j\Delta\mathscr{P}_j + \mathscr{P}_j\Delta\mathscr{S}_j)\| \le \frac{\delta_j^2}{1 - \delta_j}, \tag{5.15}$$

with $\delta_j = \|\Delta\|/\eta_j$.

Proof: Note that

$$\|\mathscr{M}(z)\| \le \|\mathscr{R}(z)\| \sum_{k=2}^{\infty} \|\Delta\mathscr{R}(z)\|^k \le \frac{1}{\eta_j}\frac{\delta_j^2}{1-\delta_j}.$$

This leads to

$$\left\|\frac{1}{2\pi i}\oint_\Gamma \mathscr{M}(z)dz\right\| \le \frac{l_\Gamma}{2\pi\eta_j}\frac{\delta_j^2}{1-\delta_j} \le \frac{\delta_j^2}{1-\delta_j}$$

and thereby proves the corollary. □

It follows that

$$\|\mathscr{S}_j\Delta\mathscr{P}_j + \mathscr{P}_j\Delta\mathscr{S}_j\|^2 = \|\mathscr{S}_j\Delta\mathscr{P}_j\|^2 + \|\mathscr{P}_j\Delta\mathscr{S}_j\|^2 \le 2\|\mathscr{S}_j\|^2\|\Delta\|^2.$$

As $\|\mathscr{S}_j\| \le (2\eta_j)^{-1}$, we conclude that

$$\|\mathscr{S}_j\Delta\mathscr{P}_j + \mathscr{P}_j\Delta\mathscr{S}_j\| \le \frac{\delta_j}{\sqrt{2}}. \tag{5.16}$$

Consequently, (5.15) implies that $\mathscr{S}_j\Delta\mathscr{P}_j + \mathscr{P}_j\Delta\mathscr{S}_j$ represents a type of first-order approximation for $\tilde{\mathscr{P}}_j - \mathscr{P}_j$. In addition, combining (5.15) and (5.16) gives

$$\|\tilde{\mathscr{P}}_j - \mathscr{P}_j\| \le \frac{\delta_j}{\sqrt{2}} + \frac{\delta_j^2}{1-\delta_j} \le \frac{\delta_j}{1-\delta_j}. \tag{5.17}$$

The nonasymptotic nature of these results allows us to potentially obtain useful bounds for multiple projection spaces simultaneously provided that $\|\Delta\|$ is small relative to η_j.

Our results concerning projections can be used to develop bounds for the differences between eigenvalues and eigenvectors of $\mathscr{T}$ and $\tilde{\mathscr{T}}$. First consider the eigenvalues.

Theorem 5.1.6 *Let λ_j be an eigenvalue of the compact, self-adjoint operator $\mathscr{T}$ and let $\mathscr{P}_j$ be the projection operator for the eigenspace. Suppose that $\tilde{\mathscr{T}}$ is another compact, self-adjoint operator with associated projection operator $\tilde{\mathscr{P}}_j$ as in (5.10). Then,*

$$\tilde{\mathscr{P}}_j\tilde{\mathscr{T}}\tilde{\mathscr{P}}_j - \lambda_j\tilde{\mathscr{P}}_j = \tilde{\mathscr{P}}_j\Delta\tilde{\mathscr{P}}_j + (\tilde{\mathscr{P}}_j - \mathscr{P}_j)(\mathscr{T} - \lambda_j I)(\tilde{\mathscr{P}}_j - \mathscr{P}_j) \tag{5.18}$$

and, consequently, for each $k = N_{j-1}+1, \ldots, N_j$,

$$\tilde{\lambda}_k - \lambda_j = \langle \Delta\tilde{e}_k, \tilde{e}_k\rangle + \langle (\tilde{\mathscr{P}}_j - \mathscr{P}_j)(\mathscr{T} - \lambda_j I)(\tilde{\mathscr{P}}_j - \mathscr{P}_j)\tilde{e}_k, \tilde{e}_k\rangle. \tag{5.19}$$

Proof: Write

$$\tilde{\mathscr{P}}_j\tilde{\mathscr{T}}\tilde{\mathscr{P}}_j - \lambda_j\tilde{\mathscr{P}}_j = \tilde{\mathscr{P}}_j\Delta\tilde{\mathscr{P}}_j + \tilde{\mathscr{P}}_j(\mathscr{T} - \lambda_j I)\tilde{\mathscr{P}}_j.$$

As $\mathscr{P}_j(\mathscr{T} - \lambda_j I) = 0$, we have

$$\tilde{\mathscr{P}}_j(\mathscr{T} - \lambda_j I)\tilde{\mathscr{P}}_j = (\tilde{\mathscr{P}}_j - \mathscr{P}_j)(\mathscr{T} - \lambda_j I)(\tilde{\mathscr{P}}_j - \mathscr{P}_j)$$

and (5.18) is proved. For (5.19), write

$$\tilde{\mathscr{P}}_j\tilde{\mathscr{T}}\tilde{\mathscr{P}}_j = \sum_{k=N_{j-1}+1}^{N_j} \tilde{\lambda}_k\tilde{e}_k \otimes \tilde{e}_k$$

and take inner products. □

As

$$\|(\tilde{\mathscr{P}}_j - \mathscr{P}_j)(\mathscr{T} - \lambda_j I)(\tilde{\mathscr{P}}_j - \mathscr{P}_j)\| \leq \|\mathscr{T}\|\|(\tilde{\mathscr{P}}_j - \mathscr{P}_j)\|^2, \tag{5.20}$$

(5.19) shows that the first-order approximation to $\tilde{\lambda}_{jk} - \lambda_j$ is $\langle\Delta\tilde{e}_{jk}, \tilde{e}_{jk}\rangle$. The remainder can be dealt with using inequalities such as (5.17).

Suppose that both $\mathscr{P}_j$ and $\tilde{\mathscr{P}}_j$ are of dimension one. Then, arguing as in the proof of Theorem 5.1.6, we see that

$$\tilde{\lambda}_{\tilde{j}}\mathscr{P}_j - \mathscr{P}_j\mathscr{T}\mathscr{P}_j = \mathscr{P}_j\Delta\mathscr{P}_j + (\tilde{\mathscr{P}}_j - \mathscr{P}_j)(\tilde{\mathscr{T}} - \tilde{\lambda}_{\tilde{j}} I)(\tilde{\mathscr{P}}_j - \mathscr{P}_j)$$

for $\tilde{j} = N_{j-1} + 1$. Thus, in this case,

$$\tilde{\lambda}_{\tilde{j}} - \lambda_j = \langle\Delta e_j, e_j\rangle + \langle(\tilde{\mathscr{P}}_{\tilde{j}} - \mathscr{P}_j)(\tilde{\mathscr{T}} - \tilde{\lambda}_{\tilde{j}} I)(\tilde{\mathscr{P}}_{\tilde{j}} - \mathscr{P}_j)e_j, e_j\rangle. \tag{5.21}$$

Furthermore, if $\mathscr{P}_j$ and $\tilde{\mathscr{P}}_j$ are of dimension one for all j then (5.21) takes the form

$$\tilde{\lambda}_j - \lambda_j = \langle\Delta e_j, e_j\rangle + \langle(\tilde{\mathscr{P}}_j - \mathscr{P}_j)(\tilde{\mathscr{T}} - \tilde{\lambda}_j I)(\tilde{\mathscr{P}}_j - \mathscr{P}_j)e_j, e_j\rangle. \tag{5.22}$$

Next we consider eigenvectors. In that regard, we first give a basic result that shows differences between eigenvectors are intimately related to differences between the corresponding projection operators.

Lemma 5.1.7 *Let $\mathscr{P} = e \otimes e$ and $\tilde{\mathscr{P}} = \tilde{e} \otimes \tilde{e}$ with $\|e\| = \|\tilde{e}\| = 1$. Then,*

1. *the eigenvalues of $\mathscr{P} - \tilde{\mathscr{P}}$ are $\pm(1 - \langle e, \tilde{e}\rangle^2)^{1/2}$,*
2. *$\|e - \tilde{e}\|^2 = 2[1 - (1 - \|\mathscr{P} - \tilde{\mathscr{P}}\|^2)^{1/2}]$ if $\langle e, \tilde{e}\rangle \geq 0$ and*
3. *$\|\mathscr{P} - \tilde{\mathscr{P}}\|_{HS}^2 = 2(1 - \langle e, \tilde{e}\rangle^2) = 2\|\mathscr{P} - \tilde{\mathscr{P}}\|^2$.*

Proof: It is clear that the image space of $\mathscr{P} - \tilde{\mathscr{P}}$ is at most of dimension two. The eigenvalues can be computed from the equivalent matrix eigen problem

$$\begin{bmatrix} 1 & \langle e, \tilde{e} \rangle \\ -\langle e, \tilde{e} \rangle & -1 \end{bmatrix} \begin{bmatrix} a \\ b \end{bmatrix} = \lambda \begin{bmatrix} a \\ b \end{bmatrix}.$$

So, part 1 follows trivially.

By Theorem 4.3.4 and part 1 of the lemma,

$$\|\mathscr{P} - \tilde{\mathscr{P}}\|^2 = 1 - \langle \tilde{e}, e \rangle^2. \tag{5.23}$$

Thus, part 2 of the lemma is a consequence of the identity

$$\|e - \tilde{e}\|^2 = 2(1 - \langle \tilde{e}, e \rangle)$$

while part 3 is due to part 1 and (5.23). □

Observe that for $\langle e, \tilde{e} \rangle \geq 0$ the inequality

$$1 - (1 - x)^{1/2} \leq \frac{x}{2} + \frac{x^2}{4(1 - x)}$$

that holds for $x \in [0, 1]$ can be used with result 2 of Lemma 5.1.7 to see that

$$\|e - \tilde{e}\|^2 \leq \|\mathscr{P} - \tilde{\mathscr{P}}\|^2 + \frac{1}{2} \frac{\|\mathscr{P} - \tilde{\mathscr{P}}\|^4}{1 - \|\mathscr{P} - \tilde{\mathscr{P}}\|^2}. \tag{5.24}$$

Moreover, as $(1 + x)^{1/2} \leq 1 + x/2$ for $x \in [0, 1]$, we obtain

$$\|e - \tilde{e}\| \leq \|\mathscr{P} - \tilde{\mathscr{P}}\| + \frac{1}{4} \frac{\|\mathscr{P} - \tilde{\mathscr{P}}\|^3}{1 - \|\mathscr{P} - \tilde{\mathscr{P}}\|^2}. \tag{5.25}$$

Our perturbation result for eigenvectors can now be stated as follows.

Theorem 5.1.8 *Let λ_j be an eigenvalue of $\mathscr{T}$ with multiplicity one and corresponding eigenvector e_j. The multiplicities of other eigenvalues are not restricted to be one. Let $\tilde{\mathscr{T}}$ be an approximating operator and suppose $(\tilde{\lambda}_j, \tilde{e}_j)$ correspond to (λ_j, e_j) with $\langle e_j, \tilde{e}_j \rangle \geq 0$. If $\inf_{k \neq j} |\tilde{\lambda}_j - \lambda_k| > 0$,*

$$\tilde{e}_j - e_j = \sum_{k \neq j} (\tilde{\lambda}_j - \lambda_k)^{-1} \mathscr{P}_k \Delta \tilde{e}_j + \mathscr{P}_j(\tilde{e}_j - e_j). \tag{5.26}$$

Let $\eta_j = (1/2) \inf_{k \neq j} |\lambda_j - \lambda_k|$ and $\delta_j = \|\Delta\|/\eta_j$. If $\delta_j < 1$ then

$$\|e_j - \tilde{e}_j - \mathscr{S}_j \Delta e_j\| \leq \psi(\delta_j)\delta_j^2 \tag{5.27}$$

for some finite function ψ, where $\psi(\delta) \downarrow 1$ as $\delta \downarrow 0$.

Proof: It is easy to verify that for $k \neq j$

$$(\tilde{\lambda}_j - \lambda_k)\mathscr{P}_k(\tilde{e}_j - e_j) = (\tilde{\lambda}_j - \lambda_k)\mathscr{P}_k\tilde{e}_j = \mathscr{P}_k\Delta\tilde{e}_j.$$

So, if $\tilde{\lambda}_j \neq \lambda_k$ for $k \neq j$,

$$\tilde{e}_j - e_j = \sum_{k=1}^{\infty} \mathscr{P}_k(\tilde{e}_j - e_j) = \sum_{k \neq j} \frac{1}{\tilde{\lambda}_j - \lambda_k}\mathscr{P}_k\Delta\tilde{e}_j + \mathscr{P}_j(\tilde{e}_j - e_j),$$

which is (5.26).

Theorem 4.2.8 gives $|\tilde{\lambda}_j - \lambda_j| \leq \|\Delta\| < \inf_{k \neq j}|\lambda_j - \lambda_k|$ so that

$$\begin{aligned}
\inf_{k \neq j} |\tilde{\lambda}_j - \lambda_k| &= \inf_{k \neq j} |\tilde{\lambda}_j - \lambda_j + \lambda_j - \lambda_k| \\
&\geq \inf_{k \neq j} |\lambda_j - \lambda_k| - |\tilde{\lambda}_j - \lambda_j| \\
&\geq \inf_{k \neq j} |\lambda_j - \lambda_k| - \|\Delta\| > 0
\end{aligned} \tag{5.28}$$

and the condition we need to use the expansion in (5.26) is met. As $|\lambda_j - \tilde{\lambda}_j| < |\lambda_j - \lambda_k|$, we can write

$$(\tilde{\lambda}_j - \lambda_k)^{-1} = (\lambda_j - \lambda_k)^{-1}\left(1 - \frac{\lambda_j - \tilde{\lambda}_j}{\lambda_j - \lambda_k}\right)^{-1} = \sum_{s=0}^{\infty} \frac{(\lambda_j - \tilde{\lambda}_j)^s}{(\lambda_j - \lambda_k)^{s+1}}.$$

Using this in conjunction with (5.26) produces

$$\begin{aligned}
\tilde{e}_j - e_j &= \sum_{k \neq j} (\lambda_j - \lambda_k)^{-1}\mathscr{P}_k\Delta\tilde{e}_j + \sum_{k \neq j}\sum_{s=1}^{\infty} \frac{(\lambda_j - \tilde{\lambda}_j)^s}{(\lambda_j - \lambda_k)^{s+1}}\mathscr{P}_k\Delta\tilde{e}_j \\
&\quad + \mathscr{P}_j(\tilde{e}_j - e_j) \\
&= \sum_{k \neq j} (\lambda_j - \lambda_k)^{-1}\mathscr{P}_k\Delta e_j + \sum_{k \neq j} (\lambda_j - \lambda_k)^{-1}\mathscr{P}_k\Delta(\tilde{e}_j - e_j) \\
&\quad + \sum_{k \neq j}\sum_{s=1}^{\infty} \frac{(\lambda_j - \tilde{\lambda}_j)^s}{(\lambda_j - \lambda_k)^{s+1}}\mathscr{P}_k\Delta\tilde{e}_j + \mathscr{P}_j(\tilde{e}_j - e_j).
\end{aligned}$$

The proof will be complete once we obtain bounds for the last three terms that arise in the last relation.

From Bessel's inequality, we see that

$$\left\| \sum_{k \neq j} (\lambda_j - \lambda_k)^{-1}\mathscr{P}_k\Delta(\tilde{e}_j - e_j) \right\| \leq \frac{\|\Delta\| \|\tilde{e}_j - e_j\|}{2\eta_j} = \frac{\delta_j}{2}\|\tilde{e}_j - e_j\|.$$

Similarly,

$$
\begin{aligned}
\left\| \sum_{k \neq j} \sum_{s=1}^{\infty} \frac{(\lambda_j - \tilde{\lambda}_j)^s}{(\lambda_j - \lambda_k)^{s+1}} \mathscr{P}_k \Delta \tilde{e}_j \right\|^2 &= \sum_{k \neq j} \left(\sum_{s=1}^{\infty} \frac{(\lambda_j - \tilde{\lambda}_j)^s}{(\lambda_j - \lambda_k)^{s+1}} \right)^2 \| \mathscr{P}_k \Delta \tilde{e}_j \|^2 \\
&= \sum_{k \neq j} \frac{(\lambda_j - \tilde{\lambda}_j)^2}{(\lambda_j - \lambda_k)^2 (\tilde{\lambda}_j - \lambda_k)^2} \| \mathscr{P}_k \Delta \tilde{e}_j \|^2 \\
&\leq \frac{\|\Delta\|^4}{4\eta_j^2 (2\eta_j - \|\Delta\|)^2} \leq \frac{\delta_j^4}{4}
\end{aligned}
$$

as a result of (5.28) and Theorem 4.2.8. Finally,

$$
\|\mathscr{P}_j(\tilde{e}_j - e_j)\| = (1 - \langle \tilde{e}_j, e_j \rangle) = \frac{1}{2} \|\tilde{e}_j - e_j\|^2.
$$

Assertion (5.27) now follows from (5.17), (5.24), and (5.25). □

5.2 Perturbation of general compact operators

To this point, we have dealt only with the case of eigenvalues and eigenvectors for a self-adjoint, compact operator $\mathscr{T}$. It is also of interest to have similar results that can be used more generally for any compact operator between two Hilbert spaces. This means that we now need to assess the effect of perturbation on singular values and vectors. That is the subject addressed in this section.

Let $\mathbb{H}_1$ and $\mathbb{H}_2$ be separable Hilbert spaces with $\mathscr{T}, \tilde{\mathscr{T}} \in \mathfrak{B}(\mathbb{H}_1, \mathbb{H}_2)$ where, again, we think of $\tilde{\mathscr{T}}$ as an approximating operator to $\mathscr{T}$. As singular values and vectors are obtained from the eigenvalues and eigenvectors of $\mathscr{T}^*\mathscr{T}$ and $\mathscr{T}\mathscr{T}^*$, we have in some sense already addressed how the singular values and vectors of $\tilde{\mathscr{T}}$ approximate those of $\mathscr{T}$. However, the goal is to assess the effect of perturbations of $\mathscr{T}$ on its singular values and the theory we have developed would, strictly speaking, only be directly applicable to perturbations that were made to $\mathscr{T}^*\mathscr{T}$ or $\mathscr{T}\mathscr{T}^*$. Thus, it it worthwhile to explore this issue in somewhat more detail.

We will assume that all the nonzero singular values are of unit multiplicity. As will be evident from our proofs, this is by no means necessary but appreciably simplifies our presentation. Let $\mathscr{T}$ and $\tilde{\mathscr{T}}$ have singular systems $\{(\lambda_j, f_{1j}, f_{2j})\}_{j=1}^{\infty}$ and $\{(\tilde{\lambda}_j, \tilde{f}_{1j}, \tilde{f}_{2j})\}_{j=1}^{\infty}$. Without loss, we will assume that $\{f_{1j}\}$ and $\{f_{2j}\}$ provide CONSs for $\mathbb{H}_1$ and $\mathbb{H}_2$. Then, for $k, \ell = 1, 2$, we will have need for the operators

$$
\begin{aligned}
\mathscr{P}_{k\ell j} &= f_{kj} \otimes f_{\ell j}, \\
\tilde{\mathscr{P}}_{k\ell j} &= \tilde{f}_{kj} \otimes \tilde{f}_{\ell j};
\end{aligned}
$$

i.e.,

$$\mathscr{P}_{k\ell j}f = \langle f_{kj}, f\rangle_k f_{\ell j},$$
$$\tilde{\mathscr{P}}_{k\ell j}f = \langle \tilde{f}_{kj}, f\rangle_k \tilde{f}_{\ell j}$$

for $f \in \mathbb{H}_k$. In particular, $\mathscr{P}_{11j}$ and $\mathscr{P}_{22j}$ are the projection operators for span$\{f_{1j}\}$ and span$\{f_{2j}\}$.

Now take

$$\Delta = \tilde{\mathscr{T}} - \mathscr{T}$$

and let

$$\zeta_j = \left(\frac{1}{2}\min_{k \neq j}|\lambda_k^2 - \lambda_j^2|\right)^{-1}(\|\mathscr{T}\| + \|\Delta\|)\|\Delta\|.$$

Our first objective is to establish analogs of (5.21) and (5.27) that are applicable to the present setting.

Theorem 5.2.1 *Let ϵ be any number in $(0, 1)$ and assume that $\zeta_j < 1 - \epsilon$. Then, there exists a constant $C = C_\epsilon \in (0, \infty)$ such that*

$$\|(\tilde{\lambda}_j\mathscr{P}_{12j} - \mathscr{P}_{22j}\mathscr{T}\mathscr{P}_{11j}) - \mathscr{P}_{22j}\Delta\mathscr{P}_{11j}\| \leq C\max(\|\mathscr{T}\|, 1)\zeta_j^2 \quad (5.29)$$

and

$$|(\tilde{\lambda}_j - \lambda_j) - \langle f_{2j}, \Delta f_{1j}\rangle_2| \leq C\max(\|\mathscr{T}\|, 1)\zeta_j^2. \quad (5.30)$$

Our applications are usually for situations with fixed $\mathscr{T}$ and j with $\tilde{\mathscr{T}} = \mathscr{T}_n$ for some approximating sequence of operators $\{\mathscr{T}_n\}$. In such instances, $\zeta_j = O(\|\Delta\|)$ with the consequence that the bounds in (5.29) and (5.30) are of order $\|\Delta\|^2$.

Proof: Relation (5.30) follows from (5.29) by taking inner products. Thus, we need only establish the latter result. A first step in this direction involves showing that

$$\begin{aligned}
&\tilde{\lambda}_j\mathscr{P}_{12j} - \mathscr{P}_{22j}\mathscr{T}\mathscr{P}_{11j} \\
&\quad = \mathscr{P}_{22j}\Delta\mathscr{P}_{11j} + (\tilde{\mathscr{P}}_{22j} - \mathscr{P}_{22j})(\tilde{\mathscr{T}} - \tilde{\lambda}_j\tilde{\mathscr{P}}_{12j})(\tilde{\mathscr{P}}_{11j} - \mathscr{P}_{11j}) \\
&\quad + \tilde{\lambda}_j\tilde{\mathscr{P}}_{22j}(\mathscr{P}_{12j} - \tilde{\mathscr{P}}_{12j})\tilde{\mathscr{P}}_{11j}.
\end{aligned}$$

To verify this identity observe that as $\mathscr{P}_{22j}\mathscr{P}_{12j}\mathscr{P}_{11j} = \mathscr{P}_{12j}$

$$\begin{aligned}
&\tilde{\lambda}_j\mathscr{P}_{12j} - \mathscr{P}_{22j}\mathscr{T}\mathscr{P}_{11j} \\
&\quad = \mathscr{P}_{22j}\Delta\mathscr{P}_{11j} - \mathscr{P}_{22j}(\tilde{\mathscr{T}} - \tilde{\lambda}_j\mathscr{P}_{12j})\mathscr{P}_{11j} \\
&\quad = \mathscr{P}_{22j}\Delta\mathscr{P}_{11j} - \mathscr{P}_{22j}(\tilde{\mathscr{T}} - \tilde{\lambda}_j\tilde{\mathscr{P}}_{12j})\mathscr{P}_{11j} - \tilde{\lambda}_j\mathscr{P}_{22j}(\tilde{\mathscr{P}}_{12j} - \mathscr{P}_{12j})\mathscr{P}_{11j}.
\end{aligned}$$

The second term on the right-hand side of the last relation is

$$-(\tilde{\mathscr{P}}_{22j} - \mathscr{P}_{22j})(\tilde{\mathscr{T}} - \tilde{\lambda}_j \tilde{\mathscr{P}}_{12j})(\tilde{\mathscr{P}}_{11j} - \mathscr{P}_{11j})$$

because $\tilde{\mathscr{P}}_{22j}(\tilde{\mathscr{T}} - \tilde{\lambda}_j \tilde{\mathscr{P}}_{12j}) = 0$ and $(\tilde{\mathscr{T}} - \tilde{\lambda}_j \tilde{\mathscr{P}}_{12j})\tilde{\mathscr{P}}_{11j} = 0$.

It remains to obtain norm bounds for the remainder terms $(\tilde{\mathscr{P}}_{22j} - \mathscr{P}_{22j})(\tilde{\mathscr{T}} - \tilde{\lambda}_j \tilde{\mathscr{P}}_{12j})(\tilde{\mathscr{P}}_{11j} - \mathscr{P}_{11j})$ and $\tilde{\lambda}_j \mathscr{P}_{22j}(\tilde{\mathscr{P}}_{12j} - \mathscr{P}_{12j})\mathscr{P}_{11j}$. To do so begin by noting that

$$\left(\frac{1}{2}\min_{k\neq j}|\lambda_k^2 - \lambda_j^2|\right)^{-1} \|\tilde{\mathscr{T}}^*\tilde{\mathscr{T}} - \mathscr{T}^*\mathscr{T}\| \leq \zeta_j.$$

Then, from (5.17), we have

$$\max(\|\tilde{\mathscr{P}}_{11j} - \mathscr{P}_{11j}\|, \|\tilde{\mathscr{P}}_{22j} - \mathscr{P}_{22j}\|) \leq \frac{\zeta_j}{1-\zeta_j} \leq C\zeta_j \tag{5.31}$$

with $C = \epsilon^{-1}$ and we can conclude that

$$\begin{aligned}
&\|(\tilde{\mathscr{P}}_{22j} - \mathscr{P}_{22j})(\tilde{\mathscr{T}} - \tilde{\lambda}_j \tilde{\mathscr{P}}_{12j})(\tilde{\mathscr{P}}_{11j} - \mathscr{P}_{11j})\| \\
&\quad \leq \|\tilde{\mathscr{P}}_{11j} - \mathscr{P}_{11j}\|\|\tilde{\mathscr{P}}_{22j} - \mathscr{P}_{22j}\|\|\tilde{\mathscr{T}}\| \\
&\quad \leq C\|\mathscr{T}\|\zeta_j^2.
\end{aligned}$$

Next,

$$\begin{aligned}
\|\mathscr{P}_{22j}(\tilde{\mathscr{P}}_{12j} - \mathscr{P}_{12j})\mathscr{P}_{11j}\| &= 1 - \langle \tilde{f}_{1j}, f_{1j}\rangle_1 \langle \tilde{f}_{2j}, f_{2j}\rangle_2 \\
&= 1 - \langle \tilde{f}_{2j}, f_{2j}\rangle_2 + (1 - \langle \tilde{f}_{1j}, f_{1j}\rangle_1)\langle \tilde{f}_{2j}, f_{2j}\rangle_2,
\end{aligned}$$

where we assume without loss of generality that $\langle \tilde{f}_{1j}, f_{1j}\rangle_1$ and $\langle \tilde{f}_{2j}, f_{2j}\rangle_2$ are nonnegative. Applying (5.24) and (5.31) and using the identity

$$1 - \langle \tilde{f}_{\ell j}, f_{\ell j}\rangle_\ell = \|\tilde{f}_{\ell j} - f_{\ell j}\|_\ell^2/2$$

leads to

$$\|\mathscr{P}_{22j}(\tilde{\mathscr{P}}_{12j} - \mathscr{P}_{12j})\mathscr{P}_{11j}\| \leq C\zeta_j^2$$

for some suitable C and completes the proof. □

The singular vector version of Theorem 5.1.8 now takes the following form.

Theorem 5.2.2 *Let ϵ be any number in $(0, 1)$ and assume that $\zeta_j < 1 - \epsilon$. Then, there exists a constant $C = C_\epsilon \in (0, \infty)$ such that*

$$\left\| (\tilde{f}_{1j} - f_{1j}) - \sum_{k \neq j} \frac{\lambda_k \langle f_{2k}, \Delta f_{1j} \rangle_2 + \lambda_j \langle f_{2j}, \Delta f_{1k} \rangle_2}{\lambda_k^2 - \lambda_j^2} f_{1k} \right\|_1 \leq C \gamma_j$$

with

$$\gamma_j = \zeta_j \left(\zeta_j + \frac{\|\Delta\|}{\|\Delta\| + \|\mathscr{T}\|} \right).$$

Proof: If $\zeta_j < 1$ then, by Theorem 5.1.8,

$$\|(\tilde{f}_{1j} - f_{1j}) - \mathscr{S}_{11j}(\tilde{\mathscr{T}}^* \tilde{\mathscr{T}} - \mathscr{T}^* \mathscr{T}) f_{1j}\|_1 \leq \psi(\zeta_j)\zeta_j^2,$$

where

$$\mathscr{S}_{11j} = \sum_{k \neq j} (\lambda_k^2 - \lambda_j^2)^{-1} \mathscr{P}_{11k}.$$

Now

$$\mathscr{S}_{11j}(\tilde{\mathscr{T}}^* \tilde{\mathscr{T}} - \tilde{\mathscr{T}}^* \tilde{\mathscr{T}}) f_{1j} = \sum_{k \neq j} \frac{\langle f_{1k}, (\tilde{\mathscr{T}}^* \tilde{\mathscr{T}} - \mathscr{T}^* \mathscr{T}) f_{1j} \rangle_1}{\lambda_k^2 - \lambda_j^2} f_{1k}$$

and we can write

$$\tilde{\mathscr{T}}^* \tilde{\mathscr{T}} - \mathscr{T}^* \mathscr{T} = \mathscr{T}^*(\tilde{\mathscr{T}} - \mathscr{T}) + (\tilde{\mathscr{T}}^* - \mathscr{T}^*)\mathscr{T} + (\tilde{\mathscr{T}}^* - \mathscr{T}^*)(\tilde{\mathscr{T}} - \mathscr{T}).$$

Thus,

$$\begin{aligned}
&\langle f_{1k}, (\tilde{\mathscr{T}}^* \tilde{\mathscr{T}} - \mathscr{T}^* \mathscr{T}) f_{1j} \rangle_1 \\
&\quad = \langle f_{1k}, \mathscr{T}^*(\tilde{\mathscr{T}} - \mathscr{T}) f_{1j} \rangle_1 + \langle f_{1k}, (\tilde{\mathscr{T}}^* - \mathscr{T}^*)\mathscr{T} f_{1j} \rangle_1 \\
&\qquad + \langle f_{1k}, (\tilde{\mathscr{T}}^* - \mathscr{T}^*)(\tilde{\mathscr{T}} - \mathscr{T}) f_{1j} \rangle_1 \\
&\quad = \lambda_k \langle f_{2k}, \Delta f_{1j} \rangle_2 + \lambda_j \langle f_{2j}, \Delta f_{1k} \rangle_2 + \langle f_{1k}, \Delta^* \Delta f_{1j} \rangle_1
\end{aligned}$$

and

$$\left\| \sum_{k \neq j} \frac{\langle f_{1k}, \Delta^* \Delta f_{1j} \rangle_1}{\lambda_k^2 - \lambda_j^2} f_{1k} \right\|_1 \leq \frac{1}{\min_{k \neq j} |\lambda_j^2 - \lambda_k^2|} \|\Delta\|^2 = \frac{\|\Delta\|}{2(\|\Delta\| + \|\mathscr{T}\|)} \zeta_j.$$

□

We conclude this chapter with an application of the developments in this section to an fda relevant problem from numerical analysis.

Example 5.2.3 *Consider the* $\mathbb{L}^2[0,1]$ *integral operator*

$$(\mathscr{T}f)(t) = \int_0^1 f(s)K(t,s)ds$$

for some continuous kernel function K *on* $[0,1]\times[0,1]$. *We do not presume that the kernel is symmetric. So,* $\mathscr{T}$ *is compact but not necessarily self-adjoint. An application of (4.15) in this instance allows us to express the operator as*

$$\mathscr{T} = \sum_{j=1}^{\infty} \lambda_j f_{1j} \otimes f_{2j}$$

with $\{(\lambda_j, f_{1j}, f_{2j})\}_{j=1}^{\infty}$ *its associated singular system. We will approximate the components of this system using a Rayleigh-Ritz motivated scheme proposed by Hansen (1988). The goal is to assess the error incurred by this approach in approximating the singular values and functions for* $\mathscr{T}$.

Let $\{e_{ij}\}_{j=1}^{\infty}, i = 1,2,$ *be two CONSs for* $\mathbb{L}^2[0,1]$. *The idea is that we approximate* $\mathscr{T}$ *by*

$$\mathscr{T}_n = \sum_{i=1}^{n}\sum_{j=1}^{n} \langle e_{2j}, \mathscr{T}e_{1i}\rangle e_{1i} \otimes e_{2j} \tag{5.32}$$

and then use its singular values and functions to approximate those for $\mathscr{T}$.

Note that under our assumptions $\mathscr{T}$ *is an HS operator. This is because* $\mathscr{T}^*\mathscr{T}$ *is an integral operator with the symmetric and continuous kernel*

$$Q(s,t) = \int_0^1 K(u,t)K(u,s)du.$$

It is therefore trace class and its eigenvalues λ_j^2 *are summable. As a result, it is easy to assess the size of* $\Delta_n = \mathscr{T}_n - \mathscr{T}$ *because*

$$\|\Delta_n\|^2 \le \|\Delta_n\|_{HS}^2 = \sum_{i=n+1}^{\infty}\sum_{j=n+1}^{\infty} \langle e_{2j}, \mathscr{T}e_{1i}\rangle^2.$$

This bound necessarily decays to zero as n diverges.

It is not difficult to see that the singular values and functions for $\mathscr{T}_n$ *derive from the singular values and vectors of the matrix* $\mathscr{A} = \{\langle e_{2j}, \mathscr{T}e_{1i}\rangle\}_{i,j=1:n}$. *Using Corollary 4.3.2, we can write*

$$\mathscr{A} = \sum_{j=1}^{n} \lambda_j^{(n)} u_j v_j^T$$

for orthonormal n-vectors $v_j = (v_{1j}, \ldots, v_{nj})^T$ *and* $u_j = (u_{1j}, \ldots, u_{nj})^T$. *The sve for* $\mathscr{T}_n$ *is now given explicitly by*

$$\mathscr{T}_n = \sum_{j=1}^{n} \lambda_j^{(n)} f_{1j}^{(n)} \otimes f_{2j}^{(n)}$$

with

$$f_{1j}^{(n)} = \sum_{i=1}^{n} v_{ij} e_{1i}$$

and

$$f_{2j}^{(n)} = \sum_{i=1}^{n} u_{ij} e_{2i}.$$

Theorem 5.2.1 now implies that for any particular nonrepeated singular value λ_j *we will have*

$$\lambda_j^{(n)} = \lambda_j + \langle f_{2j}, \Delta_n f_{1j} \rangle + O(\|\Delta_n\|^2).$$

Similarly, for a specific singular function f_{1j}, *Theorem 5.2.2 produces*

$$f_{1j}^{(n)} = f_{1j} + \sum_{k \neq j} \frac{\lambda_k \langle f_{2k}, \Delta_n f_{1j} \rangle + \lambda_j \langle f_{2j}, \Delta_n f_{1k} \rangle}{\lambda_k^2 - \lambda_j^2} f_{1k} + O(\|\Delta_n\|^2).$$

6

Smoothing and regularization

As stated in Chapter 1, our view of functional data analysis involves the statistical analysis of sample paths that arise from one or more stochastic processes. The processes themselves are presumed to be random elements of some Hilbert space such as $\mathbb{L}^2[0, 1]$ in a sense that will be precisely defined in Chapter 7. For now, it suffices to merely think of the collected data as being discretized readings from a sample of curves.

Discretization entails some loss of information. This may be sufficiently problematic in some instances to require remedial measures to recover some of what was lost. In addition, there may be contamination or distortions of the actual sample path values by noise or other sources of error. In such cases, it may be worthwhile to perform some preprocessing to filter out artifacts in the data that have arisen from extraneous sources. Such problems are not unique to fda and arise in a variety of statistical contexts. The methods that have evolved for their solution are generally referred to as *smoothing* or *nonparametric smoothing* techniques that will be the focus of this chapter.

6.1 Functional linear model

A conceptually simple smoothing problem arises from nonparametric regression analysis. In that setting, we have a real valued mean or regression function m on $[0, 1]$ that is discretely observed with additive random noise; i.e., the realized data takes the form $(t_i, Y_i), i = 1, \ldots, n$, with

$$Y_i = m(t_i) + \varepsilon_i, \tag{6.1}$$

for "time" ordinates $0 \le t_1 < \cdots < t_n \le 1$ and $\varepsilon_1, \ldots, \varepsilon_n$ zero mean, uncorrelated random variables with some common variance σ^2. The objective is estimation of m.

Theoretical Foundations of Functional Data Analysis, with an Introduction to Linear Operators, First Edition. Tailen Hsing and Randall Eubank.

An estimator of m serves a similar purpose to a sample mean in that it gives a summary of how the responses behave on the average as a function of the independent or time variable, T. However, some restrictions are needed on m in order to make the summarization process fruitful; if m is totally arbitrary the best summary statistic is just the original response values $Y_1, \ldots, Y_n$.

When m is smooth in the sense of being continuous or differentiable, the true value of m at any time point $t \in [0, 1]$ may be estimated by suitably averaging the Y_i's whose corresponding T ordinates are close to t. The process of replacing groups of response values by their average has the effect of reducing the local influence of individual responses with the outcome being smoother or less variable than what was present in the original data. This is what has motivated terminology such as *data smoothing*, *scatter plot smoothing*, and *nonparametric smoothing* for such averaging procedures. The word *nonparametric* is used here to emphasize the fact that the function m that is being estimated is presumed to derive from an infinite-dimensional space of candidate functions rather than some finite dimensional parametric family as has classically been the case in, e.g., linear regression analysis.

In this chapter, we take a broad view of nonparametric function estimation that subsumes nonparametric regression as well as smoothing problems that arise from other contexts such as fda. The formulation assumes that we have two Hilbert spaces $\mathbb{Y}$ and $\mathbb{H}$ with norms and inner products $\|\cdot\|_{\mathbb{Y}}, \langle\cdot,\cdot\rangle_{\mathbb{Y}}$, $\|\cdot\|_{\mathbb{H}}$ and $\langle\cdot,\cdot\rangle_{\mathbb{H}}$. Then, given $\mathscr{T} \in \mathfrak{B}(\mathbb{H}, \mathbb{Y})$, we observe

$$Y = \mathscr{T}m + \varepsilon, \tag{6.2}$$

for $Y, \varepsilon \in \mathbb{Y}$, and $m \in \mathbb{H}$. Here and hereafter, ε will be used to represent a generic error term that has zero mean. Although we treat $\mathscr{T}$ as being known, we should perhaps note that in practice $\mathscr{T}$ may not be known exactly and/or may be random as in Chapter 11.

We will refer to (6.2) as the *(functional) linear model*. As with nonparametric regression, our goal is to use the observed value of Y to estimate m. That is the topic that will be addressed in the following section. First, however, it will be useful to look at some specific examples of model (6.2) to gain some insight into the potential applications we have in mind.

Example 6.1.1 *Our starting point was nonparametric regression. So, let us begin by seeing how it can be treated using model (6.2). One approach would be to take* $\mathbb{H}$ *as an RKHS containing functions defined on* $[0, 1]$ *with* $\mathbb{Y} = \mathbb{R}^n$. *Then, for* $g \in \mathbb{H}$, *we define* $\mathscr{T}$ *by*

$$\mathscr{T}g = (g(t_1), \ldots, g(t_n))^T.$$

As evaluation functionals are bounded and linear in an RKHS (Example 3.2.2), $\mathscr{T} \in \mathfrak{B}(\mathbb{H}, \mathbb{Y})$.

Example 6.1.2 *Let $\{X(t) : t \in [0, 1]\}$ be a stochastic process on $[0, 1]$ with mean function $m(\cdot)$ and suppose that we observe n independent copies $X_1, \ldots, X_n$ of $X(\cdot)$ whose sample paths fall in an RKHS $\mathbb{H}$. The information for the ith sample path is digitized with readings being recorded at time ordinates $0 \le t_{i1} < \cdots < t_{ir_i} \le 1$. The realized, digitized data then takes the form $(t_{ij}, Y_{ij}), i = 1, \ldots, n, j = 1, \ldots, r_i$ for*

$$\begin{aligned} Y_{ij} &= x_i(t_{ij}) + \tilde{\varepsilon}_{ij} \\ &= m(t_{ij}) + \{x_i(t_{ij}) - m(t_{ij})\} + \tilde{\varepsilon}_{ij} \end{aligned}$$

with random errors $\tilde{\varepsilon}_{ij}$. Thus, let

$$\begin{aligned} Y &= (Y_{11}, \ldots, Y_{1r_1}, \ldots, Y_{n1}, \ldots, Y_{nr_n})^T, \\ \mathcal{T}g &= (g(t_{11}), \ldots, g(t_{1r_1}), \ldots, g(t_{n1}), \ldots, g(t_{nr_n}))^T \end{aligned}$$

and set $\varepsilon_{ij} = x_i(t_{ij}) - m(t_{ij}) + \tilde{\varepsilon}_{ij}$ to return to the (6.2) formulation. Problems of estimation in this setting have been investigated by, e.g., Rice and Silverman (1991).

Example 6.1.3 *Let R be a known, continuous bivariate function defined on $[0, 1] \times [0, 1]$. Suppose we observe $(t_i, Y_i), i = 1, \ldots, n$, where*

$$Y_i = \int_0^1 R(t_i, u)m(u)du + \varepsilon_i,$$

with $m \in \mathbb{L}^2[0, 1]$. In this case,

$$\begin{aligned} Y &= (Y_1, \ldots, Y_n)^T, \\ \mathcal{T}g &= \left(\int_0^1 R(t_1, u)g(u)du, \ldots, \int_0^1 R(t_n, u)g(u)du \right)^T. \end{aligned}$$

Estimation of m using data of this variety has been considered by, e.g., Nychka and Cox (1989).

Example 6.1.4 *Assume that $m \in \mathbb{L}^2[0, 1]$ and that $\{X(t) : t \in [0, 1]\}$ is a stochastic process from which we observe n independent copies $X_1, \ldots, X_n$. Suppose that the data at our disposal takes the form of (X_i, Y_i) and $(t_{ij}, X_i(t_{ij}))$ for $i = 1, \ldots, n$ and $j = 1, \ldots, r_i$, where*

$$Y_i = \int_0^1 m(u)X_i(u)du + \varepsilon_i.$$

This has the form of model (6.2) under the choice of

$$Y = (Y_1, \ldots, Y_n)^T,$$
$$\mathscr{T}g = \left(\int_0^1 g(u)X_1(u)du, \ldots, \int_0^1 g(u)X_n(u)du \right)^T.$$

Note that in this instance $\mathscr{T}$ is random and is not observed directly but must instead be evaluate using the observed X sample paths. If, for each given i, the t_{ij}'s are uniformly distributed, then the values produced by the $\mathscr{T}$ operator can be approximated by

$$\tilde{\mathscr{T}}g = \left(\frac{1}{r_1} \sum_{j=1}^{r_1} g(t_{1j})X_1(t_{1j}), \ldots, \frac{1}{r_n} \sum_{j=1}^{r_n} g(t_{nj})X_n(t_{nj}) \right)^T$$

or, more generally, by some other quadrature type method.

6.2 Penalized least squares estimators

We now wish to develop an estimation criterion that can be used to accompany the functional regression model (6.2). As a first step in that direction, let us return momentarily to the prototypical nonparametric regression problem corresponding to model (6.1). In that setting, an initial thought might be to obtain a least-squares estimator $\hat{m}$ of m from

$$\hat{m} = \mathrm{argmin}_{g \in \mathbb{H}} \sum_{i=1}^{n} (Y_i - g(t_i))^2 / n. \tag{6.3}$$

However, if we were to take, e.g., $\mathbb{H} = \mathbb{W}_2[0, 1]$ from Section 2.8, there would be infinitely many choices for $\hat{m}$ in (6.3); any function $g \in \mathbb{W}_2[0, 1]$ that interpolates the Y data by satisfying $g(t_i) = Y_i$ will minimize the least-squares criterion. Perhaps more to the point is that an estimator that interpolates the responses would generally be rejected a priori as being too slavish to the data as well as ineffective as a summary device.

Some additional restrictions are therefore needed to make a least-square type criterion meaningful when $\mathbb{H}$ is infinite dimensional. One way to accomplish this for nonparametric regression when $\mathbb{H} = \mathbb{W}_2[0, 1]$ is to instead consider the estimator defined by

$$\hat{m} = \mathrm{argmin}_{g \in \mathbb{H}_\delta} \sum_{i=1}^{n} (Y_i - g(t_i))^2 / n, \tag{6.4}$$

where

$$\mathbb{H}_\delta := \left\{ g \in \mathbb{W}_2[0, 1] : \int_0^1 |g^{(2)}(t)|^2 dt \le \delta \right\} \tag{6.5}$$

for some user specified $\delta > 0$. One justification for this approach derives from linear regression analysis. Specifically, if g is a line, $\int_0^1 |g^{(2)}(t)|^2 dt$ is zero and the minimizer of the least-squares criterion in (6.3) over all such functions is the simple linear regression estimator of m. With that in mind, δ in (6.5) now has the interpretation as a measure of departure from linearity. Thus, (6.4) produces a constrained least-squares estimator with bounds being enforced on how far the solution can stray from the simple linear fit.

Criterion (6.4) does not lend itself to simple practical implementation. Instead, it is better to work with the equivalent problem of finding the solution to

$$m_\eta = \text{argmin}_{g \in \mathbb{W}_2[0,1]} \left(n^{-1} \sum_{i=1}^{n} (Y_i - g(t_i))^2 + \eta \int_0^1 |g^{(2)}(t)|^2 dt \right) \tag{6.6}$$

with $\eta > 0$ now being some value that can be shown to be uniquely determined by δ in the previous formulation. The interpretation to be placed on (6.6) is that we wish to fit the response data with a smooth function (in the sense of being in $\mathbb{W}_2[0, 1]$). The first component of the estimation criterion is the (average) residual sum of squares $n^{-1} \sum_{i=1}^{n} (Y_i - g(t_i))^2$ that evaluates the fidelity of the fit to the observations. The $\int_0^1 |g^{(2)}(t)|^2 dt$ term measures the smoothness of g in terms of its average squared curvature. The relationship between the two components is that smaller values of the latter for m_η correspond to larger values of the former and conversely. So, by choosing a large value of η, we place a premium on smoothness as assessed by $\int_0^1 |g^{(2)}(t)|^2 dt$ with the consequence that the resulting estimator will behave more like a simple linear regression estimator than would be true had we opted for a smaller value. On the other hand, larger values of η deemphasize smoothness thereby allowing more flexible function forms that are capable of adhering more closely to the response values. The *smoothing parameter* η in (6.6) can now be viewed as a tuning devise that regulates the relative importance one places on fit to the data vis-a-vis smoothness.

Now suppose we adopt either of the inner products for $\mathbb{W}_2[0, 1]$ that were introduced in Section 2.8. Then,

$$\int_0^1 |g^{(2)}(t)|^2 dt = \langle g, \mathscr{W} g \rangle_{\mathbb{H}}$$

with $\mathscr{W}$ the $\mathbb{W}_2[0, 1]$ projection operator for $\mathbb{H}_1$ in (2.42). The salient feature of $\mathscr{W}$ that we wish to take forward is that it is a nonnegative operator. In light of this fact, one possible generalization of the estimator in (6.6) that could be used for m in model (6.2) is a minimizer of

$$f(g; \mathscr{T}, \mathscr{W}, Y, \eta) := \|Y - \mathscr{T} g\|_{\mathbb{Y}}^2 + \eta \langle g, \mathscr{W} g \rangle_{\mathbb{H}} \tag{6.7}$$

over $g \in \mathbb{H}$ with $\mathscr{W}$ a nonnegative element of $\mathfrak{B}(\mathbb{H})$ and $\eta \in (0, \infty)$. In the context of this, more general problem η is frequently called a *regularization parameter* and the associated estimators of m are referred to as *method of regularization estimators*. We will use "smoothing parameter" and "regularization parameter" interchangeably when referring to η. For the moment, we will treat the value of η as being given. Methods for adaptively choosing its value from data are discussed in Section 6.5.

The following result provides a characterization of the method of regularization estimator.

Theorem 6.2.1 *Assume that $(\mathscr{T}^*\mathscr{T} + \mathscr{W})$ is invertible. Then, for $\eta \in (0, \infty)$*

$$m_\eta := (\mathscr{T}^*\mathscr{T} + \eta\mathscr{W})^{-1}\mathscr{T}^*Y \tag{6.8}$$

is the unique minimizer of (6.7) over $g \in \mathbb{H}$.

Proof: Consider any candidate solution of the form $\tilde{g} = m_\eta + g$. Then,

$$f(\tilde{g}; \mathscr{T}, \mathscr{W}, Y, \eta) = \|Y\|_{\mathbb{Y}}^2 + \|\mathscr{T}\tilde{g}\|_{\mathbb{Y}}^2 - 2\langle Y, \mathscr{T}\tilde{g}\rangle_{\mathbb{Y}} + \eta\langle \tilde{g}, \mathscr{W}\tilde{g}\rangle_{\mathbb{H}}.$$

By definition, $\mathscr{T}^*Y = (\mathscr{T}^*\mathscr{T} + \eta\mathscr{W})m_\eta$ and, hence,

$$\langle Y, \mathscr{T}\tilde{g}\rangle_{\mathbb{Y}} = \langle(\mathscr{T}^*\mathscr{T} + \eta\mathscr{W})m_\eta, \tilde{g}\rangle_{\mathbb{H}}.$$

Thus,

$$\begin{aligned} f(\tilde{g}; \mathscr{T}, \mathscr{W}, Y, \eta) &= \|Y\|_{\mathbb{Y}}^2 + \langle \mathscr{T}^*\mathscr{T}\tilde{g}, \tilde{g}\rangle_{\mathbb{H}} - 2\langle(\mathscr{T}^*\mathscr{T} + \eta\mathscr{W})m_\eta, \tilde{g}\rangle_{\mathbb{H}} \\ &\quad + \eta\langle \mathscr{W}\tilde{g}, \tilde{g}\rangle_{\mathbb{H}} \\ &= \|Y\|_{\mathbb{Y}}^2 + \langle(\mathscr{T}^*\mathscr{T} + \eta\mathscr{W})(\tilde{g} - 2m_\eta), \tilde{g}\rangle_{\mathbb{H}}, \end{aligned}$$

where

$$\begin{aligned} &\langle(\mathscr{T}^*\mathscr{T} + \eta\mathscr{W})(\tilde{g} - 2m_\eta), \tilde{g}\rangle_{\mathbb{H}} \\ &\quad = \langle(\mathscr{T}^*\mathscr{T} + \eta\mathscr{W})(\tilde{g} - m_\eta), \tilde{g}\rangle_{\mathbb{H}} - \langle(\mathscr{T}^*\mathscr{T} + \eta\mathscr{W})m_\eta, \tilde{g}\rangle_{\mathbb{H}} \\ &\quad = \langle(\mathscr{T}^*\mathscr{T} + \eta\mathscr{W})(\tilde{g} - m_\eta), \tilde{g}\rangle_{\mathbb{H}} - \langle(\mathscr{T}^*\mathscr{T} + \eta\mathscr{W})\tilde{g}, m_\eta\rangle_{\mathbb{H}} \\ &\quad = \langle(\mathscr{T}^*\mathscr{T} + \eta\mathscr{W})(\tilde{g} - m_\eta), \tilde{g} - m_\eta\rangle_{\mathbb{H}} - \langle(\mathscr{T}^*\mathscr{T} + \eta\mathscr{W})m_\eta, m_\eta\rangle_{\mathbb{H}} \\ &\quad = \langle(\mathscr{T}^*\mathscr{T} + \eta\mathscr{W})g, g\rangle_{\mathbb{H}} - \langle(\mathscr{T}^*\mathscr{T} + \eta\mathscr{W})m_\eta, m_\eta\rangle_{\mathbb{H}}. \end{aligned}$$

As $(\mathscr{T}^*\mathscr{T} + \eta\mathscr{W})$ is positive definite, the last expression is uniquely minimized at $g = 0$. □

The theorem applies to values of η in $(0, \infty)$. However, it leaves open what transpires as the smoothing parameter tends to either end of this interval. These cases are conceptually important. If we are to view smoothing as a sliding scale controlled by our choice for η, then we must understand what the two extreme values for η reflect in order to appreciate what may result from selection of a value in the interior of the interval.

We begin with the simple case of ordinary Tikhonov regularization where we chose to minimize (6.7) with $\mathscr{W} = I$: i.e.,

$$f(g; \mathscr{T}, I, Y, \eta) = \|Y - \mathscr{T}g\|_{\mathbb{Y}}^2 + \eta\|g\|_{\mathbb{H}}^2. \tag{6.9}$$

We will eventually use the results for this case to handle the general problem. Applying Theorem 6.2.1 in this instance produces the Tikhonov estimator

$$m_\eta = (\mathscr{T}^*\mathscr{T} + \eta I)^{-1}\mathscr{T}^*Y. \tag{6.10}$$

The behavior of this estimator for the two extreme values of the smoothing parameter is described in our following result.

Theorem 6.2.2 *Assume that* $Y \in \text{Dom}(\mathscr{T}^\dagger)$. *Then,* m_η *in (6.10) satisfies*

$$\lim_{\eta \to \infty} m_\eta = 0$$

and

$$\lim_{\eta \to 0} m_\eta = (\mathscr{T}^*\mathscr{T})^\dagger \mathscr{T}^*Y.$$

Proof: We know that

$$\eta\|m_\eta\|_{\mathbb{H}}^2 \le \|Y - \mathscr{T}m_\eta\|_{\mathbb{Y}}^2 + \eta\|m_\eta\|_{\mathbb{H}}^2 \le \|Y - \mathscr{T}g\|_{\mathbb{Y}}^2 + \eta\|g\|_{\mathbb{H}}^2 \tag{6.11}$$

for all $g \in \mathbb{H}$. In particular, this is true for $g = 0$, which means that

$$\eta\|m_\eta\|_{\mathbb{H}} \le \|Y\|_{\mathbb{Y}}^2$$

and m_η must therefore tend to zero as η diverges.

Now consider the case where $\eta \to 0$. In this instance, use (6.11) with $g = \mathscr{T}^\dagger Y$ to see that

$$\eta\|m_\eta\|_{\mathbb{H}}^2 \le \|Y - \mathscr{T}\mathscr{T}^\dagger Y\|_{\mathbb{Y}}^2 + \eta\|\mathscr{T}^\dagger Y\|_{\mathbb{H}}^2 = \eta\|\mathscr{T}^\dagger Y\|_{\mathbb{H}}^2.$$

The equality stems from (3.19) and the fact that $Y \in \text{Dom}(\mathscr{T}^\dagger)$. So, if η_n is any sequence of smoothing parameter values that converges to zero, we must have

$$\limsup_{n \to \infty} \|m_{\eta_n}\|_{\mathbb{H}} \le \|\mathscr{T}^\dagger Y\|_{\mathbb{H}}.$$

Theorem 3.2.11 entails that the bounded sequence m_{η_n} must have a subsequence $m_{\eta_{n_k}}$ that converges weakly to some $g \in \mathbb{H}$. Then, from Theorem 3.3.6 $\mathscr{T}m_{\eta_{n_k}}$ must converge weakly to $\mathscr{T}g$. However,

$$\|Y - \mathscr{T}m_{\eta_{n_k}}\|_{\mathbb{Y}}^2 \le \|Y - \mathscr{T}m_{\eta_{n_k}}\|_{\mathbb{Y}}^2 + \eta_{n_k}\|m_{\eta_{n_k}}\|_{\mathbb{H}}^2 \le \eta_{n_k}\|\mathscr{T}^\dagger Y\|_{\mathbb{H}}^2 \to 0.$$

Thus, $\mathscr{T}g = Y$.

Now, it must be that all the $m_{\eta_{n_k}}$ are in the closed (from Theorem 2.5.6) subspace $\mathrm{Ker}(\mathscr{T})^\perp$. To see this note that we can write every element of $\mathbb{H}$ as $g = h + v$ with $v \in \mathrm{Ker}(\mathscr{T})$ and $h \in \mathrm{Ker}(\mathscr{T})^\perp$. Thus,

$$f(g; \mathscr{T}, I, Y, \eta) \geq f(h; \mathscr{T}, I, Y, \eta)$$

and only elements of $\mathrm{Ker}(\mathscr{T})^\perp$ need be examined when searching for a minimizer. Property (3.18) now has the consequence that if $m_{\eta_{n_k}}$ converges weakly to g it must be that $\mathscr{T}^\dagger \mathscr{T} g = g = \mathscr{T}^\dagger Y$.

The same argument applies to any subsequence of m_{η_n} with the consequence that every such subsequence will contain a further subsequence that converges weakly to $\mathscr{T}^\dagger Y$. As a result, m_{η_n} must also converge weakly to $\mathscr{T}^\dagger Y$.

Now suppose that there were a subsequence having

$$\|m_{\eta_{n_k}}\|_{\mathbb{H}} \leq \|\mathscr{T}^\dagger Y\|_{\mathbb{H}} - \varepsilon$$

for some $\varepsilon > 0$ and all k. This sequence would be bounded and have a further subsequence that converges weakly to $\mathscr{T}^\dagger Y$ thereby producing a contradiction. Accordingly, we must conclude that

$$\liminf_{n\to\infty} \|m_{\eta_n}\|_{\mathbb{H}} \geq \|\mathscr{T}^\dagger Y\|_{\mathbb{H}}.$$

We have now established that m_{η_n} converges weakly to $\mathscr{T}^\dagger Y$ and that $\|m_{\eta_n}\|_{\mathbb{H}} \to \|\mathscr{T}^\dagger Y\|_{\mathbb{H}}$. As,

$$\|m_{\eta_n} - \mathscr{T}^\dagger Y\|_{\mathbb{H}}^2 = \|m_{\eta_n}\|_{\mathbb{H}}^2 - 2\langle m_{\eta_n}, \mathscr{T}^\dagger Y\rangle_{\mathbb{H}} + \|\mathscr{T}^\dagger Y\|_{\mathbb{H}}^2$$

this is sufficient to conclude that $\|m_{\eta_n} - \mathscr{T}^\dagger Y\|_{\mathbb{H}} \to 0$ and complete the proof. □

Referring back to Theorem 3.5.10, we see that when $\mathscr{W} = I$ in (6.7) and we let η tend to zero, the Tikhonov estimator converges to the best (i.e., minimum norm) least-squares (approximate) solution of

$$\mathscr{T} g = Y.$$

On the other hand, the estimator converges to zero as η grows large. While one cannot debate the smoothness of this latter choice, it seems rather simplistic from an estimation standpoint. This is remedied by considering a general choice for $\mathscr{W}$ that has a nontrivial null space. In such instances, one can argue intuitively that the estimator will come from this null space when η diverges. This allows us to view the choice of $\mathscr{W}$ from a modeling perspective wherein one can choose the $\mathscr{W}$ operator in such a way that its null space corresponds to some "ideal" form for m in (6.2). For example, in the special case of (6.6), the idealized model would have m as a line.

We still need a rigorous extension of Theorem 6.2.2 that applies to general $\mathscr{W}$. This is provided by the following result.

Theorem 6.2.3 *Assume that* $Y \in \mathrm{Dom}(\mathscr{T}^\dagger)$, $\mathscr{W}$ *is nonnegative definite and that* $\mathscr{T}^*\mathscr{T} + \eta\mathscr{W}$ *is invertible for* $\eta \in (0, \infty)$. *Then,* m_η *in (6.8) satisfies*

$$\lim_{\eta\to 0} m_\eta = \mathscr{T}^\dagger Y$$

and

$$\lim_{\eta\to \infty} m_\eta = \mathscr{P}_{\mathrm{Ker}(\mathscr{W})}\mathscr{T}^\dagger Y,$$

where $\mathscr{P}_{\mathrm{Ker}(\mathscr{W})}$ *is the projection operator for the null space of* $\mathscr{W}$.

Proof: To simplify the presentation, it will be helpful to use $\mathscr{P} = \mathscr{P}_{\mathrm{Ker}(\mathscr{W})}$ and $\mathscr{Q} = I - \mathscr{P}$ throughout the proof. With this notation, we can write any element g of $\mathbb{H}$ as $g = \mathscr{P}g + \mathscr{Q}g := v + h = \mathscr{P}v + \mathscr{Q}h$. Then,

$$\begin{aligned} f(g; \mathscr{T}, \mathscr{W}, Y, \eta) &= \|Y - \mathscr{T}\mathscr{P}v + \mathscr{T}\mathscr{Q}h\|_{\mathbb{Y}}^2 + \eta\|\mathscr{W}^{1/2}\mathscr{Q}h\|_{\mathbb{H}}^2 \\ &= \|Y - \mathscr{T}\mathscr{P}v + \tilde{\mathscr{T}}\tilde{h}\|_{\mathbb{Y}}^2 + \eta\|\tilde{h}\|_{\mathbb{H}}^2 \\ &= f(\tilde{h}; \tilde{\mathscr{T}}, I, Y - \mathscr{T}\mathscr{P}v, \eta) \end{aligned}$$

for $\tilde{\mathscr{T}} = \mathscr{T}\mathscr{Q}\mathscr{W}^{-1/2}$ and $\tilde{h} = \mathscr{W}^{1/2}\mathscr{Q}h$. The use of $\mathscr{W}^{-1/2}$ here is justified because $\tilde{h}$ is the $\mathscr{W}^{1/2}$ image of an element of $\mathbb{H}$ that has no nonzero component from the operator's null space.

An application of Theorem 6.2.1 shows that for any given v in the null space of $\mathscr{W}$, $f(\tilde{h}; \tilde{\mathscr{T}}, I, Y - \mathscr{T}\mathscr{P}v, \eta)$ is minimized by

$$\tilde{h}_\eta(v) = \left(\tilde{\mathscr{T}}^*\tilde{\mathscr{T}} + \eta I\right)^{-1}\tilde{\mathscr{T}}^*(Y - \mathscr{T}\mathscr{P}v).$$

A little algebra along with the identity

$$\left(\tilde{\mathscr{T}}^*\tilde{\mathscr{T}} + \eta I\right)^{-1}\tilde{\mathscr{T}}^*\tilde{\mathscr{T}} = I - \eta\left(\tilde{\mathscr{T}}^*\tilde{\mathscr{T}} + \eta I\right)^{-1}$$

leads to

$$\begin{aligned} &f(\tilde{h}_\eta(v); \tilde{\mathscr{T}}, I, Y - \mathscr{T}\mathscr{P}v, \eta) \\ &\quad = \left\langle Y - \mathscr{P}\mathscr{T}v, (I - \tilde{\mathscr{T}}\mathscr{G}(\eta)^{-1}\tilde{\mathscr{T}}^*)(Y - \mathscr{T}\mathscr{P}v)\right\rangle_{\mathbb{Y}} \end{aligned}$$

with $\mathscr{G}(\eta) = \tilde{\mathscr{T}}^*\tilde{\mathscr{T}} + \eta I$. Now use the Sherman-Morrison–Woodbury formula from (3.14) to write

$$I - \tilde{\mathscr{T}}\mathscr{G}(\eta)^{-1}\tilde{\mathscr{T}}^* = \eta(\eta I + \tilde{\mathscr{T}}\tilde{\mathscr{T}}^*)^{-1},$$

thereby obtaining

$$f(\tilde{h}_\eta(v), \tilde{\mathscr{T}}, I, Y - \mathscr{T}\mathscr{P}v, \eta) = \eta\left\|\left(\eta I + \tilde{\mathscr{T}}\tilde{\mathscr{T}}^*\right)^{-1/2}(Y - \mathscr{T}\mathscr{P}v)\right\|_{\mathbb{Y}}^2.$$

This latter expression is minimized as a function of v by any solution to the normal equations

$$\mathscr{P}\mathscr{T}^*\left(\eta I + \tilde{\mathscr{T}}\tilde{\mathscr{T}}^*\right)^{-1}\mathscr{T}\mathscr{P}v = \mathscr{P}\mathscr{T}^*\left(\eta I + \tilde{\mathscr{T}}\tilde{\mathscr{T}}^*\right)^{-1}Y.$$

In particular, we can use the best approximate solution

$$\left[\left(\eta I + \tilde{\mathscr{T}}\tilde{\mathscr{T}}^*\right)^{-1/2}\mathscr{T}\mathscr{P}\right]^{\dagger}\left(\eta I + \tilde{\mathscr{T}}\tilde{\mathscr{T}}^*\right)^{-1/2}Y = \mathscr{P}\mathscr{T}^{\dagger}Y$$

as a result of (3.21) and (3.23).

Putting all the pieces together gives us

$$\begin{aligned} m_\eta &= \mathscr{P}\mathscr{T}^{\dagger}Y + \mathscr{Q}\mathscr{W}^{-1/2}\tilde{h}_\eta(\mathscr{P}\mathscr{T}^{\dagger}Y) \\ &= \mathscr{P}\mathscr{T}^{\dagger}Y + \mathscr{Q}\mathscr{W}^{-1/2}\left(\tilde{\mathscr{T}}^*\tilde{\mathscr{T}} + \eta I\right)^{-1}\tilde{\mathscr{T}}^*\left(Y - \mathscr{T}\mathscr{P}\mathscr{T}^{\dagger}Y\right). \end{aligned}$$

Arguing as in the proof of Theorem 6.2.2,

$$\eta\|\tilde{h}_\eta(\mathscr{P}\mathscr{T}^{\dagger}Y)\|^2_{\mathbb{H}} \le \|Y - \mathscr{T}\mathscr{P}\mathscr{T}^{\dagger}Y\|^2_{\mathbb{Y}}$$

and it follows that $\lim_{\eta\to\infty} m_\eta = \mathscr{P}\mathscr{T}^{\dagger}Y$. An application of Theorem 6.2.2 establishes that

$$\begin{aligned} \lim_{\eta\to 0}\mathscr{Q}\mathscr{W}^{-1/2}\left(\tilde{\mathscr{T}}^*\tilde{\mathscr{T}} + \eta I\right)^{-1}\tilde{\mathscr{T}}^* &= \mathscr{Q}\mathscr{W}^{-1/2}\tilde{\mathscr{T}}^{\dagger} \\ &= \mathscr{Q}\mathscr{T}^{\dagger}. \end{aligned}$$

Thus,

$$\lim_{\eta\to 0} m_\eta = \mathscr{T}^{\dagger}Y - \mathscr{Q}\mathscr{T}^{\dagger}\mathscr{T}\mathscr{P}\mathscr{T}^{\dagger}Y.$$

However, property (3.18) means that $\mathscr{T}^{\dagger}\mathscr{T} = I - \mathscr{P}_{\mathrm{Ker}(\mathscr{T})}$ with $\mathrm{Ker}(\mathscr{T})$ the null space for $\mathscr{T}$. Hence,

$$\begin{aligned} \mathscr{Q}\mathscr{T}^{\dagger}\mathscr{T}\mathscr{P}\mathscr{T}^{\dagger}Y &= (I - \mathscr{P})(I - \mathscr{P}_{\mathrm{Ker}(\mathscr{T})})\mathscr{P}\mathscr{T}^{\dagger}Y \\ &= (I - \mathscr{P})\mathscr{P}\mathscr{T}^{\dagger}Y - (I - \mathscr{P})\mathscr{P}_{\mathrm{Ker}(\mathscr{T})}\mathscr{P}\mathscr{T}^{\dagger}Y. \end{aligned}$$

It is always true that $\mathscr{P}(I - \mathscr{P}) = 0$. The condition that $\mathscr{T}^*\mathscr{T} + \eta\mathscr{W}$ be invertible entails that $\mathrm{Ker}(\mathscr{T}) \cap \mathrm{Ker}(\mathscr{W}) = 0$ which implies that $\mathscr{P}_{\mathrm{Ker}(\mathscr{T})}\mathscr{P} = 0$ and the theorem is proved. □

Example 6.2.4 *Let us return to the nonparametric regression problem from model (6.1) where* $\mathbb{H} = \mathbb{W}_2[0,1]$, $\mathbb{Y} = \mathbb{R}^n$ *and we estimate* m *by* m_η *in (6.6). In this instance,*

$$\mathscr{T}g = \begin{bmatrix} g(t_1) \\ \vdots \\ g(t_n) \end{bmatrix} = \begin{bmatrix} \langle g, K(\cdot, t_1)\rangle_{\mathbb{H}} \\ \vdots \\ \langle g, K(\cdot, t_n)\rangle_{\mathbb{H}} \end{bmatrix}$$

with K the rk for $\mathbb{W}_2[0,1]$ *in, e.g., (2.43). Then* $\mathscr{T}^\dagger Y$ *is the minimum norm solution of* $\mathscr{T}g = Y$*. This solution must come from*

$$\mathbb{M} = \text{span}\{K(\cdot, t_1), \ldots, K(\cdot, t_n)\}$$

as adding anything orthogonal to this space will only increase its norm. As a result, $\mathscr{T}^\dagger Y = \sum_{j=1}^n c_j K(\cdot, t_j)$*, where* $c = (c_1, \ldots, c_n)^T$ *is the unique solution of*

$$\mathscr{K} c = Y$$

with

$$\mathscr{K} = \{K(t_i, t_j)\}_{i,j=1.n}.$$

6.3 Bias and variance

Let m_η be the estimator for m in model (6.1) that was provided by Theorem 6.2.1. Then, an associated predictor for a future observation Y_{new} can be obtained from $\mathscr{T}m_\eta$. To assess the statistical performance of our penalized least-squares estimator, we might then use its prediction mean squared error $\mathbb{E}\|Y_{\text{new}} - \mathscr{T}m_\eta\|_{\mathbb{Y}}^2$. However, if Y_{new} is uncorrelated with the data that was used to construct the estimator, this is tantamount to consideration of

$$\text{Risk}(\eta) = \text{Var}(\eta) + \text{Bias}^2(\eta)$$

with

$$\text{Bias}^2(\eta) := \|\mathscr{T}(\mathbb{E}m_\eta - m)\|_{\mathbb{Y}}^2 \tag{6.12}$$

the squared bias of the estimator and

$$\text{Var}(\eta) := \mathbb{E}\|\mathscr{T}(m_\eta - \mathbb{E}m_\eta)\|_{\mathbb{Y}}^2 \tag{6.13}$$

its variance. Some general conclusions that can be drawn about these two quantities are provided in the following result.

Theorem 6.3.1 *The squared bias (6.12) and variance (6.13) for* $\mathscr{T}m_\eta$ *satisfy*

$$\text{Bias}^2(\eta) \le \eta\langle m, \mathscr{W}m\rangle_{\mathbb{H}}$$

and

$$\text{Var}(\eta) = \mathbb{E}\|\mathscr{T}(\mathscr{T}^*\mathscr{T} + \eta\mathscr{W})^{-1}\mathscr{T}^*\varepsilon\|_{\mathbb{Y}}^2.$$

Proof: First consider $\text{Bias}^2(\eta)$ and note that

$$\mathbb{E}m_\eta = (\mathscr{T}^*\mathscr{T} + \eta\mathscr{W})^{-1}\mathscr{T}^*\mathscr{T}m,$$

is the minimizer of

$$\|\mathscr{T}m - \mathscr{T}g\|_{\mathbb{Y}}^2 + \eta\langle g, \mathscr{W}g\rangle_{\mathbb{H}}. \tag{6.14}$$

As a result,

$$\begin{aligned}\|\mathscr{T}(\mathbb{E}m_\eta - m)\|_{\mathbb{Y}}^2 &\le \|\mathscr{T}(\mathbb{E}m_\eta - m)\|_{\mathbb{Y}}^2 + \eta\langle \mathbb{E}m_\eta, \mathscr{W}\mathbb{E}m_\eta\rangle_{\mathbb{H}} \\ &\le \|\mathscr{T}(g - m)\|_{\mathbb{Y}}^2 + \eta\langle g, \mathscr{W}g\rangle_{\mathbb{H}}\end{aligned}$$

for all $g \in \mathbb{H}$, and taking $g = m$ leads to obtain the stated result. For Var(η), one need only observe that $m_\eta - \mathbb{E}m_\eta = (\mathscr{T}^*\mathscr{T} + \eta\mathscr{W})^{-1}\mathscr{T}^*\varepsilon$. □

A useful application of the theorem is the following corollary.

Corollary 6.3.2 *Let $\mathbb{Y} = \mathbb{R}^n$ with squared norm*

$$\|Y\|_{\mathbb{Y}}^2 = n^{-1}\sum_{j=1}^n Y_j^2$$

for $Y = (Y_1, \dots, Y_n)^T \in \mathbb{Y}$. If ε contains uncorrelated random variables with mean 0 and variance σ^2 and $\mathscr{W}$ is invertible

$$\mathbb{E}\|\mathscr{T}(m_\eta - \mathbb{E}m_\eta)\|_{\mathbb{Y}}^2 = \frac{\sigma^2}{n}\sum_{j=1}^n\left(\frac{\gamma_i}{\gamma_i + \eta}\right)^2,$$

where the γ_i are the eigenvalues of the compact operator $\mathscr{W}^{-1/2}\mathscr{T}^*\mathscr{T}\mathscr{W}^{-1/2}$.

Proof: As the range of $\mathscr{T}$ is finite dimensional, it is necessarily compact. Thus, $\tilde{\mathscr{T}} := \mathscr{T}\mathscr{W}^{-1/2}$ is also compact and will have at most n nonzero singular values $\gamma_1 \ge \gamma_2 \ge \cdots \ge \gamma_n \ge 0$. The result now follows from

$$\mathscr{T}(\mathscr{T}^*\mathscr{T} + \eta\mathscr{W})^{-1}\mathscr{T}^* = \tilde{\mathscr{T}}(\tilde{\mathscr{T}}^*\tilde{\mathscr{T}} + \eta I)^{-1}\tilde{\mathscr{T}}^*$$

and the fact that $\mathbb{E}\langle\varepsilon, e\rangle_{\mathbb{Y}}^2 = \sigma^2 e^T e/n^2$ for any n-vector e. □

6.4 A computational formula

In this section, we consider a special but common setting where it is possible to give an expression for m_η that lends itself to explicit computation of the estimator. For this purpose, we take $\mathbb{Y} = \mathbb{R}^n$ and suppose that

$$\mathbb{H} = \mathbb{H}_0 \oplus \mathbb{H}_1,$$

where $\mathbb{H}_0$ and $\mathbb{H}_1$ are subspaces of $\mathbb{H}$ with $\mathbb{H}_0$ having dimension $q < \infty$. Let $\mathscr{W} = \mathscr{P}_1$ be the projection onto $\mathbb{H}_1$ and then consider the optimization of

$$f(g;\mathscr{T},\mathscr{P}_1,Y,\eta) = \|Y - \mathscr{T}g\|^2_{\mathbb{Y}} + \eta\langle g, \mathscr{P}_1 g\rangle_{\mathbb{H}} \tag{6.15}$$

over $g \in \mathbb{H}$.

Let $\{\phi_1, \ldots, \phi_q\}$ be a basis for $\mathbb{H}_0$ and express $\mathscr{T}$ as

$$\mathscr{T}g = (\mathscr{T}_1 g, \ldots, \mathscr{T}_n g)^T, \tag{6.16}$$

for linear functionals $\mathscr{T}_1, \ldots, \mathscr{T}_n$. Denote the representer for $\mathscr{T}_i$ by τ_i with ξ_i being the restriction of τ_i to $\mathbb{H}_1$: i.e.,

$$\mathscr{T}_i g = \langle \tau_i, g\rangle_{\mathbb{H}}$$

and

$$\xi_i = \mathscr{P}_1 \tau_i.$$

Theorem 6.4.1 *Define the matrices*

$$\Phi = \{\mathscr{T}_i \phi_j\}_{i=1:n, j=1:q}$$

and

$$\Psi = \{\mathscr{T}_i \xi_j\}_{i,j=1:n} = \{\langle \xi_i, \xi_j\rangle_{\mathbb{H}}\}_{i,j=1:n}$$

with

$$\mathscr{U} = (\Phi, \Psi)$$

and

$$\mathscr{V} = \begin{pmatrix} 0_{q\times q} & 0_{q\times n} \\ 0_{n\times q} & \Psi_{n\times n} \end{pmatrix}.$$

If $\mathscr{U}^T\mathscr{U} + \eta\mathscr{V}$ is invertible, the unique minimizer of (6.15) is

$$m_\eta = \sum_{j=1}^q \hat{c}_j \phi_j + \sum_{i=1}^n \hat{b}_i \xi_i,$$

where

$$\begin{pmatrix} \hat{c} \\ \hat{b} \end{pmatrix} = (\mathscr{U}^T\mathscr{U} + \eta\mathscr{V})^{-1}\mathscr{U}^T Y.$$

Proof: Any function in $\mathbb{H}$ can be written as

$$g = \sum_{j=1}^q c_j \phi_j + \sum_{j=1}^n b_j \xi_j + h$$

for coefficients c_j and b_j and some $h \in \text{span}\{\phi_1, \ldots, \phi_q, \xi_1, \ldots, \xi_n\}^{\perp}$. Now,

$$\mathscr{T}_i h = \langle \tau_i, h \rangle_{\mathbb{H}} = \langle \tau_i - \xi_i, h \rangle_{\mathbb{H}} + \langle \xi_i, h \rangle_{\mathbb{H}} = 0$$

since $\tau_i - \xi_i \in \mathbb{H}_0 = \text{span}\{\phi_1, \ldots, \phi_q\}$. Thus,

$$\mathscr{T}_i \left(\sum_{j=1}^{q} c_j \phi_j + \sum_{j=1}^{n} b_j \xi_j + h \right) = \mathscr{T}_i \left(\sum_{j=1}^{q} c_j \phi_j + \sum_{j=1}^{n} b_j \xi_j \right).$$

In view of (6.15) and (6.16), $h = 0$ and $m_\eta \in \text{span}\{\phi_1, \ldots, \phi_q, \xi_1, \ldots, \xi_n\}$.

We now know that any viable minimizer must have the form $g = \sum_{j=1}^{q} c_j \phi_j + \sum_{j=1}^{n} b_j \xi_j$. This entails that

$$\mathscr{T} g = \Phi c + \Psi b$$

and

$$\langle g, \mathscr{P}_1 g \rangle_{\mathbb{H}} = b^T \Psi b$$

for $c = (c_1, \ldots, c_q)^T$ and $b = (b_1, \ldots, b_n)^T$. Hence, the penalized least-squares criterion function in (6.15) can be expressed as

$$\left\| Y - \mathscr{U} \begin{bmatrix} c \\ b \end{bmatrix} \right\|_{\mathbb{Y}}^2 + \eta \begin{bmatrix} c^T & b^T \end{bmatrix} \mathscr{V} \begin{bmatrix} c \\ b \end{bmatrix} \tag{6.17}$$

and the conclusion of the theorem follows immediately from Theorem 6.2.1. □

In the special case where $\mathbb{H}_0$, $\mathbb{H}_1$ and $\mathbb{H}$ are RKHSs, the following result is readily established using the reproducing property (cf. Definition 2.7.1).

Theorem 6.4.2 *Let* $\mathbb{H}_0 = \mathbb{H}(K_0)$, $\mathbb{H}_1 = \mathbb{H}(K_1)$ *and* $\mathbb{H} = \mathbb{H}(K)$ *where* K_0 *and* K_1 *are rks and* $K = K_0 + K_1$. *Assume also that* $\mathbb{H}_0 \cap \mathbb{H}_1 = \{0\}$. *Then* $\mathbb{H} = \mathbb{H}_0 \oplus \mathbb{H}_1$, *and* $\tau_i(t) = \mathscr{T}_{i(\cdot)} K(t, \cdot)$ *and* $\xi_i(t) = \mathscr{T}_{i(\cdot)} K_1(t, \cdot)$, *where* $\mathscr{T}_{i(\cdot)}$ *means that* $\mathscr{T}_i$ *is applied to what follows as a function of* $(\cdot)$.

Proof: If $\mathbb{H}_0 \cap \mathbb{H}_1 = \{0\}$ then Theorem 2.7.10 entails that $\mathbb{H} = \mathbb{H}_0 \oplus \mathbb{H}_1$. By the reproducing property,

$$\tau_i(t) = \langle \tau_i, K(t, \cdot) \rangle = \mathscr{T}_{i(\cdot)} K(t, \cdot).$$

Similarly,

$$\begin{aligned} \xi_i(t) &= \langle \xi_i, K(t, \cdot) \rangle = \langle \mathscr{P}_1 \tau_i, K(t, \cdot) \rangle = \langle \tau_i, \mathscr{P}_1 K(t, \cdot) \rangle \\ &= \langle \tau_i, \mathscr{P}_1 (K_0(t, \cdot) + K_1(t, \cdot)) \rangle = \langle \tau_i, K_1(t, \cdot) \rangle \\ &= \mathscr{T}_{i(\cdot)} K_1(t, \cdot), \end{aligned}$$

□

An important application of Theorem 6.4.2 is to the Sobolev space $\mathbb{H} = \mathbb{W}_q[0,1]$, introduced in Section 2.8, for which K_0, K_1 are defined as in Theorem 2.8.1.

6.5 Regularization parameter selection

We have to this point avoided the question of how to choose the parameter η that appears in the penalized least-squares criterion. There are several data adaptive techniques that are typically used for this purpose. We will describe three of them that are appropriate for the scenario that was examined in the previous section: namely, where $\mathbb{Y} = \mathbb{R}^n$ and $\mathbb{H} = \mathbb{H}_0 \oplus \mathbb{H}_1$ for $\mathbb{H}_0$ finite dimensional with basis $\{\phi_1, \ldots, \phi_q\}$. As before the squared norm for $Y \in \mathbb{Y}$ is taken to be

$$\|Y\|_{\mathbb{Y}}^2 = n^{-1}\sum_{j=1}^{n} Y_j^2.$$

We begin with ordinary cross validation. Let $m_\eta^{[k]}$ be the minimizer of

$$n^{-1}\sum_{\substack{i=1\\ i\neq k}}^{n} (Y_i - \mathscr{T}_i g)^2 + \eta\langle g, \mathscr{P}_1 g\rangle_{\mathbb{H}}$$

and define

$$\mathrm{CV}(\eta) = n^{-1}\sum_{k=1}^{n} (Y_k - \mathscr{T}_k m_\eta^{[k]})^2.$$

Ordinary cross validation then picks η to minimize $\mathrm{CV}(\eta)$.

The intuition behind ordinary cross validation is that any good choice for η should endow the estimator with good predictive ability for future response values. If new data is not forthcoming, our only resource is to assess predictive ability by reusing the data that we currently have on hand. We then “predict” each response using an estimator that was computed without its contribution to the fit and average the squared prediction errors to give an indication of the estimators's performance for that particular choice for η.

Brute force evaluation of $\mathrm{CV}(\eta)$ is generally prohibitively time consuming. Fortunately, with some clever algebra, the task can be somewhat simplified using Wahba's “Leave-One-Out” lemma that we state and prove in the following.

For any constant y, let $m_\eta[k, y]$ be the minimizer of

$$n^{-1}(y - \mathscr{T}_k g)^2 + n^{-1}\sum_{\substack{i=1\\ i\neq k}}^{n} (Y_i - \mathscr{T}_i g)^2 + \eta\langle g, \mathscr{P}_1 g\rangle_{\mathbb{H}}$$

over $g \in \mathbb{H}$. Thus, for example, $m_\eta = m_\eta[k, Y_k]$ for each k. Somewhat beyond this simple realization is the following result.

Lemma 6.5.1 $m_\eta\left[k, \mathscr{T}_k m_\eta^{[k]}\right] = m_\eta^{[k]}$.

Proof: For any $g \in \mathbb{H}$, the definition of $m_\eta^{[k]}$ tells us that

$$n^{-1}(\mathscr{T}_k m_\eta^{[k]} - \mathscr{T}_k m_\eta^{[k]})^2 + n^{-1}\sum_{\substack{i=1\\ i\neq k}}^{n}(Y_i - \mathscr{T}_i m_\eta^{[k]})^2 + \eta\langle m_\eta^{[k]}, \mathscr{P}_1 m_\eta^{[k]}\rangle_{\mathbb{H}}$$

$$= n^{-1}\sum_{\substack{i=1\\ i\neq k}}^{n}(Y_i - \mathscr{T}_i m_\eta^{[k]})^2 + \eta\langle m_\eta^{[k]}, \mathscr{P}_1 m_\eta^{[k]}\rangle_{\mathbb{H}}$$

$$\leq n^{-1}\sum_{\substack{i=1\\ i\neq k}}^{n}(Y_i - \mathscr{T}_i g)^2 + \eta\langle g, \mathscr{P}_1 g\rangle_{\mathbb{H}}$$

$$\leq n^{-1}(\mathscr{T}_k m_\eta^{[k]} - \mathscr{T}_k g)^2 + n^{-1}\sum_{\substack{i=1\\ i\neq k}}^{n}(Y_i - \mathscr{T}_i g)^2 + \eta\langle g, \mathscr{P}_1 g\rangle_{\mathbb{H}}.$$

□

Now define

$$a_{kk}(\eta) = \frac{\mathscr{T}_k m_\eta - \mathscr{T}_k m_\eta^{[k]}}{Y_k - \mathscr{T}_k m_\eta^{[k]}}$$

and write

$$Y_k - \mathscr{T}_k m_\eta^{[k]} = \frac{Y_k - \mathscr{T}_k m_\eta}{1 - a_{kk}(\eta)}.$$

Thus,

$$\mathrm{CV}(\eta) = n^{-1}\sum_{k=1}^{n}\frac{(Y_k - \mathscr{T}_k m_\eta)^2}{[1 - a_{kk}(\eta)]^2}.$$

Letting $Y_k' = \mathscr{T}_k m_\eta^{[k]}$, it follows from Lemma 6.5.1 that

$$a_{kk}(\eta) = \frac{\mathscr{T}_k m_\eta[k, Y_k] - \mathscr{T}_k m_\eta[k, Y_k']}{Y_k - Y_k'}.$$

However,

$$m_\eta[k, y] = (\phi_1, \ldots, \phi_q)c(y) + (\xi_1, \ldots, \xi_n)b(y),$$

where $c(y), b(y)$ are the coefficient vectors from Theorem 6.4.1 with the kth element of the response vector being replaced by y. Thus, $\mathscr{T}_k m_\eta[k, y]$ is linear in y and

$$a_{kk}(\eta) = \left.\frac{\partial \mathscr{T}_k m_\eta[k, y]}{\partial y}\right|_{y=Y_k}$$

which happens to be the kth diagonal element of the "hat matrix" $\mathscr{A}(\eta)$ defined by

$$\begin{pmatrix} \mathscr{T}_1 m_\eta \\ \vdots \\ \mathscr{T}_n m_\eta \end{pmatrix} = \mathscr{A}(\eta)Y.$$

Specifically,

$$\mathscr{A}(\eta) = \mathscr{U}(\mathscr{U}^T\mathscr{U} + \eta\mathscr{V})^{-1}\mathscr{U}^T.$$

The conclusion is that for any specified η, $\mathrm{CV}(\eta)$ can be evaluated directly from the residuals $Y - \mathscr{T}m_\eta$ and the diagonal elements of $\mathscr{A}(\eta)$.

Another popular method for choosing η is generalized cross validation. The smoothing parameter value is chosen to minimize its associated criterion function defined by

$$\begin{aligned} \mathrm{GCV}(\eta) &= \frac{\|Y - \mathscr{T}m_\eta\|_{\mathbb{Y}}^2}{\left(1 - n^{-1}\mathrm{trace}\ \mathscr{A}(\eta)\right)^2} \\ &= n^{-1}\sum_{k=1}^{n}\left(Y_k - \mathscr{T}_k m_\eta^{[k]}\right)^2 w_{kk}(\eta), \end{aligned}$$

where

$$w_{kk} = \left(\frac{1 - a_{kk}(\eta)}{1 - n^{-1}\mathrm{trace}\ \mathscr{A}(\eta)}\right)^2.$$

From the last expression, one can conclude that $\mathrm{GCV}(\eta)$ is a variant of $\mathrm{CV}(\eta)$ wherein the individual weights for the squared deleted residuals are replaced by a common value using the average of the diagonal elements of $\mathscr{A}(\eta)$.

Assume now that the errors $\varepsilon = (\varepsilon_1 \ldots, \varepsilon)^T$ consist of uncorrelated random variables having zero means and common variance σ^2. Let

$$\begin{aligned} \mu &:= \mathbb{E}Y \\ &= (\mathscr{T}_1 m, \ldots, \mathscr{T}_n m)^T \end{aligned}$$

and consider the squared-error risk associated with the estimator m_η as defined by

$$\mathrm{Risk}(\eta) := \mathbb{E}\|\mathscr{A}(\eta)Y - \mu\|_{\mathbb{Y}}^2 = n^{-1}\mathbb{E}\left[(\mathscr{A}(\eta)Y - \mu)^T(\mathscr{A}(\eta)Y - \mu)\right].$$

A value of η that minimizes this quantity would be optimal in the sense of providing an estimator of μ that performed well on the average. As $\mathbb{E}\left[\mathscr{A}(\eta)Y - \mu\right] = (A(\eta) - I)\mu$ and $\mathrm{Var}(\mathscr{A}(\eta)Y) = \sigma^2\mathscr{A}(\eta)\mathscr{A}(\eta)^T$ an explicit form for the risk is seen to be

$$\mathrm{Risk}(\eta) = \frac{1}{n}\mu^T(I - \mathscr{A}(\eta))^T(I - \mathscr{A}(\eta))\mu + \frac{\sigma^2}{n}\mathrm{trace}\ \mathscr{A}(\eta)\mathscr{A}(\eta)^T.$$

Now let

$$\mathrm{MSE}(\eta) = \|(I - \mathscr{A}(\eta))Y\|_{\mathbb{Y}}^2$$

and observe that

$$\mathrm{GCV}(\eta) = \frac{\mathrm{MSE}(\eta)}{\left(1 - n^{-1}\mathrm{trace}\ \mathscr{A}(\eta)\right)^2}.$$

As before,

$$\begin{aligned}\mathbb{E}\left[\mathrm{MSE}(\eta)\right] &= \frac{1}{n}\mathbb{E}(Y^T(I - \mathscr{A}(\eta))^T(I - \mathscr{A}(\eta))Y)\\ &= \frac{1}{n}\mu^T(I - \mathscr{A}(\eta))^T(I - \mathscr{A}(\eta))\mu\\ &\quad + \frac{\sigma^2}{n}\mathrm{trace}\ (I - \mathscr{A}(\eta))^T(I - \mathscr{A}(\eta))\\ &= \mathrm{Risk}(\eta) + \sigma^2 - \frac{2\sigma^2}{n}\mathrm{trace}\ \mathscr{A}(\eta).\end{aligned} \tag{6.18}$$

From this relation, we can establish the so-called GCV theorem.

Theorem 6.5.2 *Let* $\tau_j(\eta) = \mathrm{trace}\ \mathscr{A}^j(\eta)/n, j = 1, 2,$ *and define*

$$g(\eta) = \frac{2\tau_1(\eta) + \tau_1^2(\eta)/\tau_2(\eta)}{(1 - \tau_1(\eta))^2}.$$

Then,

$$\frac{|\mathbb{E}\mathrm{GCV}(\eta) - \sigma^2 - \mathrm{Risk}(\eta)|}{\mathrm{Risk}(\eta)} \le g(\eta).$$

Proof: Using our expression for $\mathbb{E}\left[\mathrm{MSE}(\eta)\right]$, a little algebra reveals that

$$\mathbb{E}\mathrm{GCV}(\eta) - \sigma^2 - \mathrm{Risk}(\eta) = \frac{\tau_1(\eta)(2 - \tau_1(\eta))}{(1 - \tau_1(\eta))^2}\mathrm{Risk}(\eta) - \sigma^2\frac{\tau_1^2(\eta)}{(1 - \tau_1(\eta))^2}.$$

This gives

$$\begin{aligned}\frac{|\mathbb{E}\mathrm{GCV}(\eta)-\sigma^2-\mathrm{Risk}(\eta)|}{\mathrm{Risk}(\eta)} &= \left|\frac{\tau_1(\eta)(2-\tau_1(\eta))}{(1-\tau_1(\eta))^2}-\sigma^2\frac{\tau_1^2(\eta)}{\mathrm{Risk}(\eta)(1-\tau_1(\eta))^2}\right| \\ &\le \frac{\tau_1(\eta)(2-\tau_1(\eta))}{(1-\tau_1(\eta))^2}+\frac{\tau_1^2(\eta)/\tau_2(\eta)}{(1-\tau_1(\eta))^2} \\ &\le g(\eta)\end{aligned}$$

because $\sigma^2/\mathrm{Risk}(\eta) \le 1/\tau_2(\eta)$. □

The theorem can be interpreted as saying that when $g(\eta)$ is small $\mathrm{GCV}(\eta)$ provides a nearly unbiased estimator of $\sigma^2 + \mathrm{Risk}(\eta)$ in terms of the inherent estimation scale defined by $\mathrm{Risk}(\eta)$.

A somewhat more immediate consequence of (6.18) stems from the realization that

$$\mathrm{MSE}(\eta) - \sigma^2 + \frac{2\sigma^2}{n}\mathrm{trace}\ \mathscr{A}(\eta)$$

is an unbiased estimator of $\mathrm{Risk}(\eta)$. This leads us to Mallows' C_L criterion function defined as

$$C_L(\eta) = \mathrm{MSE}(\eta) + \frac{2\sigma^2}{n}\mathrm{trace}\ \mathscr{A}(\eta).$$

Again, one minimizes this criterion function to find a value for η.

Both $\mathrm{GCV}(\eta)$ and $C_L(\eta)$ provide estimators of $\mathrm{Risk}(\eta) + \sigma^2$. The latter one is unbiased. However, it will generally require an estimator of σ^2 that may introduce some level of bias back into the selection criterion. The generalized cross-validation criterion, although a priori biased, has the advantage of circumventing the problem of estimating σ^2.

6.6 Splines

The motivation for the method of regularization estimator in Theorem 6.2.1 and its associated estimation criterion derived from the penalized least-squares estimator (6.6) for the nonparametric regression problem. Thus, it seems fitting that we end this chapter by giving a detailed treatment of the root problem posed by (6.6). This entails a foray into the topic of polynomial splines, which is of independent interest for its importance in approximation theory. References on splines include de Boor (1978) and Schumaker (1981).

The term spline comes from drafting where splines were flexible strips used to draw curves connecting points. Here, though, splines mean piecewise polynomials that satisfy a number of continuity constraints.

Definition 6.6.1 *A spline of order $q \geq 1$, with knots at $0 < t_1 < \cdots < t_J < 1$, is any function of the form*

$$g(t) = \sum_{i=0}^{q-1} \theta_i t^i + \sum_{j=1}^{J} \delta_j (t - t_j)_+^{q-1} \tag{6.19}$$

for constants $\theta_0, \ldots, \theta_{q-1}, \delta_1, \ldots, \delta_J \in \mathbb{R}$.

Immediate consequences of the definition are that

1. g is a piecewise polynomial of order q on any subinterval $[t_i, t_{i+1})$,
2. for $q \geq 2$, g has $q - 2$ continuous derivatives, and
3. the $(q - 1)$st derivative of g is a step function with jumps at $t_1, \ldots, t_J$.

Conversely, it is not difficult to see that any piecewise polynomial on $[t_1, t_J]$ satisfying these three conditions can be written in the form (6.19) and must be a spline.

For a given q, J and a set of knots $t_1 < \cdots < t_J$, consider the vector space $S^q(t_1, \ldots, t_J)$ of splines of order q. It is easy to verify that $1, t, \ldots, t^{q-1}$, $(t - t_1)_+^{q-1}, \ldots, (t - t_J)_+^{q-1}$ are linearly independent and hence constitute a basis for $S^q(t_1, \ldots, t_J)$. Thus,

$$\dim(S^q(t_1, \ldots, t_J)) = q + J.$$

For computational purposes, it is desirable to have available an easily evaluated, nearly-orthogonal basis. For spline functions, this comes in the form of the Anselone-Laurent–Reinsch or B-spline basis that we now develop. For this purpose, it will be useful to take $t_0 = 0, t_{J+1} = 1$ and then define $2(q - 1)$ additional (*phantom*) knots

$$t_{-(q-1)} < t_{-(q-2)} < \cdots < t_{-1} \leq 0$$
$$1 \leq t_{J+2} < t_{J+3} < \cdots < t_{J+q}.$$

The values for these knots are arbitrary in what follows. Typically, one merely takes $t_{-(q-1)} = \cdots = t_{-1} = 0$ and $t_{J+2} = \cdots = t_{J+q} = 1$. B-splines then derive from the concept of divided differences as we spell out in the following definition.

Definition 6.6.2 *The q-th order divided difference of a function g at $\{t_i, \ldots, t_{i+q}\}$ is*

$$[t_i, \ldots, t_{i+q}]g = \frac{[t_{i+1}, \ldots, t_{i+q}]g - [t_i, \ldots, t_{i+q-1}]g}{t_{i+q} - t_i}$$

with $[t_i]g = g(t_i)$ being used to initiate the recursion.

For example, the first-order divided difference is

$$[t_i, t_{i+1}]g = \frac{g(t_{i+1}) - g(t_i)}{t_{i+1} - t_i}$$

and the second-order divided difference is

$$[t_i, t_{i+1}, t_{i+2}]g = \frac{[t_{i+1}, t_{i+2}]g - [t_i, t_{i+1}]g}{t_{i+2} - t_i} = \frac{\dfrac{g(t_{i+2}) - g(t_{i+1})}{t_{i+2} - t_{i+1}} - \dfrac{g(t_{i+1}) - g(t_i)}{t_{i+1} - t_i}}{t_{i+2} - t_i}$$

$$= (t_{i+2} - t_i)^{-1} \left\{(t_{i+2} - t_{i+1})^{-1} g(t_{i+2}) \right.$$
$$-[(t_{i+2} - t_{i+1})^{-1} + (t_{i+1} - t_i)^{-1}]g(t_{i+1})$$
$$\left. +(t_{i+1} - t_i)^{-1} g(t_i)\right\}.$$

Note that $[t_i, \ldots, t_{i+q}]g$ depends only on $t_i, \ldots, t_{i+q}$ and $g(t_i), \ldots, g(t_{i+q})$.

Theorem 6.6.3 *Let* $p_q(x) = \sum_{i=1}^{q} g(t_i)\ell_i(x)$ *for*

$$\ell_i(x) = \prod_{\substack{j=1 \\ j \neq i}}^{q} \frac{x - t_j}{t_i - t_j}.$$

Then,

1. p_q *is the unique qth order polynomial that agrees with* g *at* t_i *for* $i = 1,\ldots,q$ *and*
2. *for each* $q = 1, 2 \ldots$ *the coefficient of* x^q *in* p_{q+1} *is* $[t_i, \ldots, t_{i+q}]g$.

Proof: The function $\ell_i(x)$ is a polynomial of order q and vanishes at all the t_j except for t_i where it takes the value 1. So, property 1 holds.

To verify the second property, we proceed by induction. The claim is clearly true for $q = 1$. Next, let $p_q(x)$ be the polynomial of order q that agrees with g at $t_i,\ldots,t_{i+q-1}$ and take $\tilde{p}_q(x)$ to be the polynomial of order q that agrees with g at $t_{i+1},\ldots,t_{i+q}$. Then,

$$p(x) = \frac{x - t_i}{t_{i+q} - t_i}\tilde{p}_q(x) + \frac{t_{i+q} - x}{t_{i+q} - t_i}p_q(x)$$

is a polynomial of order $q + 1$ and agrees with g at $t_i,\ldots,t_{i+q}$. By the uniqueness of interpolating polynomials, we must have $p_{q+1}(x) = p(x)$ and the coefficient of x^q in p_{q+1} is

$$\frac{[t_{i+1},\ldots,t_{i+q}]g - [t_i,\ldots,t_{i+q-1}]g}{t_{i+q} - t_i}.$$

□

Corollary 6.6.4 *If g is a polynomial of order q on $[t_i, t_{i+q}]$, $[t_i, \ldots, t_{i+q}]g = 0$.*

Proof: Let p_i be as in Theorem 6.6.3 and note that $p_i = g$ for all $i \geq q$. Then, $[t_i, \ldots, t_{i+q}]g$ is the coefficient of x^q in p_{q+1}. As p_{q+1} is equal to g and g has degree less than q that coefficient must be zero. □

Denote by $\alpha_{rq}^{[i]}$ the coefficient of $g(t_{i+r})$ in $(t_{i+q} - t_i)[t_i, \ldots, t_{i+q}]g$: i.e.,

$$(t_{i+q} - t_i)[t_i, \ldots, t_{i+q}]g = \sum_{r=0}^{q} \alpha_{rq}^{[i]} g(t_{i+r}).$$

For example, if $q = 2$ then

$$\begin{aligned}(t_{i+2} - t_i)[t_i, \ldots, t_{i+2}]g = {} & (t_{i+2} - t_{i+1})^{-1} g(t_{i+2}) \\ & - [(t_{i+2} - t_{i+1})^{-1} + (t_{i+1} - t_i)^{-1}]g(t_{i+1}) \\ & + (t_{i+1} - t_i)^{-1} g(t_i),\end{aligned}$$

so that $\alpha_{02}^{[i]} = (t_{i+1} - t_i)^{-1}$, $\alpha_{22}^{[i]} = (t_{i+2} - t_{i+1})^{-1}$,

$$\alpha_{12}^{[i]} = -(t_{i+2} - t_{i+1})^{-1} - (t_{i+1} - t_i)^{-1}$$

and $\alpha_{r2}^{[i]} = 0$ for $r > 2$.

Now define

$$\begin{aligned}N_{iq}(t) &= (t_{i+q} - t_i)[t_i, \ldots, t_{i+q}](\cdot - t)_+^{q-1} \\ &= \sum_{r=0}^{q} \alpha_{rq}^{[i]} (t_{i+r} - t)_+^{q-1}\end{aligned} \tag{6.20}$$

for $i = -(q-1), \ldots, J$. These functions are called B-splines; the “B” stands for “basis”.

Theorem 6.6.5 *The collection $\{N_{iq}(\cdot)\}_{i=-(q-1)}^{J}$ satisfy*

1. *N_{iq} is a polynomial of order q on each interval (t_i, t_{i+1}),*
2. *$N_{iq}(t) = 0$ for $t \notin [t_i, t_{i+q}]$, and*
3. *if we take $N_{i1} := I_{[t_i, t_{i+1}]}(t)$, the N_{iq} may be evaluated recursively using*

$$N_{iq}(t) = \frac{t - t_i}{t_{i+q-1} - t_i} N_{i(q-1)}(t) + \frac{t_{i+q} - t}{t_{i+q} - t_{i+1}} N_{(i+1)(q-1)}(t).$$

Proof: Part 1 is obvious. We will prove only part 2. For this purpose, first note that $N_{iq}(t) = 0$ for $t > t_{i+q}$ as all functions in the sum in (6.20) vanish in this case. Now consider $t < t_i$. Then, from (6.20),

$$N_{iq}(t) = \sum_{r=0}^{q} \alpha_{rq}^{[i]} (t_{i+r} - t)_+^{q-1}$$
$$= \sum_{r=0}^{q} \alpha_{rq}^{[i]} (t_{i+r} - t)^{q-1}$$
$$= (t_{i+q} - t_i)[t_i, \ldots, t_{i+q}](\cdot - t)^{q-1}.$$

As the function $(\cdot - t)^{q-1}$ is a polynomial of order q on $[t_i, t_{i+q}]$, $N_{iq}(t) = 0$ by Corollary 6.6.4. □

Property 2 of Theorem 6.6.5 is called the *local support property* of the B-spline basis; given a set of knots $t_1, \ldots, t_J$, N_{iq} and N_{jq} have disjoint supports if $|i - j| > q - 1$. Thus, for example, when $|i - j| > q - 1$

$$\langle N_{iq}, N_{jq} \rangle_2 = \int_0^1 N_{iq}(t) N_{jq}(t) dt = 0,$$

which illustrates the "nearly orthogonal" quality that we mentioned earlier. This also serves the purpose of showing that the N_{iq} are linearly independent and therefore live up to their name by providing a basis for $S^q(t_1, \ldots, t_J)$.

Regularization type problems for real data lead to the consideration of a special sort of spline that satisfies some additional boundary conditions.

Definition 6.6.6 *A spline g of order $2q$ for $q \geq 1$ with knots at $t_1 < \cdots < t_J$, is said to be a natural spline of order $2q$ if g is a polynomial of order q outside of $[t_1, t_J]$. The collection of all such splines will be denoted by $N^{2q}(t_1, \ldots, t_J)$.*

It is clear that any $g \in N^{2q}(t_1, \ldots, t_J)$ must satisfy

$$g^{(j)}(0) = g^{(j)}(1) = 0, j = q, \ldots, 2q - 1,$$

which are known as the natural boundary conditions. These impose $2q$ linear constraints on the coefficients in (6.19) with the consequence that $N^{2q}(t_1, \ldots, t_J)$ has dimension $2q + J - 2q = J$. The following result stems from this fact.

Theorem 6.6.7 *Given $0 \leq t_1 < \cdots < t_J \leq 1$ and specified constants $z_1, \ldots, z_J$, there is a unique natural spline g of order $2q$ with knots $t_1, \ldots, t_J$ and $g(t_j) = z_j, j = 1, \ldots, J$.*

The natural spline whose existence is guaranteed by Theorem 6.6.7 is called a *natural interpolating spline*. It turns out to be optimal in a way that is very important in the context of (6.6) and related problems.

Theorem 6.6.8 *For knots* $0 < t_1 < \cdots < t_J < 1$ *let* $\tilde{g}$ *be the natural spline of order* $2q$ *that interpolates* $z_1, z_2, \ldots, z_J$. *Then, for all functions* $g \in \mathbb{W}_q[0,1]$ *that interpolate the* z_j,

$$\int_0^1 |\tilde{g}^{(q)}(t)|^2 dt \le \int_0^1 |g^{(q)}(t)|^2 dt. \tag{6.21}$$

Equality holds in (6.21) if $g = \tilde{g} + h$ *for* h *a* q*th order polynomial that vanishes at* $t_1, \ldots, t_J$. *If* $J \ge q$, *then* $h = 0$ *on* $[0,1]$.

Proof: Let g be a function in $\mathbb{W}_q[0,1]$ that interpolates the z_j and set $h = g - \tilde{g}$. Then, $h(t_j) = 0$ for all j. Using integration by parts and the natural boundary conditions, we find that

$$\begin{aligned}\int_0^1 \tilde{g}^{(q)}(t)h^{(q)}(t)dt &= \tilde{g}^{(q+1)}(t)h^{(q)}(t)|_0^1 - \int_0^1 \tilde{g}^{(q+1)}(t)h^{(q-1)}(t)dt \\ &= -\int_0^1 \tilde{g}^{(q+1)}(t)h^{(q-1)}(t)dt.\end{aligned}$$

This suggests the way to proceed and by continuing to integrate by parts we eventually arrive at

$$\begin{aligned}\int_0^1 \tilde{g}^{(q)}(t)h^{(q)}(t)dt &= (-1)^{q-1}\int_0^1 \tilde{g}^{(2q-1)}(t)h^{(1)}(t)dt \\ &= (-1)^{q-1}\sum_{j=1}^{J-1} \tilde{g}^{(2q-1)}(t_j^+)[h(t_{j+1}) - h(t_j)] = 0,\end{aligned}$$

where we have used the property that $\tilde{g}^{(2q-1)}$ vanishes on $(0, t_1)$ and $(t_J, 1)$, while being constant on each interval (t_j, t_{j+1}) with value $\tilde{g}^{(2q-1)}(t_j^+)$. Thus,

$$\begin{aligned}\int_0^1 |g^{(q)}(t)|^2 dt &= \int_0^1 \left(\tilde{g}^{(q)}(t) + h^{(q)}(t)\right)^2 dt \\ &= \int_0^1 |\tilde{g}^{(q)}(t)|^2 dt + 2\int_0^1 \tilde{g}^{(q)}(t)h^{(q)}(t)dt + \int_0^1 |h^{(q)}(t)|^2 dt \\ &= \int_0^1 |\tilde{g}^{(q)}(t)|^2 dt + \int_0^1 |h^{(q)}(t)|^2 dt \ge \int_0^1 |\tilde{g}^{(q)}(t)|^2 dt.\end{aligned}$$

From the developments so far, we see that equality will be attained in (6.21) if and only if $\int_0^1 |h^{(q)}(t)|^2 dt = 0$: namely, $h^{(q)} = 0$ a.e. Recall that both g and $\tilde{g}$ are members of $\mathbb{W}_q[0,1]$ so that $h = g - \tilde{g} \in \mathbb{W}_q[0,1]$. As

$$h(t) = \sum_{i=0}^{q-1} \frac{h^{(i)}(0)}{i!} t^i + \int_0^1 \frac{(t-u)_+^{q-1}}{(q-1)!} h^{(q)}(u)du,$$

we conclude that

$$h(t) = \sum_{i=0}^{q-1} \frac{h^{(i)}(0)}{i!} t^i.$$

Note that any qth order polynomial can have at most $q-1$ roots unless it is identically equal to 0. The final statement in the theorem is a consequence of that fact as $h(t_j) = 0$ for $j = 1, \ldots, J$. □

As promised, the preceding result has a very powerful consequence for regularization-related problems.

Theorem 6.6.9 *Suppose that $n > q$ and $\eta \in (0, \infty)$. Let L be a mapping from $\mathbb{R}^n$ to $[0, \infty)$. Then, if*

$$L(g(t_1), \ldots, g(t_n)) + \eta \int_0^1 |g^{(q)}(t)|^2 dt \tag{6.22}$$

has a minimizer in $\mathbb{W}_q[0,1]$, it must be an element of $N^{2q}(t_1, \ldots, t_n)$.

Proof: Let $\hat{g}$ be any minimizer of (6.22). If $\tilde{g}$ is the element of $N^{2q}(t_1, \ldots, t_n)$ that interpolates $\hat{g}$ at $t_1, \ldots, t_J$

$$L(\hat{g}(t_1), \ldots, \hat{g}(t_n)) = L(\tilde{g}(t_1), \ldots, \tilde{g}(t_n)).$$

The result is now a consequence of Theorem 6.6.8. □

Theorem 6.6.9 has the remarkable implication that for any optimization problem corresponding to a criterion of the form (6.22), we need only look for minimizers over the finite dimensional space of natural splines of order $2q$ rather than all of $\mathbb{W}_q[0,1]$. Thus, we have traded an infinite dimensional problem for one that has only finite-dimensional complexity.

Now let us return to the nonparametric regression estimator (6.6) and, for that purpose, take $\mathbb{Y} = \mathbb{R}^n$ with

$$\|Y\|_{\mathbb{Y}}^2 = n^{-1} \sum_{j=1}^{n} Y_j^2$$

for $Y \in \mathbb{R}^n$. Then, $\mathbb{H} = \mathbb{W}_q[0, 1] = \mathbb{H}_0 \oplus \mathbb{H}_1$ and $\mathscr{T} g = (g(t_1), \ldots, g(t_n))^T$. If $\mathscr{P}_1$ is the projection operator for $\mathbb{H}_1$, our estimation criterion is

$$\|Y - \mathscr{T} g\|_{\mathbb{Y}}^2 + \eta \langle g, \mathscr{P}_1 g \rangle_{\mathbb{H}} = n^{-1} \sum_{j=1}^{n} (Y_j - g(t_j))^2 + \eta \int_0^1 |g^{(q)}(t)|^2 dt.$$

When $q = 2$, this gives us the estimator in (6.6).

For $j = 1, \ldots, n$ let g_j be the element of $N^{2q}(t_1, \ldots, t_n)$ that interpolates $z_1 = \cdots = z_{j-1} = 0, z_j = 1, z_{j+1} = \cdots = z_n = 0$. Then, any $g \in N^{2q}(t_1, \ldots, t_n)$ can be written as

$$g(\cdot) = \sum_{j=1}^{n} g(t_j) g_j(\cdot).$$

In particular, by combining this fact with Theorem 6.6.9, we see that minimization of our estimation criterion over $\mathbb{H} = \mathbb{W}_q[0, 1]$ is equivalent to finding the vector $g = (g(t_1), \ldots, g(t_n))^T$ that minimizes

$$\|Y - g\|_{\mathbb{Y}}^2 + \eta g^T \mathscr{W} g$$

for

$$\mathscr{W} = \left\{ \int_0^1 g_i^{(q)}(t) g_j^{(q)}(t) dt \right\}_{i,j=1,n}.$$

The optimal coefficient vector is

$$\begin{aligned} m_\eta &= (m_\eta(t_1), \ldots, m_\eta(t_n))^T \\ &= (I + \eta \mathscr{W}) Y \end{aligned} \tag{6.23}$$

giving

$$m_\eta(\cdot) = \sum_{j=1}^{n} m_\eta(t_j) g_j(\cdot). \tag{6.24}$$

Thus, the nonparametric regression estimator determined by (6.23) and (6.24) is a natural spline of order $2q$ that is called a *smoothing spline*. In particular, when $q = 2$, we get the cubic smoothing spline estimator (6.6).

To conclude this chapter, we provide a result concerning the bias and variance properties of smoothing spline estimators of m in model (6.1).

Theorem 6.6.10 *Assume that* $\mathbb{E}[\varepsilon\varepsilon^T] = \sigma^2 I, t_j = (2j-1)/2n, j = 1, \ldots, n$ *and let* $m = (m(t_1), \ldots, m(t_n))^T$. *For any* n *and* $\eta > 0$,

$$\|\mathbb{E}m_\eta - m\|_{\mathbb{Y}}^2 \leq \frac{\eta}{4}\left(m^T m + m^T \mathscr{W} m\right)$$

and

$$\mathbb{E}\|m_\eta - \mathbb{E}m_\eta\|_{\mathbb{R}^n}^2 \leq \frac{q\sigma^2}{n} + \frac{C\sigma^2}{n}\eta^{-1/(2q)}$$

for a constant C *that depends only on* q.

Proof: Let $\gamma_1 \leq \gamma_2 \leq \cdots \leq \gamma_n$ be the eigenvalues of $\mathscr{W}$ with $e_1, \ldots, e_n$ the corresponding eigenvector. Utreras (1983) shows that in this case $\gamma_j = 0, j = 1, \ldots, q$ and

$$C_1(j-q)^{2q} \leq \gamma_j \leq C_2(j-q)^{2q}, j = q+1, \ldots, n,$$

where $0 < C_1 \leq C_2 < \infty$ depend only on q. Clearly, then

$$\begin{aligned}
\|\mathbb{E}(m_\eta - m)\|_{\mathbb{Y}}^2 &= \|(I + \eta\mathscr{W})^{-1}m - m\|_{\mathbb{Y}}^2 \\
&= n^{-1}\sum_{j=1}^n \left(\frac{\eta\gamma_j}{1+\eta\gamma_j}\right)^2 (m^T e_j)^2 \\
&= n^{-1}\sum_{j=1}^n \left[(1+\gamma_j)\left(m^T e_j\right)^2\right](1+\gamma_j)^{-1}\left(\frac{\eta\gamma_j}{1+\eta\gamma_j}\right)^2.
\end{aligned}$$

Now note that

$$\sum_{j=1}^n (1+\gamma_j)(m^T e_j)^2 = m^T m + m^T \mathscr{W} m$$

and

$$\begin{aligned}
\sup_{q+1\leq j\leq n} (1+\gamma_j)^{-1}\left(\frac{\eta\gamma_j}{1+\eta\gamma_j}\right)^2 &\leq \sup_{u>0}(1+u\eta^{-1})^{-1}\left(\frac{u}{1+u}\right)^2 \\
&= \eta \sup_{u>0}(\eta+u)^{-1}\left(\frac{u}{1+u}\right)^2 \\
&\leq \eta \sup_{u>0}\frac{u}{(1+u)^2} = \frac{\eta}{4}.
\end{aligned}$$

For the variance term

$$\begin{aligned}\mathbb{E}\|m_\eta - \mathbb{E}m_\eta\|_{\mathbb{Y}}^2 = n^{-1}\mathbb{E}\left[\varepsilon^T(I+\eta\mathscr{W})^{-2}\varepsilon\right] &= \frac{\sigma^2}{n}\mathbb{E}\left[\text{trace }(I+\eta\mathscr{W})^{-2}\varepsilon\varepsilon^T\right]\\ &= \frac{\sigma^2}{n}\sum_{j=1}^{n}(1+\eta\gamma_j)^{-2}.\end{aligned}$$

Approximating the sum by an integral gives

$$\begin{aligned}\sum_{j=1}^{J}(1+\eta\gamma_j)^{-2} &\le q + \int_0^\infty (1+\theta x^{2q})^{-2}dx\\ &= q + \theta^{-1/(2q)}\int_1^\infty u^{-2}(u-1)^{1/(2q)-1}du/2q,\end{aligned}$$

where $\theta = C_1\eta$. □

7

Random elements in a Hilbert space

In this chapter, we give an overview of various foundational issues that arise in fda and related settings. There are two somewhat different perspectives of functional data. The first one is that functional data are realizations of random variables that take values in a Hilbert space; for convenience, we call this the *random element perspective*. The second view is that functional data are the sample paths of a (typically continuous time) stochastic process with smooth mean and covariance functions; we will refer to this as the *stochastic process perspective*. The differences between the two perspectives are subtle and worth exploring from a theoretical standpoint.

To develop the random element perspective, we need to lay a rigorous foundation for the study of Hilbert space valued "random variables" so that we can develop concepts for the mean and covariance in that abstract environment. The stochastic process perspective uses the covariance function of the stochastic process as the fundamental tool for assessing the variability. The classical theorem by Karhunen and Lòeve appears in this context.

For the purpose of developing the notions of prediction and canonical correlations, we define the concept of closed linear span corresponding to both the random element and stochastic process perspective and explore the properties of various associated congruence relations. This work will be useful in Chapter 10. Finally, we present some large sample theory for the case where one acquires (many) independent realizations of some random element of interest.

Theoretical Foundations of Functional Data Analysis, with an Introduction to Linear Operators, First Edition. Tailen Hsing and Randall Eubank.

7.1 Probability measures on a Hilbert space

Throughout this chapter, we work within the context of a separable Hilbert space $\mathbb{H}$ with associated norm and inner product $\|\cdot\|$ and $\langle\cdot,\cdot\rangle$. Topologically, we can view $\mathbb{H}$ as a metric space with metric

$$d(f,g) = \|f-g\| = \langle f-g, f-g\rangle^{1/2}.$$

As usual, the Borel σ-field of any topological space $\mathbb{X}$ is the smallest σ-field containing all the open (relative to the norm-based metric) subsets of $\mathbb{X}$ and will be denoted by $\mathscr{B}(\mathbb{X})$. The smallest σ-field containing a class $\mathscr{C}$ of sets is also indicated by the standard notation $\sigma(\mathscr{C})$.

Let $\mathscr{M}$ be the class of all sets of the form $\{x \in \mathbb{H} : \langle x,f\rangle \in B\}$ for $f \in \mathbb{H}$ and B an open subset of $\mathbb{R}$.

Theorem 7.1.1 *The σ-fields $\sigma(\mathscr{M})$ and $\mathscr{B}(\mathbb{H})$ are identical.*

Proof: The proof is accomplished directly by establishing that $\sigma(\mathscr{M}) \subset \mathscr{B}(\mathbb{H})$ and $\mathscr{B}(\mathbb{H}) \subset \sigma(\mathscr{M})$. With this as our aim, first note that each set in $\mathscr{M}$ is either open or closed since $x \mapsto \langle x,f\rangle$ is a continuous mapping from $\mathbb{H}$ into $\mathbb{R}$. Hence, $\mathscr{M} \subset \mathscr{B}(\mathbb{H})$ and we must have $\sigma(\mathscr{M}) \subset \mathscr{B}(\mathbb{H})$.

Next, recall that $\mathscr{B}(\mathbb{H})$ is the smallest σ-field that contains all the open sets $\{x \in \mathbb{H} : \|x-f\| > r\}, f \in \mathbb{H}, r \in (0,\infty)$. To show $\mathscr{B}(\mathbb{H}) \subset \sigma(\mathscr{M})$, it suffices to show that these open sets are in $\sigma(\mathscr{M})$. Theorem 2.4.14 ensures that there exists a CONS $\{e_j\}_{j=1}^{\infty}$ in $\mathbb{H}$ and, for any $r \in (0,\infty)$, we have

$$\{x \in \mathbb{H} : \|x\| > r\} = \cup_{j=1}^{\infty} \left\{ x \in \mathbb{H} : \sum_{k=1}^{j} \langle x, e_k\rangle^2 > r^2 \right\}. \tag{7.1}$$

Initially, we can observe that

$$\begin{aligned}
&\{x \in \mathbb{H} : \langle x, e_1\rangle^2 + \langle x, e_2\rangle^2 > r^2\} \\
&\quad = \cup_{q\in\mathbb{Q}} \{x \in \mathbb{H} : |\langle x, e_1\rangle|^2 > q > r^2 - |\langle x, e_2\rangle|^2\} \\
&\quad = \cup_{q\in\mathbb{Q}} \left(\{x \in \mathbb{H} : |\langle x, e_1\rangle|^2 > q\} \cap \{x \in \mathbb{H} : |\langle x, e_2\rangle|^2 > r^2 - q\}\right)
\end{aligned}$$

for $\mathbb{Q}$ the set of rational numbers. As both $\{x \in \mathbb{H} : |\langle x, e_1\rangle|^2 > q\}$ and $\{x \in \mathbb{H} : |\langle x, e_2\rangle|^2 > r^2 - q\}$ are in $\mathscr{M}$, $\{x \in \mathbb{H} : \langle x, e_1\rangle^2 + \langle x, e_2\rangle^2 > r^2\} \in \sigma(\mathscr{M})$. Induction using (7.1) allows us to conclude that

$$\{x \in \mathbb{H} : \|x\| > r\} \in \sigma(\mathscr{M}). \tag{7.2}$$

To complete the proof, let f be any element of $\mathbb{H}$ and write

$$\begin{aligned}&\{x \in \mathbb{H} : \|x - f\| > r\}\\&= \{x \in \mathbb{H} : \|x\|^2 > r^2 + 2\langle x,f\rangle - \|f\|^2\}\\&= \cup_{q\in\mathbb{Q}}(\{x \in \mathbb{H} : \|x\|^2 > q\} \cap \{x \in \mathbb{H} : 2\langle x,f\rangle < q - r^2 + \|f\|^2\}).\end{aligned}$$

By (7.2) and the above-mentioned argument, the last expression is a set in $\sigma(\mathscr{M})$ and, hence, $\mathscr{B}(\mathbb{H}) \subset \sigma(\mathscr{M})$. □

The following result provides a Hilbert space specific characterization of measurability that we will need in subsequent sections of this chapter.

Theorem 7.1.2 *Let χ be a mapping from some probability space $(\Omega, \mathscr{F}, \mathbb{P})$ into $(\mathbb{H}, \mathscr{B}(\mathbb{H}))$. Then,*

1. *χ is measurable if $\langle \chi,f\rangle$ is measurable for all $f \in \mathbb{H}$ and*
2. *if χ is measurable, its distribution is uniquely determined by the (marginal) distributions of $\langle \chi,f\rangle$ over $f \in \mathbb{H}$.*

Proof: If χ is measurable $\langle \chi,f\rangle$ is measurable as $x \mapsto \langle x,f\rangle$ is continuous for all f. Conversely, if $\langle \chi,f\rangle$ is measurable for all f, for any open $B \subset \mathbb{R}$, the set $\{x \in \mathbb{H} : \langle x,f\rangle \in B\} \in \mathscr{M}$ satisfies

$$\chi^{-1}(\{x \in \mathbb{H} : \langle x,f\rangle \in B\}) = \{\omega \in \Omega : \langle \chi(\omega),f\rangle \in B\} \in \mathscr{B}.$$

Thus, Theorem 7.1.1 tells us that χ is measurable.

Next, let $\check{\mathscr{M}}$ be the class containing finite intersections of sets in $\mathscr{M}$. Clearly, $\sigma(\check{\mathscr{M}}) = \sigma(\mathscr{M}) = \mathscr{B}(\mathbb{H})$ due to Theorem 7.1.1. We first establish that $\check{\mathscr{M}}$ is a determining class; namely, if two probability measure μ_1, μ_2 agree on $\check{\mathscr{M}}$ it means that they must also agree on $\mathscr{B}(\mathbb{H})$.

Define

$$G = \{A \in \mathscr{B}(\mathbb{H}) : \mu_1(A) = \mu_2(A)\}.$$

It is easy to verify that G is a λ system. As $\check{\mathscr{M}}$ is a π-system and $\check{\mathscr{M}} \subset G$, the π-λ theorem (e.g., 1995, Billingsley) implies that $\sigma(\check{\mathscr{M}}) \subseteq G$ and, hence, that $\check{\mathscr{M}}$ is indeed determining.

Now, let $\mathscr{M}_i = \{x \in \mathbb{H} : \langle x,f_i\rangle \in B_i\}$ for some $f_i \in \mathbb{H}$, open sets $B_i \subset \mathbb{R}$ and $i = 1, \ldots, n$. Then,

$$\mathbb{P}\circ\chi^{-1}(\cap_{i=1}^n M_i) = \mathbb{P}(\langle \chi,f_i\rangle \in B_i, 1 \le i \le n).$$

The fact that $\check{\mathscr{M}}$ is determining means that probabilities of the form $\mathbb{P}(\langle \chi, f_i \rangle \in B_i, 1 \le i \le n)$ uniquely determine $\mathbb{P} \circ \chi^{-1}$ while these quantities are, in turn, determined by the joint distributions of $\langle \chi, f_i \rangle$, $1 \le i \le n$. The Cramér–Wold Theorem entails that these joint distributions are determined by the marginal distributions of $\langle \chi, f \rangle, f \in \mathbb{H}$. □

In subsequent discussions, we will refer to measurable functions that take values in a Hilbert space $\mathbb{H}$ as being *random elements* of $\mathbb{H}$. As such, they represent a useful abstraction of the random variable concept that is well suited for our intended direction of study.

7.2 Mean and covariance of a random element of a Hilbert space

Let χ be a random element of a separable Hilbert space $\mathbb{H}$ defined on a probability space $(\Omega, \mathscr{F}, \mathbb{P})$. Theorem 2.6.5 suggests one notion of a mean that can be used here.

Definition 7.2.1 *If* $\mathbb{E}(\|\chi\|) < \infty$, *the mean element of* χ, *or simply the mean of* χ, *is defined as the Bochner integral*

$$m = \mathbb{E}(\chi) := \int_{\Omega} \chi \, d\mathbb{P}. \tag{7.3}$$

This definition provides the natural extension of the mean of a random variable to the case of random elements. It is, roughly speaking, a "weighted sum" of the possible realizations of χ that returns another, nonrandom, element of $\mathbb{H}$.

An alternative way to define the mean element in general is to observe that the assumption $\mathbb{E}(\|\chi\|) < \infty$ implies that the functional $f \mapsto \mathbb{E}[\langle \chi, f \rangle]$ is in $\mathfrak{B}(\mathbb{H}, \mathbb{R})$ and then define m as its representer (cf. Theorem 3.2.1). An application of Theorem 3.1.7 reveals that either definition gives

$$\langle m, f \rangle = \mathbb{E}[\langle \chi, f \rangle] \tag{7.4}$$

for $f \in \mathbb{H}$. The linear functional approach corresponds to the Gelfand–Pettis integral, which is more general than the Bochner integral. In view of (7.4), the two integrals are identical in this setting.

Assuming the expectation exists, one might take $\mathbb{E}\|\chi - m\|^2$ as a variance type measure associated with a random element χ. In this regard, we have the following analog of a familiar variance identity.

Theorem 7.2.2 *Assume that* $\mathbb{E}\|\chi\|^2 < \infty$. *Then*

$$\mathbb{E}\|\chi - m\|^2 = \mathbb{E}\|\chi\|^2 - \|m\|^2,$$

where m *is the mean of* χ.

Proof: Write

$$\mathbb{E}\|\chi - m\|^2 = \mathbb{E}\|\chi\|^2 - 2\mathbb{E}\langle\chi, m\rangle + \|m\|^2.$$

The result follows from Theorem 3.1.7 that allows us to interchange the expectation and inner product operations. □

The next step is to develop a concept of covariance for χ. In that regard, one might recall that the covariance for a random p-vector X is

$$\mathbb{E}[(X - \mathbb{E}X)(X - \mathbb{E}X)^T] = \mathbb{E}[(X - \mathbb{E}X) \otimes (X - \mathbb{E}X)].$$

This is a $p \times p$ matrix and therefore an element of $\mathfrak{B}(\mathbb{R}^p)$. Our Hilbert space formulation builds on this idea. Specifically, if χ is a random element of a Hilbert space $\mathbb{H}$, we define the corresponding covariance operator as follows.

Definition 7.2.3 *Assume that* $\mathbb{E}\|\chi\|^2 < \infty$. *Then, the covariance operator for* χ *is the element of* $\mathfrak{B}_{HS}(\mathbb{H})$ *given by the Bochner integral*

$$\mathscr{K} = \mathbb{E}\left[(\chi - m) \otimes (\chi - m)\right] := \int_\Omega (\chi - m) \otimes (\chi - m)\, d\mathbb{P}. \tag{7.5}$$

If $\chi(\omega)$ is in $\mathbb{H}$, a direct calculation shows that $(\chi(\omega) - m) \otimes (\chi(\omega) - m)$ is a Hilbert–Schmidt operator with norm $\|\chi(\omega) - m\|^2$. As a result, $(\chi - m) \otimes (\chi - m)$ is a random element of the Hilbert space $\mathfrak{B}_{HS}(\mathbb{H})$. From Theorem 7.2.2, the assumption that $\mathbb{E}\|\chi\|^2 < \infty$ implies that the expectation of the HS norm of $(\chi - m) \otimes (\chi - m)$ is finite. As $\mathfrak{B}_{HS}(\mathbb{H})$ is a separable Hilbert space (Section 4.4), it follows from Theorem 2.6.5 that $\mathscr{K}$ is well defined as the Bochner integral (7.5) that returns a nonrandom element of $\mathfrak{B}_{HS}(\mathbb{H})$.

The following result is the extension of a familiar covariance identity for finite dimensions.

Theorem 7.2.4 $\mathbb{E}\left[(\chi - m) \otimes (\chi - m)\right] = \mathbb{E}(\chi \otimes \chi) - m \otimes m.$

Proof: All three of $\chi \otimes m, m \otimes \chi$, and $m \otimes m$ are HS operators. The latter of the three is constant on Ω while

$$\mathbb{E}\|\chi \otimes m\|_{HS} = \mathbb{E}\|m \otimes \chi\|_{HS} = \|m\|\mathbb{E}\|\chi\|.$$

Thus, the result holds if for all $f \in \mathbb{H}$

$$\mathbb{E}(m \otimes \chi)f = \mathbb{E}(\chi \otimes m)f = (m \otimes m)f = \langle m, f\rangle m,$$

which is immediate from (4.30). □

For simplicity of notation, we will generally assume that $m = 0$ unless stated otherwise. In this instance,

$$\mathscr{K} = \mathbb{E}(\chi \otimes \chi) := \int_{\Omega} (\chi \otimes \chi)\, d\mathbb{P}.$$

Some properties of the covariance operator are listed in our following result.

Theorem 7.2.5 *Suppose that* $m = 0$ *and* $\mathbb{E}(\|\chi\|^2) < \infty$. *For* $f, g \in \mathbb{H}$

1. $\langle \mathscr{K}f, g\rangle = \mathbb{E}\left[\langle \chi, f\rangle\langle \chi, g\rangle\right]$,
2. $\mathscr{K}$ *is a nonnegative-definite, trace-class operator with*

$$\|\mathscr{K}\|_{TR} = \mathbb{E}\|\chi\|^2,$$

and

3. $\mathbb{P}\left(\chi \in \overline{\mathrm{Im}(\mathscr{K})}\right) = 1$.

Proof: To prove part 1 first note that as $\mathscr{K} \in \mathfrak{B}_{HS}(\mathbb{H})$ we can use (4.30) to obtain

$$\mathscr{K}f = \left(\int_{\Omega} \chi \otimes \chi\, d\mathbb{P}\right) f = \int_{\Omega} \chi\langle \chi, f\rangle\, d\mathbb{P}$$

for any $f \in \mathbb{H}$. As a result, we can express $\mathscr{K}f$ as $\int_{\Omega} \chi\langle \chi, f\rangle\, d\mathbb{P}$. Then, an application of Theorem 3.1.7 to the linear functional $Tf := \langle f, g\rangle$ gives the result.

From part 1, we can see that $\mathscr{K}$ is clearly nonnegative definite. To show that $\mathscr{K}$ is trace class, let $\{e_j\}$ be any CONS for $\mathbb{H}$ and observe that

$$\|\mathscr{K}\|_{TR} = \sum_{j=1}^{\infty} \langle \mathscr{K}e_j, e_j\rangle = \sum_{j=1}^{\infty} \mathbb{E}\langle \chi, e_j\rangle^2 = \mathbb{E}\|\chi\|^2 < \infty.$$

Finally, from part 4 of Theorem 3.3.7,

$$(\mathrm{Im}(\mathscr{K}))^{\perp} = \mathrm{Ker}\ (\mathscr{K}^*) = \mathrm{Ker}\ (\mathscr{K})$$

as $\mathscr{K}$ is self-adjoint. Hence, for any $f \in (\mathrm{Im}(\mathscr{K}))^{\perp}$,

$$\mathbb{E}[\langle \chi, f\rangle^2] = \langle \mathscr{K}f, f\rangle = 0.$$

This implies that, with probability one, χ is orthogonal to any function in $(\mathrm{Im}(\mathscr{K}))^{\perp}$. Thus, with probability one,

$$\chi \in (\mathrm{Im}(\mathscr{K}))^{\perp\perp} = \overline{\mathrm{Im}(\mathscr{K})}$$

by part 3 of Theorem 2.5.6. □

The following result is an immediate consequence of Theorem 4.2.4 and part 2 of Theorem 7.2.5.

Theorem 7.2.6 *The operator $\mathscr{K}$ admits the eigen decomposition*

$$\mathscr{K} = \sum_{j=1}^{\infty} \lambda_j e_j \otimes e_j. \tag{7.6}$$

The eigenfunctions $\{e_j\}_{j=1}^{\infty}$ form a CONS for $\overline{\mathrm{Im}(\mathscr{K})}$ while the eigenvalues are nonnegative with the set $\{\lambda_j\}_{j=1}^{\infty}$ being either finite or consisting of a sequence that tends to zero. Each nonzero eigenvalue has finite multiplicity.

Relation (7.6) extends the spectral decomposition of a variance–covariance matrix for a random vector. Indeed, it gives (1.3) as a special case when χ is a random vector in $\mathbb{R}^p$. The combination of Theorems 7.2.5 and 7.2.6 provide the corresponding extension of the principal component decomposition (1.4) that we state formally as follows.

Theorem 7.2.7 *Suppose that $\mathscr{K}$ has the eigen decomposition (7.6). Then, with probability one,*

$$\chi = \sum_{j=1}^{\infty} \langle \chi, e_j \rangle e_j, \tag{7.7}$$

where $\langle \chi, e_j \rangle, j \geq 1$, are uncorrelated random variables with mean zero and variances λ_j.

The decomposition in (7.7) has various optimality properties. For example, it is straightforward to extend Theorem 1.1.1 to this context. Another possibility is provided by our following result.

Theorem 7.2.8 *If $\{f_j\}_{j=1}^{\infty}$ is any CONS for $\mathbb{H}$,*

$$\mathbb{E}\left\| \chi - \sum_{j=1}^{n} \langle \chi, f_j \rangle f_j \right\|^2 = \mathbb{E}\|\chi\|^2 - \sum_{j=1}^{n} \langle \mathscr{K} f_j, f_j \rangle,$$

which is minimized by taking $f_j = e_j, 1 \leq j \leq n$.

Proof: We know that

$$\mathbb{E}\left\|\chi - \sum_{j=1}^{n}\langle\chi,f_j\rangle f_j\right\|^2 = \mathbb{E}\|\chi\|^2 + \mathbb{E}\left\|\sum_{j=1}^{n}\langle\chi,f_j\rangle f_j\right\|^2 - 2\mathbb{E}\left\langle\chi,\ \sum_{j=1}^{n}\langle\chi,f_j\rangle f_j\right\rangle$$

with

$$\mathbb{E}\left\|\sum_{j=1}^{n}\langle\chi,f_j\rangle f_j\right\|^2 = \mathbb{E}\left\langle\chi,\ \sum_{j=1}^{n}\langle\chi,f_j\rangle f_j\right\rangle = \sum_{j=1}^{n}\mathbb{E}\langle\chi,f_j\rangle^2 = \sum_{j=1}^{n}\langle\mathscr{K}f_j,f_j\rangle.$$

So,

$$\mathbb{E}\left\|\chi - \sum_{j=1}^{n}\langle\chi,f_j\rangle f_j\right\|^2 = \mathbb{E}\|\chi\|^2 - \sum_{j=1}^{n}\langle\mathscr{K}f_j,f_j\rangle.$$

Part 2 of Theorem 7.2.5 gives $\|\mathscr{K}\|_{TR} = \mathbb{E}\|\chi\|^2$ and we can now use relation (4.35) along with Theorem 4.2.5 to complete the proof. □

As seen from the developments in Section 1.1, principle component decompositions tell us only a part of the story about the relationships that may be present among variables. It can also be of interest to examine the dependence between two different groups of variables using, e.g., canonical correlation analysis (cca). We will present a general treatment of abstract cca in Chapter 10 and therefore postpone further discussion of the topic until that point. However, in preparation for that work, we need to have at our disposal some general Hilbert space version of the finite-dimensional concept of a cross-covariance matrix in (1.8). The remainder of the section is devoted to establishing the existence and properties of this particular operator.

Suppose now that we have two random elements χ_1, χ_2 defined on the same probability space $(\Omega, \mathscr{F}, \mathbb{P})$ but taking values in two separable Hilbert spaces $\mathbb{H}_1$ and $\mathbb{H}_2$, respectively. Assume that, for $i = 1, 2$,

$$\mathbb{E}\|\chi_i\|_i^2 < \infty$$

and, for simplicity, take $\mathbb{E}\chi_1 = \mathbb{E}\chi_2 = 0$. Then, the cross-covariance operator for $\chi_1, \chi_2$is defined as the Bochner integral

$$\mathscr{K}_{12} = \int_{\Omega}(\chi_2 \otimes_2 \chi_1)\, d\mathbb{P}, \tag{7.8}$$

where the tensor product $\chi_2 \otimes_2 \chi_1$ is defined as in Definition 3.4.6 to be the mapping that takes any element $f \in \mathbb{H}_2$ to $\langle \chi_2, f \rangle_2 \chi_1$. This integral exists for essentially the same reasons as we used to verify the existence of the covariance operator. In particular, Theorem 3.4.7 shows that $\chi_2(\omega) \otimes_2 \chi_1(\omega)$ is an HS operator with HS norm $\|\chi_1(\omega)\|_1 \|\chi_2(\omega)\|_2$ for any $\omega \in \Omega$ so that Theorem 2.6.5 can again be employed to see that the integral is well defined as an element of $\mathfrak{B}_{HS}(\mathbb{H}_2, \mathbb{H}_1)$.

The extension of Theorem 7.2.5 to this situation takes the following form.

Theorem 7.2.9 *Suppose that* $\mathbb{E}\chi_i = 0$ *and* $\mathbb{E}\|\chi_i\|_i^2 < \infty$ *for* $i = 1, 2$. *Then, for any* $g \in \mathbb{H}_1, f \in \mathbb{H}_2$,

1. $\langle \mathscr{K}_{12} f, g \rangle_1 = \mathbb{E}\left[\langle \chi_1, g \rangle_1 \langle \chi_2, f \rangle_2\right]$,

2. $|\langle \mathscr{K}_{12} f, g \rangle_1| \le \langle \mathscr{K}_1 g, g \rangle_1^{1/2} \langle \mathscr{K}_2 f, f \rangle_2^{1/2}$, *and*

3. *the adjoint of* $\mathscr{K}_{12}$ *is* $\mathscr{K}_{21} = \int_\Omega (\chi_1 \otimes_1 \chi_2)\, d\mathbb{P}$.

Proof: The first claim of the theorem is verified by the same technique that was used to prove part 1 of Theorem 7.2.5. The second and third claims follow immediately from the first. □

In the multivariate analysis case, canonical correlation revolves around the singular value decomposition of the generalized correlation measure provided by the matrix $\mathscr{R}_{12} = \mathscr{K}_1^{-1/2} \mathscr{K}_{12} \mathscr{K}_2^{-1/2}$ with $\mathscr{K}_1, \mathscr{K}_2$, and $\mathscr{K}_{12}$ the covariance and cross-covariance matrices for the two random variables that are the subject of the analysis (cf. Chapter 1). Unfortunately, this approach fails in the case of general Hilbert space valued random elements because the compact nature of covariance operators renders them noninvertible for anything but the finite-dimensional case (Theorem 4.1.4). Nevertheless, it is a remarkable fact that there is still an infinite-dimensional parallel of $\mathscr{R}_{12}$ that is the subject of the following theorem. The development here follows that in Baker (1970); a somewhat more general treatment can be found in Baker (1973).

Theorem 7.2.10 *There exists an operator* $\mathscr{R}_{12} \in \mathfrak{B}(\mathbb{H}_2, \mathbb{H}_1)$ *with* $\|\mathscr{R}_{12}\| \le 1$ *such that* $\mathscr{K}_{12} = \mathscr{K}_1^{1/2} \mathscr{R}_{12} \mathscr{K}_2^{1/2}$.

Proof: Let (λ_{1j}, e_{1j}) be the eigenvalues and eigenfunctions of $\mathscr{K}_1$ and $\mathscr{P}_n$ the projection in $\mathbb{H}_1$ on span$\{e_{11}, \ldots, e_{1n}\}$. Then, for every $f \in \mathbb{H}_2$,

$$\|\mathscr{P}_n \mathscr{K}_1^{-1/2} \mathscr{K}_{12} f\|_1^2 = \langle \mathscr{K}_{12} f, \mathscr{P}_n \mathscr{K}_1^{-1} \mathscr{K}_{12} f \rangle_1. \tag{7.9}$$

By part 2 of Theorem 7.2.9,

$$\begin{aligned}
&\langle \mathscr{K}_{12}f, \mathscr{P}_n\mathscr{K}_1^{-1}\mathscr{K}_{12}f\rangle_1 \\
&\le \langle \mathscr{K}_2 f, f\rangle_2^{1/2}\langle \mathscr{K}_1\mathscr{P}_n\mathscr{K}_1^{-1}\mathscr{K}_{12}f, \mathscr{P}_n\mathscr{K}_1^{-1}\mathscr{K}_{12}f\rangle_1^{1/2} \qquad (7.10)\\
&= \langle \mathscr{K}_2 f, f\rangle_2^{1/2}\|\mathscr{P}_n\mathscr{K}_1^{-1/2}\mathscr{K}_{12}f\|_1.
\end{aligned}$$

Combining (7.9) and (7.10) gives

$$\|\mathscr{P}_n\mathscr{K}_1^{-1/2}\mathscr{K}_{12}f\|_1 \le \|\mathscr{K}_2^{1/2}f\|_2$$

for any n, and consequently

$$\|\mathscr{K}_1^{-1/2}\mathscr{K}_{12}f\|_1 \le \|\mathscr{K}_2^{1/2}f\|_2$$

for every $f \in \mathbb{H}_2$. So, if $f \in \text{Im}(\mathscr{K}_2^{1/2})$ in that $f = \mathscr{K}_2^{1/2}\tilde{f}$ for some $\tilde{f} \in \mathbb{H}_2$, we have

$$\|\mathscr{K}_1^{-1/2}\mathscr{K}_{12}\mathscr{K}_2^{-1/2}f\|_1 \le \|f\|_2; \qquad (7.11)$$

i.e., $\mathscr{R}_{12} := \mathscr{K}_1^{-1/2}\mathscr{K}_{12}\mathscr{K}_2^{-1/2}$ is bounded on $\overline{\text{Im}(\mathscr{K}_2^{1/2})}$ with norm at most one. Now $\mathscr{R}_{12}$ may be extended to $\overline{\text{Im}(K_2^{1/2})}$ by application of the extension principle and the extended operator has the same norm. Finally, define $\mathscr{R}_{12}f = 0$ for $f \in \overline{\text{Im}(K_2^{1/2})}^{\perp}$ to complete the definition of $\mathscr{R}_{12}$ on the entirety of $\mathbb{H}_1$. □

The operator $\mathscr{R}_{12}$ is unique in a certain sense. If there is another operator $\mathscr{R}$ such that $\mathscr{K}_{12} = \mathscr{K}_1^{1/2}\mathscr{R}\mathscr{K}_2^{1/2}$ with $\|\mathscr{R}\| \le 1$, then $\mathscr{K}_1^{1/2}(\mathscr{R}_{12} - \mathscr{R})\mathscr{K}_2^{1/2}f = 0$ for all $f \in \mathbb{H}_2$. Thus, $\mathscr{R}$ and $\mathscr{R}_{12}$ must coincide when viewed as linear mappings from $\overline{\text{Im}(\mathscr{K}_1^{1/2})}$ to $\overline{\text{Im}(\mathscr{K}_2^{1/2})}$.

As in the mva setting, the $\mathscr{R}_{12}$ operator provides an assessment of the dependence relationship between χ_1 and χ_2. In this regard, Baker (1970) shows that in the Gaussian case (i.e., when $(\langle\chi_1, g\rangle_1, \langle\chi_2, f\rangle_2)$ are bivariate normal for all $g \in \mathbb{H}_1, f \in \mathbb{H}_2$), the mutual information of χ_1 and χ_2 is finite if and only if $\mathscr{R}_{12}$ is HS and $\|\mathscr{R}_{12}\| < 1$. This has the implication that $\mathscr{R}_{12}$ is an element of $\mathfrak{B}_{HS}(\mathbb{H}_2, \mathbb{H}_1)$ with norm strictly less than one unless χ_1 is a deterministic function of χ_2.

7.3 Mean-square continuous processes and the Karhunen–Lòeve Theorem

So far, we have dealt with random elements in an abstract Hilbert space. Such a space may be a purely algebraic construction and have nothing to do with functions. In this section, we examine another important paradigm for fda.

Specifically, our interest here is in dealing with a stochastic process $X = \{X(t) : t \in E\}$ on a probability space $(\Omega, \mathscr{F}, \mathbb{P})$, where E is a general compact metric space. The index t is referred to as "time" and a continuous time process would correspond to $E = [0, 1]$, for example. In fda, and more generally in statistics, X represents a random function that can be fully or partially observed. The basic measure-theoretic assumption of a stochastic process is that $X(t)$ is a random variable, i.e., $\mathscr{F}$-measurable, for each fixed t. This alone does not imply that $X(\cdot)$ is a random element of $\mathbb{L}^2(E, \mathscr{B}(E), \mu)$, for instance; in the following section, we will give conditions that ensure this.

We note in passing that Theorem 7.1.2 has the consequence that random elements in a Hilbert space $\mathbb{H}$ may be viewed as stochastic processes with index set $E = \mathfrak{B}(\mathbb{H})$. However, we will not pursue that viewpoint as that particular abstraction takes us beyond what is relevant for this book.

The mean function of the process X is defined by

$$m(t) = \mathbb{E}[X(t)] \tag{7.12}$$

and the covariance function or covariance kernel by

$$K(s, t) = \text{Cov}(X(s), X(t)) \tag{7.13}$$

for $s, t \in E$, provided the expectations exist. Processes with well-defined mean and covariance functions are referred to as *second-order processes*. As

$$\sum_{i=1}^{n} \sum_{j=1}^{n} a_i a_j K(t_i, t_j) = \text{Var}\left(\sum_{j=1}^{n} a_j X(t_j)\right),$$

we can conclude that

Theorem 7.3.1 *The function K in (7.13) is nonnegative definite.*

This result in combination with the Moore–Aronszajn theorem (Theorem 2.7.4) guarantees that the covariance kernel for a stochastic process generates a reproducing kernel Hilbert space for which it is the rk. The implications of this fact will be explored in Section 7.6.

We will focus on second-order processes such that

$$\lim_{n\to\infty} \mathbb{E}\left[X(t_n) - X(t)\right]^2 = 0, \tag{7.14}$$

for any $t \in E$ and any sequence $\{t_n\}$ in E converging to t. Processes with this property are said to be *mean-square continuous* and they are the focal entity that underlies the Karhunen–Lòeve Theorem that will be established in this section.

The following result characterizes mean-square continuity by the the continuity of the covariance function.

Theorem 7.3.2 *Let X be a second-order process. Then, X is mean-square continuous if and only if its mean and covariance functions are continuous.*

Proof: As

$$\mathbb{E}[X(s) - X(t)]^2 = K(s,s) + K(t,t) - 2K(s,t) + (m(s) - m(t))^2,$$

the continuity of m and K implies (7.14). To show the other direction, we first see that the continuity of m follows from

$$\begin{aligned} |m(s) - m(t)| &= |\mathbb{E}[X(s) - X(t)]| \\ &\leq \left\{\mathbb{E}[X(s) - X(t)]^2\right\}^{1/2} \end{aligned}$$

and (7.14). Without loss of generality, assume that $m(t) \equiv 0$ in the following. Write

$$K(s,t) - K(s',t') = (K(s,t) - K(s',t)) + (K(s',t) - K(s',t')).$$

The Cauchy–Schwarz inequality then gives

$$|K(s,t) - K(s',t)| \leq K^{1/2}(t,t)\left(\mathbb{E}\left[X(s) - X(s')\right]^2\right)^{1/2}$$

and

$$|K(s',t) - K(s',t')| \leq K^{1/2}(s',s')\left(\mathbb{E}\left[X(t) - X(t')\right]^2\right)^{1/2}$$

so that mean-square continuity of X is seen to imply continuity of K. □

Observe that a by-product of the above-mentioned proof is the fact that if the mean function $m(t)$ is continuous, then the covariance function $K(s,t)$ is continuous at all (s,t) if and only if it is continuous at all "diagonal points" (t,t).

As a mean-square continuous process X may not be a random elements in any Hilbert space, Definition 7.2.3 cannot be applied to define the covariance operator. However, the following integral operator on $\mathbb{L}^2(E, \mathscr{B}(E), \mu)$ is well defined

$$(\mathscr{K}f)(t) = \int_E K(t,s)f(s)d\mu(s), \tag{7.15}$$

where μ is a finite measure. We refer to $\mathscr{K}$ as the covariance operator of a mean-square continuous process. Note that the measure μ is exogenous to X and is often set to be Lebesgue measure when E is an interval, or the counting measure when E is a finite (discrete) set. By Mercer's Theorem (Theorem 4.6.5),

$$K(s,t) = \sum_{j=1}^{\infty} \lambda_j e_j(s) e_j(t), \tag{7.16}$$

where the sum converges absolutely and uniformly on the support of μ, with (λ_j, e_j) the eigenvalue and (continuous) eigenfunction pairs of $\mathscr{K}$.

The next step is to define the $\mathbb{L}^2$ stochastic integral of a mean-square continuous process X. For convenience, we will assume in the rest of the section that the mean function is identically 0. For any function $f \in \mathbb{L}^2(E, \mathscr{B}(E), \mu)$, define

$$I_X(f; \{E_i, t_i : 1 \le i \le m(n)\}) = \sum_{i=1}^{m(n)} X(t_i) \int_{E_i} f(u) d\mu(u),$$

where $E_i \in \mathscr{B}(E)$ and $t_i \in E$. The total boundedness of E ensures that for each $n > 0$ there is a partition E_i, $1 \le i \le m(n)$, of E such that each E_i is an element of $\mathscr{B}(E)$ and has diameter less than $1/m(n)$; let t_i be an arbitrary point in E_i. Take two such partitions $\{E_i, t_i : 1 \le i \le m(n)\}$ and $\{E'_j, t'_j : 1 \le j \le m(n')\}$. Then,

$$\begin{aligned}
&\mathbb{E}\Big[I_X(f; \{E_i, t_i : 1 \le i \le m(n)\}) - I_X\left(f; \{E'_j, t'_j : 1 \le j \le m(n')\}\right)\Big]^2 \\
&= \sum_{i_1=1}^{m(n)} \sum_{i_2=1}^{m(n)} K(t_{i_1}, t_{i_2}) \int_{E_{i_1}} f(u) d\mu(u) \int_{E_{i_2}} f(v) d\mu(v) \\
&\quad + \sum_{j_1=1}^{m(n')} \sum_{j_2=1}^{m(n')} K(t'_{j_1}, t'_{j_2}) \int_{E'_{j_1}} f(u) d\mu(v) \int_{E'_{j_2}} f(u) d\mu(v) \\
&\quad - 2 \sum_{i=1}^{m(n)} \sum_{j=1}^{m(n')} K(t_i, t'_j) \int_{E_i} f(u) d\mu(v) \int_{E'_j} f(u) d\mu(v).
\end{aligned}$$

As K is uniformly continuous, each double sum on the right-hand side can be made arbitrarily close to $\int\int_{E \times E} K(u,v) f(u) f(v) d\mu(u) d\mu(v)$ for large enough n and n'. By the completeness of $\mathbb{L}^2(\Omega, \mathscr{F}, \mathbb{P})$, we conclude that there is a random variable $I_X(f) \in \mathbb{L}^2(\Omega, \mathscr{F}, \mathbb{P})$ to which $I_X(f; \{E_i, t_i : 1 \le i \le m(n)\})$ converges in mean square as $n \to \infty$ and the limit is independent of the choice of $\{E_i, t_i : 1 \le i \le m(n)\}$.

Theorem 7.3.3 *Let $\{X(t) : t \in E\}$ be a mean-square continuous stochastic process with mean zero. Then, for any $f, g \in \mathbb{L}^2(E, \mathscr{B}(E), \mu)$,*

1. $\mathbb{E}[I_X(f)] = 0$,
2. $\mathbb{E}[I_X(f)X(t)] = \int_E K(u,t)f(u)d\mu(u)$ *for any $t \in E$, and*
3. $\mathbb{E}[I_X(f)I_X(g)] = \int\int_{E\times E} K(u,v)f(u)g(v)d\mu(u)d\mu(v)$.

Proof: The general strategy is the same for proving all three assertions. For instance,

$$|\mathbb{E}[I_X(f)]| = |\mathbb{E}[I_X(f) - I_X(f; \{E_i, t_i : 1 \le i \le m(n)\})]|$$
$$\le \left\{\mathbb{E}[I_X(f) - I_X(f; \{E_i, t_i : 1 \le i \le m(n)\})]^2\right\}^{1/2},$$

which tends to 0 as $n \to \infty$. □

The following corollary is a simple consequence of part 3 of Theorem 7.3.3 and the orthogonality of eigenfunctions.

Corollary 7.3.4 *Let λ_i and e_i be the eigenvalues and eigenfunctions of the operator $\mathscr{K}$ in (7.15) and (7.16). Then, the $I_X(e_j)$ are mean zero random variables with*

$$\mathbb{E}[I_X(e_i)I_X(e_j)] = \delta_{ij}\lambda_i.$$

We are now ready to state the Karhunen–Lòeve Theorem .

Theorem 7.3.5 *Let $\{X(t) : t \in E\}$ be a mean-square continuous stochastic process with mean zero. Then,*

$$\lim_{n\to\infty} \sup_{t\in E} \mathbb{E}\left[X(t) - X_n(t)\right]^2 = 0, \tag{7.17}$$

where $X_n(t) := \sum_{j=1}^n I_X(e_j)e_j(t)$.

Proof: Using Theorem 7.3.3 and Corollary 7.3.4, we see that

$$\mathbb{E}[X_n(t) - X(t)]^2 = \mathbb{E}[X_n(t)]^2 - 2\mathbb{E}[X_n(t)X(t)] + \mathbb{E}[X(t)]^2$$
$$= \sum_{j=1}^n \lambda_j e_j^2(t) - 2\sum_{j=1}^n \lambda_j e_j^2(t) + K(t,t)$$
$$= K(t,t) - \sum_{j=1}^n \lambda_j e_j^2(t),$$

which tends to zero uniformly by Mercer's Theorem. □

The random variables $I_X(e_j)$ in the Karhunen–Lòeve Theorem are sometimes referred to as the principle component scores or simply scores. The partial sum $X_n(t)$ is the n-term Karhunen–Lòeve expansion and (7.17) says that the $\mathbb{L}^2(\Omega, \mathscr{F}, \mathbb{P})$ distance between $X(t)$ and its Karhunen–Lòeve expansion can be made arbitrarily small uniformly in t.

Finally, consider two mean-square continuous processes $X_1 = \{X_1(s), s \in E_1\}$ and $X_2 = \{X_2(t), t \in E_2\}$, where E_1, E_2 are both compact metric spaces. Define the auto-covariance functions

$$K_i(s,t) = \mathrm{Cov}(X_i(s), X_i(t))$$

for $s, t \in E_i$ and the cross-covariance function

$$K_{12}(s,t) = \mathrm{Cov}(X_1(s), X_2(t))$$

for $s \in E_1, t \in E_2$. Let $\mathbb{H}_i = \mathbb{L}^2(E_i, \mathscr{B}(E_i), \mu_i)$ for some arbitrary finite measure μ_i on E_i for $i = 1, 2$. In this case, the auto-covariance operators of X_1, X_2 are defined by

$$(\mathscr{K}_i f)(t) = \int_{E_i} K_i(s,t) f(s) d\mu_i(s) \tag{7.18}$$

for $f \in \mathbb{H}_i$ and cross-covariance operators by

$$\begin{aligned} (\mathscr{K}_{12} f)(t) &= \int_{E_2} K_{12}(t,s) f(s) d\mu_2(s), \\ (\mathscr{K}_{21} g)(t) &= \int_{E_1} K_{12}(s,t) g(s) d\mu_1(s) \end{aligned} \tag{7.19}$$

for $g \in \mathbb{H}_1, f \in \mathbb{H}_2$. The following result provides the stochastic process analog of Theorem 7.2.9.

Theorem 7.3.6 *For any $g \in \mathbb{H}_1, f \in \mathbb{H}_2$,*

1. $\langle \mathscr{K}_{12} f, g \rangle_1 = \int_{E_1} \int_{E_2} K_{12}(s,t) g(s) f(t) d\mu_1(s) d\mu_2(t)$,
2. $\langle \mathscr{K}_{12} f, g \rangle_1 \leq \langle \mathscr{K}_1 g, g \rangle_1^{1/2} \langle \mathscr{K}_2 f, f \rangle_2^{1/2}$, *and*
3. *the adjoint of $\mathscr{K}_{12}$ is $\mathscr{K}_{21}$*

where $\langle \cdot, \cdot \rangle_i$ denotes inner product of $\mathbb{H}_i$.

The proof of Theorem 7.2.10 is predicated on the Cauchy–Schwarz inequality in Theorem 7.2.9. In light of this fact and the parallel between Theorem 7.2.9 and Theorem 7.3.6, we see that the conclusion of Theorem 7.2.10 holds in exactly the same way for the covariance operators defined in (7.18) and (7.19) for mean-square continuous processes.

7.4 Mean-square continuous processes in $\mathbb{L}^2(E, \mathscr{B}(E), \mu)$

We have considered two different perspectives to this point: namely, random elements and second-order process. However, much of the fda research nowadays assumes that one is in a situation that involves the intersection of these two settings. The two dominant examples for this are processes that take values in RKHSs and in $\mathbb{L}^2(E, \mathscr{B}(E), \mu)$. In this section, we consider the latter space, which is more common but for which the theoretical issues are somewhat harder to handle. The RKHS setting will be discussed in Section 7.5.

Let us start with a mean-square continuous process $X = \{X(t), t \in E\}$ defined on some probability space $(\Omega, \mathscr{F}, \mathbb{P})$. Thus, $X(t)$ is $\mathscr{F}$-measurable for each $t \in E$. The first question is what additional assumption is required that will ensure that X is a random element of $\mathbb{H} = \mathbb{L}^2(E, \mathscr{B}(E), \mu)$. As usual, E is a compact metric space and μ a finite measure.

A convenient technical assumption in this regard is the joint measurability of $X(t, \omega)$ in both arguments t and ω: namely, $X(t, \omega)$ is measurable with respect to the product σ-field $\mathscr{B}(E) \times \mathscr{F}$. Joint measurability implies that for each ω, $X(\cdot, \omega)$ is a measurable function on E which places us in a position where we can consider issues such as whether $X(\cdot, \omega)$ belongs to $\mathbb{H}$ or whether X is a random element in $\mathbb{H}$, etc.

Theorem 7.4.1 *Suppose that the process X is jointly measurable and $X(\cdot, \omega) \in \mathbb{H}$ for each ω. Then, the mapping*

$$\omega \mapsto X(\cdot, \omega)$$

is measurable from Ω to $\mathbb{H}$: i.e., X is a random element of $\mathbb{H}$.

Proof: By joint measurability, $\langle X(\cdot, \omega), f\rangle$ is measurable for each $f \in \mathbb{H}$. A proof of this can be found in the standard treatment of Fubini's Theorem. The present result then follows from Theorem 7.1.2. □

In Theorem 7.4.1, the notation X is used to denote a stochastic process and a Hilbert space random element. It might be useful to emphasize a subtle distinction between the two roles for X. As mentioned earlier, the process $X(t)$ is a concrete object that can potentially be observed. The Hilbert space element X is an abstract representation of X; for instance, for any t, the symbol $X(t)$ has no meaning as elements of $\mathbb{H}$ are equivalence classes of functions. For convenience of notation, we will continue to use X to represent both the process and the Hilbert space element.

The question now arises what might represent an easily understood condition that would imply joint measurability. The following result provides one possible answer.

Theorem 7.4.2 *Assume that for each t, $X(t, \cdot)$ is measurable and that $X(\cdot, \omega)$ is continuous for each $\omega \in \Omega$. Then, $X(t, \omega)$ is jointly measurable and hence X is a random element of $\mathbb{H}$. In this case, the distribution of X is uniquely determined by the (finite-dimensional) distributions of $(X(t_1, \cdot), \ldots, X(t_n, \cdot))$ for all $t_1, \ldots, t_n \in E$ and all n.*

Proof: Consider any bivariate function of the form $g(t, \omega) := \sum_{i=1}^{k} I_{E_i}(t) f_i(\omega)$, where the E_i are disjoint sets in $\mathscr{B}(E)$ and the $f_i(\omega)$ are measurable functions. For any Borel set B of the real line,

$$g^{-1}(B) = \cup_{i=1}^{k} \left(E_i \times f_i^{-1}(B) \right),$$

which is in the product σ-field $\mathscr{B}(E) \times \mathscr{F}$. Thus, g is jointly measurable.

Let $\{E_i, t_i : 1 \le i \le m(n)\}$ be a partition defined as in Section 7.3 and define the jointly measurable function

$$X_n(t, \omega) = \sum_{i=1}^{m(n)} I_{E_i}(t) X(t_i, \omega).$$

By the uniform continuity of $X(t, \omega)$ in t for any fixed ω, we have $X_n(t, \omega) \to X(t, \omega)$ uniformly in t for any fixed ω. Thus, $X(t, \omega)$ is jointly measurable. The rest of the proof is straightforward due to Theorem 7.4.1 and the construction of X_n. □

Theorem 7.4.2 leads us to the following obvious question of how to verify the sample path continuity of a stochastic process. There is a large literature that considers this problem. One sufficient condition for continuity is the Kolmogorov criterion (see, e.g., Bass 2011), which we now describe.

For two processes X and Y with $E = [0, 1]$ as their index set, we say they are modifications of each other if

$$\mathbb{P}(X(t) = Y(t)) = 1 \tag{7.20}$$

for all $t \in [0, 1]$. By combining a countable number of zero-measure sets in Ω, (7.20) implies that

$$\mathbb{P}(X(t) = Y(t) \text{ for all } t \in \mathscr{C}) = 1 \tag{7.21}$$

for any countable sets $\mathscr{C}$. Note, however, that we cannot conclude from (7.20) alone that (7.21) holds for $\mathscr{C} = [0, 1]$, a notion referred to as indistinguishability. Processes that are modifications of each other have the same finite-dimensional distributions. Kolmogorov's criterion says that if there are finite, positive constants α, β, C such that

$$\mathbb{E}|X(t_1) - X(t_2)|^{\alpha} \le C|t_1 - t_2|^{1+\beta} \tag{7.22}$$

for all $t_1, t_2 \in [0, 1]$, X has a modification Y for which all realizations are continuous functions on $[0, 1]$. This allows us to verify continuity by calculations using the moments of the process.

To illustrate the use of (7.22), consider the case where X corresponds to the Brownian motion process on the interval $[0, 1]$. This means that all the $X(t)$ are normally distributed with mean zero and

$$\mathbb{E}[X(t_1)X(t_2)] = \min(t_1, t_2). \tag{7.23}$$

One can conclude from this that Brownian motion has uncorrelated and, hence, independent increments. Using this property in conjunction with the recurrence,

$$\begin{aligned}&\mathbb{E}[X^k(t_2)X^r(t_1)]\\&\quad = \mathbb{E}[(X(t_2) - X(t_1))\, X^{k-1}(t_2)X^r(t_1)] + \mathbb{E}[X^{k-1}(t_2)X^{r+1}(t_1)]\end{aligned}$$

reveals that

$$\mathbb{E}|X(t_2) - X(t_1)|^4 = 3|t_2 - t_1|^2.$$

So, Brownian motion satisfies (7.22) with $\alpha = 4, \beta = 1$ and $C = 3$.

The following result shows what happens if a mean-square continuous process is also a random element of $\mathbb{H}$.

Theorem 7.4.3 *Let $X = \{X(t) : t \in E\}$ be a mean-square continuous process that is jointly measurable. Then*

1. *the mean function m belongs to $\mathbb{H}$ and coincides with mean element of X in $\mathbb{H}$,*
2. *the covariance operator $\mathbb{E}(X \otimes X)$ is defined and coincides with the operator $\mathscr{K}$ in (7.15),*
3. *for any $f \in \mathbb{H}$*

$$\begin{aligned}I_X(f) &= \int_E X(t)f(t)d\mu(t)\\&= \langle X, f\rangle.\end{aligned}$$

Proof: By Fubini's Theorem,

$$\mathbb{E}\left(\int_E X(t)f(t)d\mu(t)\right) = \int_E m(t)f(t)d\mu(t),$$

for $f \in \mathbb{H}$ and therefore m is the mean element of X as a result of (7.4). Similarly, for any $f, g \in \mathbb{H}$, Fubini's Theorem allows us to write

$$\mathbb{E}\left(\iint_{E\times E}[X(s)-m(s)][X(t)-m(t)]f(s)g(t)d\mu(s)d\mu(t)\right)$$
$$=\iint_{E\times E}K(s,t)f(s)g(t)d\mu(s)d\mu(t).$$

The left-hand side of this last expression is $\mathbb{E}(\langle X-m,f\rangle\langle X-m,g\rangle)$ while the right-hand side is $\langle \mathscr{K}f, g\rangle$. Thus, part 1 of Theorem 7.2.5 shows that $\mathscr{K}$ is the covariance operator of X.

To show part 3, let $\{E_i, t_i : 1 \le i \le m(n)\}$ be a partition defined as in Section 7.3. Then, by Fubini's Theorem,

$$\mathbb{E}\left(I_X(f;\{E_i,t_i : 1\le i\le m(n)\}) - \int_E X(t)f(t)d\mu(t)\right)^2$$
$$=\sum_{i_1=1}^{m(n)}\sum_{i_2=1}^{m(n)}K(t_{i_1},t_{i_2})\int_{E_{i_1}}f(u)d\mu(u)\int_{E_{i_2}}f(v)d\mu(v)$$
$$+\iint_{E\times E}K(u,v)f(u)f(v)d\mu(u)d\mu(v)$$
$$-2\sum_{i=1}^{m(n)}\int_{E_i}\int_E K(t_i,v)f(u)f(v)d\mu(u)d\mu(v),$$

where the right-hand side tends to zero as $n\to\infty$. This verifies part 3 and concludes the proof. □

One immediate implication of Theorem 7.4.3 is that when X is mean-square continuous and jointly measurable, the scores $I_X(e_j)$ in Theorem 7.3.5 can be represented as

$$\begin{aligned} I_X(e_j) &= \int_E X(t)e_j(t)d\mu(t) \\ &= \langle X, e_j\rangle. \end{aligned} \tag{7.24}$$

Our last result in this section shows that even if the mean-square continuous process X is not a random element of $\mathbb{H} = \mathbb{L}^2(E, \mathscr{B}(E), \mu)$, one can speak about an abstract random element χ of $\mathbb{H}$ that plays the role of X in (7.24).

Theorem 7.4.4 *Let $X = \{X(t) : t \in E\}$ be a mean-square continuous stochastic process with mean zero. Then, there is a random element χ of $\mathbb{H}$ whose covariance operator is given by (7.15).*

Proof: For any choice of $\{E_i, t_i : 1 \le i \le m(n)\}$, define the process

$$\chi(t, \omega; \{E_i, t_i : 1 \le i \le m(n)\}) = \sum_{i=1}^{m(n)} I_{E_i}(t) X(t_i, \omega).$$

Now view this as an element of $\mathbb{H}_0 := \mathbb{L}^2(E \times \Omega, \mathscr{F}(E) \times \mathscr{B}, \mu \times \mathbb{P})$, where the measurability is checked as in Theorem 7.4.2. Observe that

$$\left\| \chi(t, \omega; \{E_i, t_i : 1 \le i \le m(n)\}) - \chi\left(t, \omega; \{E'_j, t'_j : 1 \le j \le m(n')\}\right) \right\|_{\mathbb{H}_0}^2$$
$$= \mathbb{E}\left[I_X(f; \{E_i, t_i : 1 \le i \le m(n)\}) - I_X\left(f; \{E'_j, t'_j : 1 \le j \le m(n')\}\right) \right]^2,$$

with $f(t) \equiv 1$. By completeness, there exists an element χ in $\mathbb{H}_0$ such that

$$\| \chi(t, \omega; \{E_i, t_i : 1 \le i \le m(n)\}) - \chi(t, \omega) \|_{\mathbb{H}_0} \to 0 \tag{7.25}$$

as $n \to \infty$. Note that the event

$$\left\{ \omega \in \Omega : \int_E \chi^2(t, \omega) d\mu(t) = \infty \right\}$$

has probability zero. Thus, if necessary we can modify χ and define a new element in $\mathbb{H}_0$, also denoted by χ for convenience, such that $\int_E \chi^2(t, \omega) d\mu(t) < \infty$ for all ω. Then Theorem 7.4.1 can be invoked to conclude that $\chi(\cdot, \omega)$ is a random element of $\mathbb{L}^2(E, \mathscr{B}(E), \mu)$.

Let $\tilde{\mathscr{K}} = \mathbb{E}(\chi \otimes \chi)$ be the covariance operator of χ. From part 1 of Theorem 7.2.5, $\langle \tilde{\mathscr{K}} f, g \rangle = \mathbb{E}(\langle \chi, f \rangle \langle \chi, g \rangle)$ for any $f, g \in \mathbb{L}^2(E, \mathscr{B}(E), \mu)$, and (7.25) now has the consequence that $\mathbb{E}(\langle \chi, f \rangle \langle \chi, g \rangle)$ is the limit of

$$\mathbb{E}(I_X(f; \{E_i, t_i : 1 \le i \le m(n)\}) I_X(g; \{E_i, t_i : 1 \le i \le m(n)\}))$$
$$= \sum_{i_1=1}^{m(n)} \sum_{i_2=1}^{m(n)} K(t_{i_1}, t_{i_2}) \int_{E_{i_1}} f(u) d\mu(u) \int_{E_{i_2}} g(v) d\mu(v)$$

as $n \to \infty$. On the other hand, the continuity of K implies that the last expression tends to $\langle \mathscr{K} f, g \rangle$, from which we conclude that $\tilde{\mathscr{K}}$ and $\mathscr{K}$ are the same operator. □

7.5 RKHS valued processes

In Section 7.4, we considered the setting where a mean-square continuous stochastic process $\{X(t), t \in E\}$ is also a random element of $\mathbb{H} = \mathbb{L}^2(E, \mathscr{B}(E), \mu)$. Here we consider what transpires when the Hilbert space $\mathbb{H}$ is an RKHS $\mathbb{H}(R)$ where the rk R is a continuous function defined on $E \times E$.

In contrast to the $\mathbb{L}^2(E, \mathscr{B}(E), \mu)$ functional space in Section 7.4, as the elements of $\mathbb{H}(R)$ are functions, there is no need to distinguish between the process X and the potential Hilbert space element X. As such, the theoretical development at least in the fda context becomes a bit cleaner. For example, measurability issues are more straightforward as the following result shows. As usual, we say that X is a stochastic process if $X(t)$ is a random variable for each fixed t and we say that X is a random element of $\mathbb{H}(R)$ if X is a measurable mapping from the probability space into $\mathbb{H}(R)$.

Theorem 7.5.1 *A random element X of $\mathbb{H}(R)$ is a stochastic process. Conversely, a stochastic process X taking values in $\mathbb{H}(R)$ is a random element of $\mathbb{H}(R)$.*

Proof: Denote the probability space under discussion by $(\Omega, \mathscr{F}, \mathbb{P})$. First assume that X is a random element of $\mathbb{H}$. By the reproducing property, $X(t) = \langle X, R(\cdot, t)\rangle$. As the composition of two measurable transformations, $X(t)$ is necessarily a random variable.

To go the other direction, let X be an $\mathbb{H}(R)$ valued process. For any $g \in \mathbb{H}(R)$, we may find a sequence of the form $g_n(\cdot) = \sum_{i=1}^{n} a_i R(\cdot, t_i)$ that converges to g for which $\langle X(\cdot, \omega), g_n\rangle = \sum_{i=1}^{n} a_i X(t_i, \omega)$. Thus, by the continuity of the inner product, we see that $\langle X(\cdot, \omega), g\rangle$ is the point-wise limit of a sequence of measurable function in $\mathbb{H}(R)$ and must aslo be measurable. The result is now a consequence of Theorem 7.1.2. □

With the aid of the previous theorem, we can obtain expressions for the mean and covariance functions of an RKHS valued process.

Theorem 7.5.2 *Let X be a random element of $\mathbb{H}(R)$ with $\mathbb{E}\|X\|^2 < \infty$. Then, X is a mean-square continuous process on E and the mean element m and covariance operator $\mathscr{K}$ are related to the mean and covariance functions $m(t)$ and $K(s, t)$ by*

$$m(t) = \mathbb{E}[X(t)] = \langle m, R(\cdot, t)\rangle \tag{7.26}$$

and

$$\begin{aligned} K(s, t) &= \mathrm{Cov}(X(t), X(s)) \\ &= \langle \mathscr{K} R(\cdot, t), R(\cdot, s)\rangle. \end{aligned} \tag{7.27}$$

Furthermore, $K \in \mathbb{H}(R) \otimes \mathbb{H}(R)$, *with*

$$\|K\|_{\mathbb{H}(R)\otimes\mathbb{H}(R)} \leq \mathbb{E}\|X - m\|^2 < \infty. \tag{7.28}$$

Proof: That assertion (7.26) follows from (7.4) as $X(t) = \langle X, R(\cdot, t)\rangle$. To prove (7.27), we need only observe that

$$\text{Cov}(X(t), X(s)) = \mathbb{E}\langle X, R(\cdot, t)\rangle\langle X, R(\cdot, s)\rangle - \mathbb{E}\langle X, R(\cdot, t)\rangle\mathbb{E}\langle X, R(\cdot, s)\rangle$$

and then apply (7.4) and Theorem 7.2.5. The continuity of K is a consequence of the continuity of R.

Finally, $X(s)X(t)$ is a bivariate stochastic process with sample paths in the direct product space $\mathbb{H}(R) \otimes \mathbb{H}(R)$, which is also an RKHS with rk $R(\cdot, s)R(\cdot, t)$ by Theorem 2.7.13. As a result, $X(s)X(t)$ is a random element of $\mathbb{H}(R) \otimes \mathbb{H}(R)$ and $\mathbb{E}\|X(s)X(t)\|_{\mathbb{H}(R)\otimes\mathbb{H}(R)} = \mathbb{E}\|X\|^2 < \infty$. Thus, the assertions on K follow from the fact that $K(s, t)$ is the mean element of the random element $\{X(s) - m(s)\}\{X(t) - m(t)\}$. □

The following result contains parallels of Mercer's Theorem and the Karhunen and Lòeve Theorem for an RKHS valued process.

Theorem 7.5.3 *Let X be a random element of $\mathbb{H}(R)$ with mean zero and $\mathbb{E}\|X\|^2 < \infty$. If the covariance operator $\mathscr{K}$ has the eigen decomposition*

$$\mathscr{K} = \sum_{j=1}^{\infty} \lambda_j e_j \otimes e_j,$$

then

$$K(s, t) = \sum_{j=1}^{\infty} \lambda_j e_j(s) e_j(t), \tag{7.29}$$

where the sum converges absolutely and uniformly, and

$$\lim_{n\to\infty} \sup_{t\in E} \mathbb{E}\left[X(t) - X_n(t)\right]^2 = 0, \tag{7.30}$$

where $X_n(t) := \sum_{j=1}^{n} \langle X, e_j\rangle e_j(t)$.

Proof: By (7.27) the covariance function of X satisfies

$$\begin{aligned} K(s, t) &= \langle \mathscr{K} R(\cdot, s), R(\cdot, t)\rangle \\ &= \sum_{j=1}^{\infty} \lambda_j \langle e_j, R(\cdot, s)\rangle\langle e_j, R(\cdot, t)\rangle \\ &= \sum_{j=1}^{\infty} \lambda_j e_j(s) e_j(t). \end{aligned}$$

As

$$K(t,t) = \langle \mathscr{K}R(\cdot,t), R(\cdot,t)\rangle,$$

we have

$$\sup_{t\in E} K(t,t) \leq \|\mathscr{K}\| \sup_{t\in E} R(t,t) < \infty.$$

Thus, uniformly in s, t,

$$\sum_{j=1}^{\infty} \lambda_j |e_j(s)e_j(t)| \leq \left(\sum_{j=1}^{\infty} \lambda_j e_j^2(s) \sum_{j=1}^{\infty} \lambda_j e_j^2(t) \right)^{1/2}$$
$$= (K(s,s)K(t,t))^{1/2} < \infty.$$

Finally, as the scores $\langle X, e_j\rangle$ are uncorrelated (Theorem 7.2.7), (7.30) can be established along similar lines as those in the proof of Theorem 7.3.5. □

Note that the components of the expansions in (7.29) and (7.30) are not the same as those in Mercer's Theorem and the Karhunen and Lòeve Theorem. The latter quantities are based on the integral covariance operator on $\mathbb{L}^2(E, \mathscr{B}(E), \mu)$.

At this point, one may begin to wonder about the conditions that are required for a process to take values in a particular RKHS. To motivate that discussion, let us again consider the Brownian motion process on $[0, 1]$. As stated in (7.23), this is a zero mean process that has covariance kernel

$$K(s,t) = \min(s,t).$$

The eigen decomposition for the corresponding covariance operator $\mathscr{K}$ in $\mathbb{L}^2[0,1]$ was the subject of Example 4.6.3. By (2.41), $\text{Im}(\mathscr{K})$ contains functions f in $\mathbb{W}_1[0,1]$ satisfying $f(0) = 0$. However, Brownian motion is the classic example of a continuous process that has nowhere differentiable sample paths, which means that $\mathbb{P}(X \in \text{Im}(\mathscr{K})) = 0$. This would, at first, seem to be a contradiction to part 3 of Theorem 7.2.5. However, that is not the case because $\text{Im}(\mathscr{K})$ is not closed in $\mathbb{L}^2[0,1]$ (Theorem 4.3.7) and the completion of $\text{Im}(\mathscr{K})$ in $\mathbb{L}^2[0,1]$ is precisely the set of functions in $\mathbb{L}^2[0,1]$ that are equal to 0 at 0 (Theorem 2.3.8). By Theorem 7.6.4 below, $\text{Im}(\mathscr{K}) \subset \mathbb{H}(K)$. Thus, the following result is an attempt to describe this phenomenon in a general way.

Theorem 7.5.4 *Suppose that X is a random element of $\mathbb{H}(K)$ with $\mathbb{E}\|X\|^2 < \infty$, where K is the covariance function of X. Then, $\mathbb{H}(K)$ must be finite dimensional.*

Proof: The assumption $\mathbb{E}\|X\|^2_{\mathbb{H}(K)} < \infty$ implies that the covariance operator $\mathscr{K}$ of X on $\mathbb{H}(K)$ is well defined and therefore must be trace class by Theorem 7.2.5. Then (7.26) of Theorem 7.5.2 has the consequence that $\mathscr{K}$

is the identity mapping of $\mathbb{H}(K)$. However, by Theorem 4.1.2, the identity of an infinite-dimensional space is not compact and hence the conclusion. □

Necessary and sufficient conditions for the sample paths of a Gaussian process to lie in an RKHS were determined by Driscoll (1973). The general case was examined by Lukíc and Beder (2001). They showed that if a process has covariance kernel K, a necessary condition for its sample paths to fall in an RKHS $\mathbb{H}(R)$ is that $\mathbb{H}(K) \subseteq \mathbb{H}(R)$ and the process has a valid covariance operator. This is also sufficient when, e.g., K is continuous.

7.6 The closed span of a process

One of the wonderful aspects of congruence mappings is that they provide us with the potential for translating a problem formulated in one space into an equivalent problem in another space where the mathematics may be more tractable. This turns out to be particularly true in the case of stochastic processes as a result of work begun in Parzen (1961). Here we describe certain aspects of this representation theory that are particularly relevant to the development of canonical correlations of functional data. As in previous sections, one can approach from the viewpoint of a stochastic process or a Hilbert space valued random element. We start with the first perspective.

Let $\{X(t), t \in E\}$ be a second-order process with mean zero and a continuous covariance function $K(s,t)$. For the purpose of statistical inference, a space of particular interest is the completion of the set of all random variables of the form

$$\sum_{i=1}^{n} a_i X(t_i), \ a_i \in \mathbb{R}, t_i \in E, n = 1, 2, \ldots, \tag{7.31}$$

in the space $\mathbb{L}^2(\Omega, \mathscr{F}, \mathbb{P})$. We denote the completed space by $\mathbb{L}^2(X)$, i.e.,

$$\mathbb{L}^2(X) = \overline{\left\{ \sum_{j=1}^{n} a_j X(t_j), t_j \in E, a_j \in \mathbb{R}, n = 1, 2, \ldots \right\}}, \tag{7.32}$$

and call it the *closed linear span* of X.

Recall from Section 2.7 that $\mathbb{H}(K)$ denotes the RKHS with rk K; $\mathbb{H}(K)$ is the completion of the space that contains functions of the form $\sum_{i=1}^{n} a_i K(\cdot, t_i)$ with $a_i \in \mathbb{R}, t_i \in E, n \in \mathbb{Z}^+$ and with inner product determined by

$$\langle K(\cdot, s), K(\cdot, t) \rangle_{\mathbb{H}(K)} = K(s,t).$$

The following result was first discussed by Lòeve.

Theorem 7.6.1 *$\mathbb{L}^2(X)$ is congruent to $\mathbb{H}(K)$.*

Proof: The proof is accomplished by verifying that the correspondence

$$\sum_{i=1}^{\infty} a_i X(t_i) \longleftrightarrow \sum_{i=1}^{\infty} a_i K(\cdot, t_i) \tag{7.33}$$

determines a congruence between $\mathbb{L}^2(X)$ and $\mathbb{H}(K)$. □

For $f \in \mathbb{H}(K)$, denote by $Z(f)$, the random variable in $\mathbb{L}^2(X)$ determined by the congruence (7.33). As such,

$$\mathrm{Cov}(Z(f_1), Z(f_2)) = \langle f_1, f_2 \rangle_{\mathbb{H}(K)} \tag{7.34}$$

for $f_1, f_2 \in \mathbb{H}(K)$. We now explore the properties of the congruence beginning with the following insightful example.

Example 7.6.2 *Suppose that $E = \{t_1, \dots, t_p\}$ so that $X(\cdot)$ can be represented as the random p-vector $X = (X(t_1), \dots, X(t_p))^T$. Its covariance kernel is equivalent to the $p \times p$ matrix $\mathscr{K} = \{K(t_i, t_j)\}_{i,j=1:p}$ and the congruent RKHS $\mathbb{H}(K)$ in this instance was described in Example 2.7.8. Analogous to that development we will use $X(\cdot)$ and X to refer to the stochastic process and vector forms for the $X(\cdot)$ process and use f and $f(\cdot)$ to represent the vector and function forms of an element $f(\cdot)$ of $\mathbb{H}(K)$. We now claim that*

$$Z(f) = f^T \mathscr{K}^\dagger X \tag{7.35}$$

with $\mathscr{K}^\dagger$ the Moore–Penrose generalized inverse of $\mathscr{K}$. To verify this, we need only observe that

$$\mathrm{Var}\left(f^T \mathscr{K}^\dagger X\right) = f^T \mathscr{K}^\dagger f = \|f(\cdot)\|^2_{\mathbb{H}(K)}.$$

The previous example provides an important case where $Z(f)$ can actually be evaluated. As both $\mathbb{L}^2(X)$ and $\mathbb{H}(K)$ are built up from finite-dimensional subspaces, it provides us with a prescription for connecting the essential ingredients for the two spaces. The following result leverages that idea and gives a generalization of Example 7.6.2.

Theorem 7.6.3 *A function f is in $\mathbb{H}(K)$ if there exists $Y \in \mathbb{L}^2(X)$ such that $f(\cdot) = \mathbb{E}[YX(\cdot)]$, in which case $Y = Z(f)$.*

Proof: Let Y be the limit of a sequence $Y_n = \sum_{i=1}^{n} a_{ni} X(t_{ni})$ in $\mathbb{L}^2(X)$. If $f \in \mathbb{H}(K)$ is the congruent image of Y, then $f_n(\cdot) = \sum_{i=1}^{n} a_{ni} K(\cdot, t_{ni})$ must converge to f as a result of the isometry. For each $t \in E$,

$$f_n(t) = \sum_{i=1}^{n} a_{ni}\mathbb{E}[X(t)X(t_{ni})] = \mathbb{E}[Y_n X(t)]$$

and continuity of the $\mathbb{L}^2(X)$ inner product ensures that $\mathbb{E}[Y_n X(t)] \to \mathbb{E}[YX(t)]$ and $Y = Z(f)$.

Conversely, suppose that f is in $\mathbb{H}(K)$. Then, f corresponds to $Z(f)$ in $\mathbb{L}^2(X)$ and by the reproducing property $f(t) = \langle f, K(t,\cdot)\rangle_{\mathbb{H}(\mathscr{K})} = \mathbb{E}[Z(f)X(t)]$. □

Although Theorem 7.6.3 is insightful, it falls well short of the type of relation we were able to produce in Example 7.6.2. Such explicit characterizations are of paramount importance because if we are to solve optimization problems deriving from $\mathbb{L}^2(X)$ in $\mathbb{H}(K)$, we need to have a formula that tells us how to return to the space of random variables that is actually of interest. Our following approach is an attempt in that regard.

Suppose now that E is a compact metric space. Let $\mathscr{K}$ denote the covariance operator on $\mathbb{H} = \mathbb{L}^2(E, \mathscr{B}(E), \mu)$ defined by (7.15) and (λ_j, e_j) the eigenvalue–eigenfunction pairs of $\mathscr{K}$. Denote by $\mathbb{G}(\mathscr{K})$, the Hilbert space of functions in the range of $\mathscr{K}^{1/2}$, equipped with the inner product

$$\langle \mathscr{K}^{1/2} f, \mathscr{K}^{1/2} g\rangle_{\mathbb{G}(\mathscr{K})} = \langle f, g\rangle_{\mathbb{H}} \tag{7.36}$$

for $f, g \in \mathbb{H}$.

The relationship (7.36) shows that functions in $\mathbb{G}(\mathscr{K})$ can be represented as $\sum_{j=1}^{\infty} \lambda_j^{1/2} a_j e_j$ with $\sum_{j=1}^{\infty} a_j^2 < \infty$, or, equivalently, as $\sum_{j=1}^{\infty} \lambda_j a_j e_j$ with $\sum_{j=1}^{\infty} \lambda_j a_j^2 < \infty$. We will primarily use the latter representation in the following. To summarize,

$$\mathbb{G}(\mathscr{K}) = \left\{ \sum_{j=1}^{\infty} \lambda_j a_j e_j : \sum_{j=1}^{\infty} \lambda_j a_j^2 < \infty \right\} \tag{7.37}$$

and

$$\left\langle \sum_{j=1}^{\infty} \lambda_j a_j e_j, \sum_{j=1}^{\infty} \lambda_j b_j e_j \right\rangle_{\mathbb{G}(\mathscr{K})} = \sum_{j=1}^{\infty} \lambda_j a_j b_j. \tag{7.38}$$

Theorem 7.6.4 $\mathbb{G}(\mathscr{K}) = \mathbb{H}(K)$.

Proof: Mercer's Theorem gives us

$$K(\cdot, t) = \sum_{j=1}^{\infty} \lambda_j e_j(t) e_j(\cdot) = \sum_{j=1}^{\infty} \lambda_j a_j e_j(\cdot) \tag{7.39}$$

for any fixed $t \in E$, where $a_j = e_j(t)$. As

$$\sum_{j=1}^{\infty} \lambda_j a_j^2 = \sum_{j=1}^{\infty} \lambda_j e_j^2(t) = K(t,t) < \infty,$$

we conclude that $K(\cdot,t) \in \mathbb{G}(\mathscr{K})$. Now, for $g = \sum_j \lambda_j b_j e_j \in \mathbb{G}(\mathscr{K})$, (7.38) and (7.39) entail that

$$\langle g, K(\cdot,t)\rangle_{\mathbb{G}(\mathscr{K})} = \sum_{j=1}^{\infty} \lambda_j a_j b_j = \sum_{j=1}^{\infty} \lambda_j b_j e_j(t) = g(t).$$

Therefore, K is an rk for $\mathbb{G}(\mathscr{K})$ and our claim has been verified due to the unicity aspect of the Moore–Aronszajn Theorem (Theorem 2.7.4). □

Theorem 7.6.4 shows that functions of $\mathbb{H}(K)$ can be represented as $\sum_{j=1}^{\infty} \lambda_j a_j e_j$ with $\sum_{j=1}^{\infty} \lambda_j a_j^2 < \infty$ and therefore the congruence between $\mathbb{L}^2(X)$ and $\mathbb{H}(K)$ can be described as

$$Z(f) = \sum_{j=1}^{\infty} a_j I_X(e_j) \tag{7.40}$$

for $f = \sum_{j=1}^{\infty} \lambda_j a_j e_j \in \mathbb{H}(K)$, where, as in the Karhunen–Lòeve Theorem (Theorem 7.3.5), $I_X(e_j)$ is the score of X that corresponds to e_j. To see why this is true we only needs to observe that

$$\mathrm{Var}\left(\sum_{j=1}^{\infty} a_j I_X(e_j)\right) = \sum_{j=1}^{\infty} \lambda_j a_j^2 = \|f\|^2_{\mathbb{H}(K)}.$$

Suppose we make the additional assumption that X is a random element of $\mathbb{H} = \mathbb{L}^2(E, \mathscr{B}(E), \mu)$. Then (7.24) shows that $Z_j = \langle X, e_j\rangle$, in which case we can write

$$Z(f) = \sum_{j=1}^{\infty} a_j \langle X, e_j\rangle \tag{7.41}$$

for $f = \sum_{j=1}^{\infty} \lambda_j a_j e_j$. With (7.38) and (7.41), it is tempting to conclude that $Z(f)$ in this case is just $\langle X, f\rangle_{\mathbb{H}(K)}$. Unfortunately, in general, that is not true as X cannot be a random element of $\mathbb{H}(K)$ except when $\mathbb{H}(K)$ is finite dimensional (Theorem 7.5.4).

Example 7.6.5 *Take* $X = \left(X(t_1), \ldots X(t_p)\right)^T$ *as in Example 7.6.2 with* $\mathbb{H} = \mathbb{R}^p$ *and*

$$\langle X(\cdot), g(\cdot)\rangle = \sum_{i=1}^{p} g(t_i) X(t_i)$$

for $g = (g(t_1), \ldots, g(t_p))^T \in \mathbb{R}^p$. *Let* $\mathscr{K} = \sum_{j=1}^{q} \lambda_j e_j e_j^T$ *for nonzero eigenvalues* $\lambda_1, \ldots, \lambda_q$ *with* $q \le p$ *and associated eigenvectors* $e_j = (e_j(t_1), \ldots, e_j(t_p))^T$. *Then,* $X = \sum_{j=1}^{q} \langle X(\cdot), e_j(\cdot)\rangle e_j$ *with probability one and if* $f(\cdot) = \sum_{j=1}^{q} \lambda_j a_j e_j(\cdot)$

$$
\begin{aligned}
Z(f) &= \sum_{j=1}^{q} a_j \langle X(\cdot), e_j(\cdot)\rangle \\
&= f^T \mathscr{K}^{\dagger} X
\end{aligned}
$$

because $\mathscr{K}^{\dagger} = \sum_{j=1}^{q} \lambda_j^{-1} e_j e_j^T$.

So far in this section, we have considered mean-square continuous processes. We next turn to the more abstract setting where χ is a random element in a Hilbert space $\mathbb{H}$. Assume that $\mathbb{E}\chi = 0$ and $\mathbb{E}\|\chi\|^2 < \infty$. As explained briefly in Section 7.3, we can view χ as a stochastic process indexed by $f \in \mathbb{H}$, i.e., $\chi = \{\langle \chi, f\rangle : f \in \mathbb{H}\}$. The covariance function of this process is

$$
K(f, g) = \mathrm{Cov}(\langle \chi, f\rangle, \langle \chi, g\rangle)
$$

and the closed linear span is

$$
\mathbb{L}^2(\chi) = \overline{\{\langle \chi, f\rangle, f \in \mathbb{H}\}},
$$

where the completion is taken in $\mathbb{L}^2(\Omega, \mathscr{F}, \mathbb{P})$. As in Theorem 7.6.1, we can define the RKHS $\mathbb{H}(K)$ and the correspondence

$$
\langle \chi, f\rangle \longleftrightarrow K(\cdot, f)
$$

that determines the congruence between $\mathbb{L}^2(\chi)$ and $\mathbb{H}(K)$. Unfortunately, it is hard to visualize functions whose argument is a Hilbert space element. Thus, it might be useful to consider a different parameterization of this congruence.

Let $\mathscr{K}$ be the covariance operator of the random element χ as defined by Definition 7.2.3 and (λ_j, e_j) its eigenvalue–eigenfunction pairs. In terms of (λ_j, e_j), we can express $\mathbb{L}^2(\chi)$ as

$$
\mathbb{L}^2(\chi) = \left\{ \sum_{j=1}^{\infty} a_j \langle \chi, e_j\rangle : \sum_{j=1}^{\infty} \lambda_j a_j^2 < \infty \right\}. \tag{7.42}
$$

Define the Hilbert space

$$
\mathbb{G}(\mathscr{K}) = \left\{ \sum_{j=1}^{\infty} \lambda_j a_j e_j : \sum_{j=1}^{\infty} \lambda_j a_j^2 < \infty \right\} \tag{7.43}
$$

with inner product

$$\left\langle \sum_{j=1}^{\infty} \lambda_j a_j e_j, \sum_{j=1}^{\infty} \lambda_j a_j e_j \right\rangle_{\mathbb{G}(\mathscr{K})} = \sum_{j=1}^{\infty} \lambda_j a_j b_j. \tag{7.44}$$

Then, $\mathbb{G}(\mathscr{K})$ can take the role of $\mathbb{H}(K)$ and the the congruence between $\mathbb{L}^2(\chi)$ and $\mathbb{H}(\mathscr{K})$ is

$$Z(f) = \sum_{j=1}^{\infty} a_j \langle \chi, e_j \rangle \tag{7.45}$$

for $f = \sum_{j=1}^{\infty} \lambda_j a_j e_j \in \mathbb{G}(\mathscr{K})$.

Observe that (7.43), (7.44), and (7.45) for the random element setting completely parallel (7.37), (7.38), and (7.40) for the process setting. The only difference is how the covariance operators are defined in the two settings and that $\mathbb{G}(\mathscr{K})$ in the former is an RKHS. In the special case of Theorem 7.4.3 where a mean-square continuous process is also a random element in $\mathbb{L}^2(E, \mathscr{B}(E), \mu)$, the space $\mathbb{G}(\mathscr{K})$ defined from the process and random element settings are identical. This will facilitate our study of canonical correlations in Chapter 10 in a unified manner for the two settings.

7.7 Large sample theory

In this section, we introduce some basic large-sample results that are useful for the inference problems that arise in fda. More generally, probability theory in Banach and Hilbert spaces is an important branch of modern probability. The interested reader is referred to Ledoux and Talagrand (2013) for a complete treatment of this topic.

Suppose that we have independent realization $X_1, \ldots, X_n$ of some real valued random variable X with $\mathbb{E}X = m$. Then, we know that under various conditions $\overline{X}_n = n^{-1} \sum_{i=1}^{n} X_i$ converges almost surely to m and, when suitably normalized, has an approximate normal distribution. Results of this nature are referred to as the strong law of large numbers and the central limit theorem, respectively. The goal of this section is to obtain analogs of these results with the X_i replaced by random elements of a Hilbert space.

Let $\chi_1, \chi_2, \ldots$ be random elements in $\mathbb{H}$ and define

$$S_n = \sum_{i=1}^{n} \chi_i.$$

Theorem 7.7.1 *If $\chi_1, \ldots, \chi_n$ are pairwise independent with mean 0, then*

$$\mathbb{E}\|S_n\|^2 = \sum_{i=1}^{n} \mathbb{E}\|\chi_i\|^2.$$

Proof: Let $\{e_k, k \geq 1\}$ be a CONS for $\mathbb{H}$. As the χ_i's are pairwise independent, for $i \neq j$, we have

$$\mathbb{E}\langle \chi_i, e_k\rangle\langle \chi_j, e_k\rangle = \mathbb{E}\langle \chi_i, e_k\rangle\mathbb{E}\langle \chi_j, e_k\rangle = \langle \mathbb{E}\chi_i, e_k\rangle\langle \mathbb{E}\chi_j, e_k\rangle = 0.$$

It then follows that

$$\begin{aligned}\mathbb{E}\|S_n\|^2 &= \sum_{k=1}^{\infty} \mathbb{E}\langle S_n, e_k\rangle^2 = \sum_{k=1}^{\infty}\sum_{i=1}^{n} \mathbb{E}\langle \chi_i, e_k\rangle^2 \\ &= \sum_{i=1}^{n}\sum_{k=1}^{\infty} \mathbb{E}\langle \chi_i, e_k\rangle^2 = \sum_{i=1}^{n} \mathbb{E}\|\chi_i\|^2.\end{aligned}$$

□

Theorem 7.7.2 *Let $\chi_1, \chi_2, \ldots$ be pairwise independent and identically distributed with $\mathbb{E}\|\chi_1\| < \infty$. Then, $\lim_{n\to\infty} n^{-1}S_n = \mathbb{E}(\chi_1)$ a.s.*

Proof: We follow the line of proof in Etemadi (1983). First define

$$\chi_i' = \chi_i I_{\{\|\chi_i\| \leq i\}}$$

and

$$S_n' = \sum_{i=1}^{n} \chi_i'.$$

Then, take $k_n = [\alpha^n]$ for $\alpha > 1$, where $[a]$ denotes the integral part of a and let $\{e_k\}_{k=1}^{\infty}$ be a CONS for $\mathbb{H}$. Applying Theorem 7.2.2, Theorem 7.7.1, and Markov's inequality, we see that for any $\epsilon > 0$

$$\begin{aligned}\sum_{n=1}^{\infty} \mathbb{P}\left(\frac{\|S_{k_n}' - \mathbb{E}S_{k_n}'\|}{k_n} > \epsilon\right) &\leq \epsilon^{-2}\sum_{n=1}^{\infty} k_n^{-2}\sum_{i=1}^{k_n} \mathbb{E}\|\chi_i'\|^2 \\ &= \epsilon^{-2}\sum_{i=1}^{\infty} \mathbb{E}\|\chi_i'\|^2 \sum_{\{n:k_n \geq i\}} k_n^{-2} \\ &\leq 4(1-\alpha^{-2})^{-1}\epsilon^{-2}\sum_{i=1}^{\infty} \mathbb{E}\|\chi_i'\|^2 i^{-2},\end{aligned}$$

where the last inequality follows from a simple calculation in Durrett (1995, page 57). However,

$$\sum_{i=1}^{\infty} \mathbb{E}\|\chi_i'\|^2 i^{-2} = \sum_{i=1}^{\infty} i^{-2} \sum_{j=0}^{i-1} \mathbb{E}\left[\|\chi_1\|^2 I_{\{j \le \|\chi_1\| \le j+1\}}\right]$$

$$= \sum_{j=0}^{\infty} \mathbb{E}\left[\|\chi_1\|^2 I_{\{j \le \|\chi_1\| \le j+1\}}\right] \sum_{i=j+1}^{\infty} i^{-2}$$

$$\le C \sum_{j=0}^{\infty} (j+1)^{-1} \mathbb{E}\left[\|\chi_1\|^2 I_{\{j \le \|\chi_1\| \le j+1\}}\right]$$

for some $C < \infty$. Noticing that the last sum is bounded by $\mathbb{E}\|\chi_1\|$, we conclude from the Borel–Cantelli Lemma that

$$\lim_{n\to\infty} \frac{S'_{k_n} - \mathbb{E}S'_{k_n}}{k_n} = 0 \quad \text{a.s.}$$

By Lebesgue's dominated convergence theorem,

$$\lim_{n\to\infty} \|\mathbb{E}\chi_n' - \mathbb{E}\chi_1\| \le \lim_{n\to\infty} \mathbb{E}\|\chi_1\| I_{\{\|\chi_1\|>n\}} = 0.$$

Hence,

$$\lim_{n\to\infty} \left\| \frac{\mathbb{E}S'_{k_n}}{k_n} - \mathbb{E}\chi_1 \right\| \le \lim_{n\to\infty} \frac{1}{k_n} \sum_{i=1}^{k_n} \|\mathbb{E}\chi_i' - \mathbb{E}\chi_1\| = 0$$

and we have

$$\lim_{n\to\infty} \frac{S'_{k_n}}{k_n} = \mathbb{E}\chi_1 \quad \text{a.s.}$$

A standard argument using the Borel–Cantelli Lemma now shows that, with probability one, $X_n' = X_n$ eventually (i.e., X_n' and X_n are tail equivalent). Thus,

$$\lim_{n\to\infty} \frac{S_{k_n}}{k_n} = \mathbb{E}\chi_1 \quad \text{a.s.} \tag{7.46}$$

The remainder of the proof deviates slightly from the Etemadi arguments we have used to this point. We now observe that for any n there exists a positive integer $m(n)$ such that

$$k_{m(n)-1} = [\alpha^{m(n)-1}] < n \le [\alpha^{m(n)}] = k_{m(n)}.$$

It follows that

$$\left\| \frac{S_n}{n} - \frac{S_{k_{m(n)}}}{k_{m(n)}} \right\| = \left\| \frac{S_{k_{m(n)}}}{n} - \frac{S_{k_{m(n)}}}{k_{m(n)}} + \frac{S_n - S_{k_{m(n)}}}{n} \right\|$$

$$\leq \left(\frac{k_{m(n)}}{n} - 1 \right) \left\| \frac{S_{k_{m(n)}}}{k_{m(n)}} \right\| + \frac{1}{n} \sum_{i=n+1}^{k_{m(n)}} \|\chi_i\|$$

$$\leq (\alpha - 1) \left\| \frac{S_{k_{m(n)}}}{k_{m(n)}} \right\| + \frac{1}{n} \sum_{i=n+1}^{k_{m(n)}} \|\chi_i\|.$$

From (7.46),

$$\lim_{n\to\infty} \left\| \frac{S_{k_{m(n)}}}{k_{m(n)}} \right\| = \|\mathbb{E}\chi_1\| \leq \mathbb{E}\|\chi_1\| \quad \text{a.s.}$$

In addition, Etemadi's strong law of large numbers for real-valued random variables ensures that $k_{m(n)}^{-1} \sum_{i=1}^{k_{m(n)}} \|\chi_i\|$ and $n^{-1} \sum_{i=1}^{n} \|\chi_i\|$ tend to $\mathbb{E}\|\chi_1\|$ with probability one. Therefore,

$$\limsup_{n\to\infty} \frac{1}{n} \sum_{i=n+1}^{k_{m(n)}} \|\chi_i\|$$

$$= \limsup_{n\to\infty} \left(\frac{k_{m(n)}}{n} \frac{1}{k_{m(n)}} \sum_{i=1}^{k_{m(n)}} \|\chi_i\| - \frac{1}{n} \sum_{i=1}^{n} \|\chi_i\| \right)$$

$$\leq (\alpha - 1)\mathbb{E}\|\chi_1\| \quad \text{a.s.}$$

This gives

$$\limsup_{n\to\infty} \left\| \frac{S_n}{n} - \frac{S_{k_{m(n)}}}{k_{m(n)}} \right\| \leq 2(\alpha - 1)\mathbb{E}\|\chi_1\|$$

and the result follows as we let $\alpha \downarrow 1$. □

The remaining concept to consider is that of convergence in distribution or law. To explore that topic, we need a suitable notion of weak convergence of probability measures that we now develop.

Let $\mathbb{P}, \mathbb{P}_n, n \geq 1$ be probability measures on $(\mathbb{H}, \mathscr{B}(\mathbb{H}))$. We say that P_n converges weakly to P, denoted by $P_n \xrightarrow{w} P$, if

$$\int_{\mathbb{H}} f(x)dP_n(x) \to \int_{\mathbb{H}} f(x)dP(x)$$

for any bounded and continuous functions f on $\mathbb{H}$; for random elements χ and $\chi_n, n \geq 1$, we say that χ_n converges in distribution to χ if $\mathbb{P}\circ\chi_n^{-1} \stackrel{w}{\longrightarrow} \mathbb{P}\circ\chi^{-1}$ and denote this relationship by $\chi_n \stackrel{d}{\longrightarrow} \chi$.

A general treatment of weak convergence can be found in Billingsley (1999). Here, we introduce an approach that will be especially effective for our applications. The idea was adapted from de Acosta (1970) and Mas (2006).

First let us recall the definition of tightness.

Definition 7.7.3 *An arbitrary set of probability measures $\{\mu_\alpha\}_{\alpha\in I}$ on $(\mathbb{H}, \mathscr{B}(\mathbb{H}))$ is tight if for any $\epsilon > 0$ there exists a compact set W such that*

$$\inf_{\alpha\in I} \mu_\alpha(W) \geq 1 - \epsilon.$$

For any $S \subset \mathbb{H}$ and any $\epsilon > 0$, let

$$S^\epsilon = \{x \in \mathbb{H} : \inf\{\|x - z\| : z \in S\} \leq \epsilon\}. \tag{7.47}$$

Theorem 7.7.4 *Let $\{\mu_\alpha\}_{\alpha\in I}$ be a family of probability measures on $(\mathbb{H}, \mathscr{B}(\mathbb{H}))$. Assume that for each $\epsilon, \delta > 0$, there exists a finite subset $\{y_1, \ldots, y_k\} \subset \mathbb{H}$ such that*

1. $\inf_{\alpha\in I} \mu_\alpha(S^\epsilon) \geq 1 - \delta$, *where* $S := \text{span}\{y_1, \ldots, y_k\}$ *and*
2. $\inf_{\alpha\in I} \mu_\alpha(\{x \in \mathbb{H} : |\langle x, y_j\rangle| \leq r, j = 1, \ldots, k\}) \geq 1 - \delta$ *for some* $r > 0$.

Then, $\{\mu_\alpha\}_{\alpha\in I}$ is tight.

Proof: Let $\epsilon = \delta$ and pick $\{y_1, \ldots, y_k\} \subset \mathbb{H}$ and r to satisfy the two conditions of the theorem. Define

$$A_\epsilon = S^\epsilon \cap \{x \in \mathbb{H} : |\langle x, y_j\rangle| \leq r, j = 1, \ldots, k\}.$$

We first show that $A_\epsilon \subset W_\epsilon^\epsilon$ for some compact set W_ϵ, where the ϵ subscript has been used to emphasize that the sets depend on ϵ.

For $x, z \in \mathbb{H}$ write $x = x_1 + x_2$ and $z = z_1 + z_2$ with x_1 and z_1 the projections of x and z onto S. Then,

$$\langle x, z\rangle = \langle x_1, z_1\rangle + \langle x_2, z_2\rangle. \tag{7.48}$$

If $x \in A_\epsilon$, we know from the definition of S^ϵ that $\|x_2\| \leq \epsilon$ and, hence,

$$|\langle x_2, z_2\rangle| \leq \epsilon\|z_2\| \leq \epsilon\|z\|. \tag{7.49}$$

Without loss of generality, assume that $y_1, \ldots, y_k$ are orthonormal. Then,

$$|\langle x_1, z_1\rangle| = \left|\sum_{j=1}^{k} \langle x, y_j\rangle\langle z, y_j\rangle\right| \leq kr\|z\| \tag{7.50}$$

as each $|\langle x, y_j\rangle| \leq r$ by the choice of A_ϵ. From (7.48)–(7.50)

$$\|x\| = \sup\{|\langle x, z\rangle| : \|z\| \leq 1\} \leq \epsilon + kr =: \nu,$$

which shows that $A_\epsilon \subset B_\nu$, where $B_\nu = \{x \in \mathbb{H} : \|x\| \leq \nu\}$. Thus,

$$A_\epsilon \subset B_\nu \cap S^\epsilon. \tag{7.51}$$

Now, if $x \in B_\nu \cap S^\epsilon$, this means that $\|x\| \leq \nu$ and there exists a $y_j \in S$ such that $\|x - y_j\| \leq \epsilon$. So, from the triangle inequality,

$$\|y_j\| \leq \|x\| + \|x - y_j\| \leq \nu + \epsilon,$$

which implies that

$$\inf_{z \in S \cap B_{\nu+\epsilon}} \|x - z\| \leq \|x - y_j\| \leq \epsilon.$$

Thus, we conclude

$$B_\nu \cap S^\epsilon \subset (S \cap B_{\nu+\epsilon})^\epsilon. \tag{7.52}$$

In combination with (7.51), this produces

$$A_\epsilon \subset (S \cap B_{\nu+\epsilon})^\epsilon.$$

For notational convenience, we will use

$$W_\epsilon = S \cap B_{\nu+\epsilon}$$

in what follows. Note that as S is finite dimensional, W_ϵ is compact.

Conditions 1 and 2 ensure that $\inf_{\alpha \in I} \mu_\alpha(A_\epsilon) \geq 1 - 2\epsilon$. Consequently, we see that $\inf_{\alpha \in I} \mu_\alpha(W_\epsilon^\epsilon) \geq 1 - 2\epsilon$; more generally, for each $\epsilon > 0, j \geq 1$,

$$\inf_{\alpha \in} \mu_\alpha \left(W_{\epsilon/2^j}^{\epsilon/2^j}\right) \geq 1 - 2\epsilon/2^j$$

and the set $W := \cap_{j=2}^{\infty} W_{\epsilon/2^j}^{\epsilon/2^j}$ satisfies $\inf_{\alpha \in I} \mu_\alpha(W) \geq 1 - \epsilon$.

To conclude the proof, it remains to verify that W is compact. Now, W is obviously closed. Thus, we need only show it is totally bounded to be able to apply the Heine–Borel theorem. Specifically, we need to show that for any $\nu > 0$ there exists a finite set $F \subset W$ such that $W \subset F^{\nu}$. For this purpose, write

$$W = \cap_{j=1}^{\infty}(W_{\epsilon/2^j}^{\epsilon/2^j} \cap W) = \cap_{j=1}^{\infty}\{(W_{\epsilon/2^j} \cap W) \cup (W_{\epsilon/2^j}^{\epsilon/2^j} \cap W - W_{\epsilon/2^j} \cap W)\}.$$

Note that, for each j, $W_{\epsilon/2^j} \cap W$ is compact as it is a closed subset of a compact set. For any $\nu > 0$, pick a large enough j such that $\epsilon/2^j < \nu/2$. By the total boundedness of $W_{\epsilon/2^j} \cap W$, there exists a finite set $F \subset W_{\epsilon/2^j} \cap W$ such that $W_{\epsilon/2^j} \cap W \subset F^{\nu/2}$. So,

$$(W_{\epsilon/2^j} \cap W) \cup (W_{\epsilon/2^j}^{\epsilon/2^j} \cap W - W_{\epsilon/2^j} \cap W) \subset F^{\nu},$$

which implies that $W \subset F^{\nu}$ and completes the proof. □

Condition 1 of Theorem 7.7.4 is sometimes referred to as *flat concentration* in the literature. We will apply Theorem 7.7.4 in establishing weak convergence mainly through the following result.

Theorem 7.7.5 *Let $\chi, \chi_n, n \geq 1$, be random elements in $(\mathbb{H}, \mathscr{B}(\mathbb{H}))$. Assume that $\langle \chi_n, f\rangle \xrightarrow{d} \langle \chi, f\rangle$ in $\mathbb{R}$ for all $f \in \mathbb{H}$ and for each $\epsilon, \delta > 0$, there exists a finite-dimensional subspace S such that*

$$\inf_{n\geq 1} \mathbb{P}(\chi_n \in S^{\epsilon}) \geq 1 - \delta \tag{7.53}$$

for S^{ϵ} defined as in (7.47). Then, $\chi_n \xrightarrow{d} \chi$.

Proof: It is easy to show that the assumptions of Theorem 7.7.4 hold for the family of probability measures $\{\mathbb{P}\circ\chi_n^{-1} : n \geq 1\}$. Therefore, $\{\mathbb{P}\circ\chi_n^{-1} : n \geq 1\}$ is tight and, by Prohorov's Theorem (e.g., Billingsley 1995), it is relatively compact. Now, let $\{\mathbb{P}\circ\chi_{n'}^{-1}\}$ and $\{\mathbb{P}\circ\chi_{n''}^{-1}\}$ be two weakly convergent subsequences; for convenience, write $\chi_{n'} \xrightarrow{d} \tilde{\chi}$ and $\chi_{n''} \xrightarrow{d} \check{\chi}$ for some random elements $\tilde{\chi}$ and $\check{\chi}$ in $\mathbb{H}$. By the continuous mapping theorem, $\langle \chi_{n'}, f\rangle \xrightarrow{d} \langle \tilde{\chi}, f\rangle$ and $\langle \chi_{n''}, f\rangle \xrightarrow{d} \langle \check{\chi}, f\rangle$ for all $f \in \mathbb{H}$. However, it follows that

$$\langle \tilde{\chi}, f\rangle \stackrel{d}{=} \langle \check{\chi}, f\rangle \stackrel{d}{=} \langle \chi, f\rangle, \quad f \in \mathbb{H},$$

where $\stackrel{d}{=}$ indicates "equal in distribution." By Theorem 7.1.2 , $\chi \stackrel{d}{=} \tilde{\chi} \stackrel{d}{=} \check{\chi}$ and so both $\chi_{n'}$ and $\chi_{n''}$ converge in distribution to χ. This completes the proof. □

We conclude by proving an elementary central limit theorem for random elements of a Hilbert space.

Theorem 7.7.6 *Let $\chi_1, \chi_2, \ldots$ be independent and identically distributed random elements in $\mathbb{H}$ with mean 0 and $\mathbb{E}\|\chi_1\|^2 < \infty$. Then*

$$\xi_n := n^{-1/2} \sum_{i=1}^{n} \chi_i \xrightarrow{d} \xi,$$

where ξ is Gaussian random element of $\mathbb{H}$ with covariance operator equal to $\mathbb{E}(\chi_1 \otimes \chi_1)$.

Proof: We need only verify the two conditions in Theorem 7.7.5. First, by the central limit theorem for real-values random variables, for any $f \in \mathbb{H}$, the distribution of $\langle \xi_n, f \rangle$ converges to $N(0, \langle \mathbb{E}(\chi_1 \otimes \chi_1), f \rangle)$, which is the distribution of $\langle \xi, f \rangle$.

To show the second condition, let $\{e_j\}$ be any CONS for $\mathbb{H}$ and take S in Theorem 7.7.5 to be $S_J = \text{span}\{e_1, \ldots, e_J\}$. Let ξ_{nJ} and ξ'_{nJ} be the projections of ξ_n on S_J and $S_J^{\perp}$, respectively, and let χ'_{iJ} be the projection of χ_i on $S_J^{\perp}$. Then, for any $\epsilon > 0$,

$$\mathbb{P}(\|\xi'_{nJ}\| \leq \epsilon) = \mathbb{P}(\xi_n \in S_J^{\epsilon}).$$

By Chebyshev's inequality,

$$\mathbb{P}(\|\xi'_{nJ}\| > \epsilon) \leq \epsilon^{-2} \mathbb{E}(\|\chi'_{1J}\|^2),$$

which will be smaller than any δ for J sufficiently large. □

8

Mean and covariance estimation

In this chapter, we study mean and covariance estimation for both the random element and stochastic process settings described in Chapter 7. For the first perspective, let $\mathbb{H}$ be a separable Hilbert space and suppose that we have a random element χ of $\mathbb{H}$ with $\mathbb{E}\|\chi\|^2 < \infty$. The mean element and covariance operator of χ from Section 7.2 are therefore well defined as

$$m = \mathbb{E}\chi$$

and

$$\mathscr{K} = \mathbb{E}(\chi - m) \otimes (\chi - m).$$

In Section 8.1, we address the base problem of estimating m and $\mathscr{K}$ using a random sample $\chi_1, \ldots, \chi_n$ from χ. The estimators of choice are the natural parallels of the sample mean and covariance that are used for real-valued random variables.

Suppose, on the other hand, that $\{X(t) : t \in E\}$ is a second-order stochastic process. Then, the mean function

$$m(t) = \mathbb{E}X(t)$$

and covariance function

$$K(s, t) = \mathbb{E}(X(s) - m(s))(X(t) - m(t))$$

are well defined for all $s, t \in E$, and they are the basic quantities of interest in our inference problem. Sections 8.2 and 8.3 focus on two scenarios that

Theoretical Foundations of Functional Data Analysis, with an Introduction to Linear Operators, First Edition. Tailen Hsing and Randall Eubank.

are more realistic from an fda standpoint. In both instances, we deal with stochastic processes that take values in a Hilbert function space. However, the sample paths are now only assumed to be observed at discrete time points with the resulting values then distorted by additive random noise. For Section 8.2, we let $\mathbb{H} = \mathbb{L}^2(E, \mathscr{B}(E), \mu)$ and estimate m and K with local linear regression type smoothers. In Section 8.3, the choice is $\mathbb{H} = \mathbb{W}_q[0, 1]$ and estimation is by penalized least squares methods.

8.1 Sample mean and covariance operator

Let $\mathbb{H}$ be a separable Hilbert space and χ a random element of $\mathbb{H}$. Assume that we observe an iid sample $\chi_1, \ldots, \chi_n$ from χ. We will then estimate m and $\mathscr{K}$ by the sample mean

$$m_n = \frac{1}{n} \sum_{i=1}^{n} \chi_i$$

and sample covariance operator

$$\mathscr{K}_n = \frac{1}{n-1} \sum_{i=1}^{n} (\chi_i - m_n) \otimes (\chi_i - m_n). \tag{8.1}$$

It is easy to verify that m_n and $\mathscr{K}_n$ are random elements of $\mathbb{H}$ and $\mathfrak{B}_{HS}(\mathbb{H})$, respectively, and they are unbiased for their population counterparts. We now apply the results in Section 7.7 to derive the asymptotic properties of m_n and $\mathscr{K}_n$ as $n \to \infty$.

For the sample mean, we have the following result whose proof follows directly from Theorems 7.7.2 and 7.7.6.

Theorem 8.1.1 *If $\mathbb{E}\|\chi_1\| < \infty$ then $m_n \xrightarrow{a.s.} m$. If $\mathbb{E}\|\chi_1\|^2 < \infty$ then*

$$\sqrt{n}(m_n - m) \xrightarrow{d} \xi$$

in $\mathbb{H}$ where ξ is a Gaussian random element with mean zero and covariance operator $\mathscr{K}$.

The basic asymptotic properties of $\mathscr{K}_n$ are established as

Theorem 8.1.2 *If $\mathbb{E}\|\chi_1\|^2 < \infty$ then $\mathscr{K}_n \xrightarrow{a.s.} \mathscr{K}$. If $\mathbb{E}\|\chi_1\|^4 < \infty$ then*

$$\sqrt{n}(\mathscr{K}_n - \mathscr{K}) \xrightarrow{d} \mathfrak{Z}$$

in $\mathscr{B}_{HS}(\mathbb{H})$ where $\mathfrak{Z}$ is a Gaussian random element with mean zero and covariance operator

$$\mathbb{E}((\chi_1 - m) \otimes (\chi_1 - m) - \mathscr{K}) \otimes_{HS} ((\chi_1 - m) \otimes (\chi_1 - m) - \mathscr{K}).$$

Proof: Write

$$\mathscr{K}_n = \frac{1}{n-1} \sum_{i=1}^{n} (\chi_i - m) \otimes (\chi_i - m) - \frac{n}{n-1} (m_n - m) \otimes (m_n - m).$$

The first conclusion follows from Theorems 7.7.2 and 8.1.1. The second conclusion will follow from Theorem 7.7.6 once we verify that

$$\mathbb{E}\|(\chi_i - m) \otimes (\chi_i - m) - \mathscr{K}\|_{HS}^2 < \infty.$$

However, from Theorem 7.2.2,

$$\begin{aligned}\mathbb{E}\|(\chi_i - m) \otimes (\chi_i - m) - \mathscr{K}\|_{HS}^2 &\leq \mathbb{E}\|(\chi_i - m) \otimes (\chi_i - m)\|_{HS}^2 \\ &= \mathbb{E}\|\chi_i - m\|^4 < \infty\end{aligned}$$

and the proof is complete. □

To illustrate the implication of this result, let $\mathscr{G}$ be any element of $\mathfrak{B}_{HS}(\mathbb{H})$ and initially take $m = 0$. Then, the limiting distribution of $\langle \sqrt{n}(\mathscr{K}_n - \mathscr{K}), \mathscr{G} \rangle_{HS}$ is that of the random variable $\langle \mathfrak{Z}, \mathscr{G} \rangle_{HS}$. This quantity is normally distributed with mean zero and, from part 1 of Theorem 7.2.5,

$$\begin{aligned}\mathrm{Var}(\langle \mathfrak{Z}, \mathscr{G} \rangle_{HS}) &= \mathbb{E}\langle \mathfrak{Z}, \mathscr{G} \rangle_{HS}^2 \\ &= \langle \mathbb{E}(\chi_1 \otimes \chi_1 - \mathscr{K}) \otimes_{HS} (\chi_1 \otimes \chi_1 - \mathscr{K}) \mathscr{G}, \mathscr{G} \rangle_{HS} \\ &= \mathbb{E}\langle (\chi_1 \otimes \chi_1 - \mathscr{K}), \mathscr{G} \rangle_{HS}^2 \\ &= \mathrm{Var}(\langle \chi_1 \otimes \chi_1, \mathscr{G} \rangle_{HS}).\end{aligned}$$

We can evaluate the HS inner product in this last expression directly. For that purpose, we choose any CONS $\{e_j\}$ for $\mathbb{H}$ whose first element in $e_1 = \chi_1/\|\chi_1\|$. Then, $(\chi_1 \otimes \chi_1)e_j = 0$ for all $j > 1$ in which case Definition 4.4.4 for the HS inner product produces

$$\begin{aligned}\langle \chi_1 \otimes \chi_1, \mathscr{G} \rangle_{HS} &= \langle (\chi_1 \otimes \chi_1)e_1, \mathscr{G} e_1 \rangle \\ &= \langle \chi_1, \mathscr{G} \chi_1 \rangle.\end{aligned}$$

So,

$$\mathrm{Var}(\langle \mathfrak{Z}, \mathscr{G} \rangle_{HS}) = \mathrm{Var}(\langle \chi_1, \mathscr{G} \chi_1 \rangle).$$

When $m \neq 0$, this becomes

$$\mathrm{Var}(\langle \mathfrak{Z}, \mathscr{G} \rangle_{HS}) = \mathrm{Var}(\langle (\chi_1 - m), \mathscr{G}(\chi_1 - m) \rangle).$$

8.2 Local linear estimation

We assume in this section that X is a stochastic process on $E = [0, 1]$ where the mean function $m(t)$ and covariance function $K(s, t)$ are both smooth. For convenience, we also assume that X can be viewed as a random element of $\mathbb{L}^2[0, 1]$. Thus, we are in the setting described in Section 7.4.

Our attention is then directed toward the use of local linear regression estimators along the lines of those in Fan and Gijbels (1996) for estimation of m and K. The basic nonparametric regression formulation that underlies such estimators can be employed directly to construct estimators for the mean function but requires a slight adjustment for estimation of covariances as we subsequently demonstrate.

Suppose that $X_1, \ldots, X_n$ are independent copies of a X. Consider the model

$$Y_{ij} = X_i(T_{ij}) + \varepsilon_{ij}, \quad j = 1, \ldots, r, i = 1, \ldots, n, \tag{8.2}$$

where the T_{ij}'s are iid sampling points and the ε_{ij} are iid errors. For data of the form (8.2), a local linear smoothing based estimator of $m(t)$ is obtained by minimization of

$$\sum_{i=1}^{n} \sum_{j=1}^{r} W_h(T_{ij} - t)(Y_{ij} - a_0 - a_1(T_{ij} - t))^2 \tag{8.3}$$

with respect to a_0, a_1. This is just a weighted sum of squared residuals corresponding to a simple linear regression fit. What makes it special and worthwhile for our situation is how the weights are chosen. This is accomplished with the kernel weight function W and associated bandwidth $h > 0$ via the relation

$$W_h(u) = h^{-1} W\left(\frac{u}{h}\right). \tag{8.4}$$

Here, W is a symmetric probability density function on $[-1, 1]$, which makes h a scale parameter that governs how concentrated the associated weights $W_h(T_{ij} - t)$ are around t. We also assume that W is of bounded variation and satisfies the moment conditions

$$\int_{-1}^{1} u^j W(u) du = \begin{cases} 1, & j = 0, \\ 0, & j = 1, \\ C \neq 0, & j = 2. \end{cases} \tag{8.5}$$

When the bandwidth is small, criterion (8.3) uses only those responses whose time ordinates are in the immediate neighborhood of the point of estimation t. First-order Taylor expansions of the $m(T_{ij})$ around t then suggest that we estimate $m(t)$ by the minimizing intercept term, $\hat{a}_0$, of (8.3); i.e., the estimator we will consider is

$$m_h(t) = \frac{S_0(t)M_2(t) - S_1(t)M_1(t)}{M_0(t)M_2(t) - M_1^2(t)}, \tag{8.6}$$

with $M_p, S_p, p = 0, 1, 2$, defined by

$$M_p(t) = \frac{1}{nr} \sum_{i=1}^{n} \sum_{j=1}^{r} W_{hp}(T_{ij} - t), \tag{8.7}$$

$$S_p(t) = \frac{1}{nr} \sum_{i=1}^{n} \sum_{j=1}^{r} W_{hp}(T_{ij} - t)Y_{ij} \tag{8.8}$$

for

$$W_{hp}(u) = \left(\frac{u}{h}\right)^p W_h(u). \tag{8.9}$$

In the spirit of typical developments in the nonparametric smoothing literature, we will analyze the behavior of

$$\sup_{t\in[0,1]} |m_h(t) - m(t)|$$

as $n \to \infty$ and h tends to zero at a rate that is some function of the sample size. What distinguishes the calculations performed here from those that arise in classical nonparametric regression problems is the necessity of dealing with an additional random (and correlated) component from our stochastic process formulation and the presence of replicates.

We now spell out the conditions that are required for our analysis. First, the T_{ij} are assumed to be a random sample from a continuous random variable T with positive density function $f(\cdot)$ having a derivative in $C[0, 1]$. We should perhaps mention at this point that the didactic assumption of an equal number of sampling points for every sample path can be relaxed (Li and Hsing, 2010). The ε_{ij} are also independent and identically distributed with $\mathbb{E}\varepsilon_{11} = 0$, $\text{Var}(\varepsilon_{11}) = \sigma^2 < \infty$ and, collectively, the X_i, T_{ij} and ε_{ij} are independent of one another. Finally, we will need

$$\mathbb{E}|\varepsilon_{11}|^q < \infty \tag{8.10}$$

and

$$\mathbb{E} \sup_{t\in[0,1]} |X(t)|^q < \infty \tag{8.11}$$

for some positive q falling in a range that will be made explicit in the following. Moment conditions such as (8.11) hold rather generally. In particular, they are satisfied for Gaussian processes with continuous sample paths (cf. Landau and Shepp 1970). To simplify subsequent notation, we will use the definitions

$$\delta_{n1}(h) = (\{1 + (hr)^{-1}\} \log n/n)^{1/2},$$
$$\delta_{n2}(h) = (\{1 + (hr)^{-1} + (hr)^{-2}\} \log n/n)^{1/2}$$

when stating our major results in the following.

Under the above-mentioned conditions, we are able to establish

Theorem 8.2.1 *Assume that m has a uniformly bounded second derivative on* $[0, 1]$, *(8.10) and (8.11) hold for some* $q \in (2, \infty)$ *and that* $h \to 0$ *as* $n \to \infty$ *in such a way that* $(h^2 + h/r)^{-1}(\log n/n)^{1-2/q} \to 0$. *Then,*

$$\sup_{t\in[0,1]} |m_h(t) - m(t)| = O(h^2 + \delta_{n1}(h)) \quad a.s. \tag{8.12}$$

Proof: Straightforward calculations give us

$$m_h(t) - m(t) = \frac{\tilde{S}_0(t)M_2(t) - \tilde{S}_1(t)M_1(t)}{M_0(t)M_2(t) - M_1^2(t)}$$

with

$$\tilde{S}_p(t) = S_p(t) - m(t)M_p(t) - hm'(t)M_{p+1}(t).$$

To establish the result, we will proceed by obtaining almost sure, uniform approximations for the $\tilde{S}_p(t)$ and $M_p(t)$ with the aid of Lemma 8.2.2 in the following text.

First, let us consider M_p. As the T_{ij} are iid and the support of W is $[-1, 1]$, uniformly for all t

$$\begin{aligned}\mathbb{E}M_p(t) &= \int_0^1 W_{hp}(s - t)f(s)ds \\ &= \int_{-t/h}^{(1-t)/h} u^p W(u)f(t + hu)du \\ &= (f(t) + O(h)) \int_{\max(-1,-t/h)}^{\min(1,(1-t)/h)} u^p W(u)du.\end{aligned}$$

Thus, for $p = 0, 1, 2$, $\mathbb{E}M_p(t)$ is uniformly bounded away from ∞, whereas for $p = 0, 2$, $\mathbb{E}M_p(t)$ is also uniformly bounded away from 0 as $h \to 0$. Applying Lemma 8.2.2 with $Z_{ij} \equiv 1$ then enables us to conclude that these statements hold for $M_p(t)$ in place of $\mathbb{E}[M_p(t)]$ with probability one. Thus, the rate of convergence for $m_h(t) - m(t)$ is determined by those of the $\tilde{S}_p$.

Observe that

$$\begin{aligned}\tilde{S}_p(t) &= \frac{1}{nr}\sum_{i=1}^{n}\sum_{j=1}^{r} W_{hp}(T_{ij}-t)(Y_{ij}-m(t)-m'(t)(T_{ij}-t)) \\ &= U_p(t) + \frac{1}{nr}\sum_{i=1}^{n}\sum_{j=1}^{r} W_{hp}(T_{ij}-t)[m(T_{ij})-m(t)-m'(t)(T_{ij}-t)]\end{aligned}$$

with

$$U_p(t) = \frac{1}{nr}\sum_{i=1}^{n}\sum_{j=1}^{r} W_{hp}(T_{ij}-t)[\varepsilon_{ij}+X_i(T_{ij})-m(T_{ij})].$$

Taylor's Theorem and the assumption that m'' is uniformly bounded then lead to

$$\begin{aligned}&\frac{1}{nr}\sum_{i=1}^{n}\sum_{j=1}^{r} W_{hp}(T_{ij}-t)[m(T_{ij})-m(t)-m'(t)(T_{ij}-t)] \\ &= \frac{1}{nr}\sum_{i=1}^{n}\sum_{j=1}^{r} W_{hp}(T_{ij}-t)O((T_{ij}-t)^2) \\ &= M_p(t)O(h^2) \quad \text{a.s.} \\ &= O(h^2) \quad \text{a.s.}\end{aligned}$$

as a result of what we previously established for M_p. Thus, with probability one,

$$\tilde{S}_p(t) = U_p(t) + O(h^2).$$

Finally, express $U_p(t)$ as the sum of two zero mean process; i.e., write

$$U_p(t) = U_{1p}(t) + U_{2p}(t)$$

with

$$U_{1p}(t) = \frac{1}{nr}\sum_{i=1}^{n}\sum_{j=1}^{r} W_{hp}(T_{ij}-t)\varepsilon_{ij}$$

and

$$U_{2p}(t) = \frac{1}{nr}\sum_{i=1}^{n}\sum_{j=1}^{r} W_{hp}(T_{ij}-t)[X_i(t_{ij})-m(T_{ij})].$$

An application of Lemma 8.2.2 to each of the U_{ip}, $i = 1, 2$ verifies (8.12) and concludes the proof. □

Lemma 8.2.2 *Let $Z_{ij}, 1 \leq i \leq n, 1 \leq j \leq r,$ be real-valued random variables satisfying*

$$\sup_{i,j} \mathbb{E}|Z_{ij}|^q < \infty \tag{8.13}$$

and

$$\frac{1}{nr}\sum_{i=1}^{n}\sum_{j=1}^{r}|Z_{ij}|^q = O(1) \quad a.s. \tag{8.14}$$

for some $q \in (2, \infty)$. Let $T_{ij}, 1 \leq i \leq n, 1 \leq j \leq r,$ be independent random variables taking values in $[0, 1]$ with probability density functions that are uniformly bounded. Assume the sets of random variables $\{Z_{ij}, T_{ij}, 1 \leq j \leq r\}, i = 1, \ldots, n,$ are mutually independent and that there is a universal constant C such that

$$\sup_{i,j,k} \mathbb{E}[|Z_{ij}Z_{ik}||T_{ij}, T_{ik}] < C \quad a.s. \tag{8.15}$$

Define

$$\overline{Z}_p(t) = \frac{1}{nr}\sum_{i=1}^{n}\sum_{j=1}^{r} W_{hp}(T_{ij} - t)Z_{ij}.$$

Set $\beta_n = h^2 + h/r$ and assume that $h \to 0$ in such a way that $\beta_n^{-1}(\log n/n)^{1-2/q} = o(1)$. Then,

$$\sup_{t\in[0,1]} \sqrt{nh^2/(\beta_n \log n)}|\overline{Z}_p(t) - \mathbb{E}\overline{Z}_p(t)| = O(1) \quad a.s. \tag{8.16}$$

Proof: As both $W(u)$ and u^p are of bounded variation, $W_{hp}(u)$ is also of bounded variation. As a result, $W_{hp}(u) = W_{hp,1}(u) - W_{hp,2}(u)$ for increasing functions $W_{hp,1}(u)$ and $W_{hp,2}(u)$; without loss of generality, assume that $W_{hp,1}(-h) = W_{hp,2}(-h) = 0$ and write

$$\begin{aligned}
\overline{Z}_p(t) &= \frac{1}{n}\sum_{i=1}^{n}\sum_{j=1}^{r} W_{hp}(T_{ij} - t)Z_{ij} \\
&= \frac{1}{nr}\sum_{i=1}^{n}\sum_{j=1}^{r} Z_{ij}I(T_{ij} - t \leq h)\int_{-h}^{T_{ij}-t} dW_{hp}(v) \\
&= \int_{-h}^{h}\frac{1}{nr}\sum_{i=1}^{n}\sum_{j=1}^{r} Z_{ij}I(v \leq T_{ij} - t \leq h)dW_{hp}(v) \\
&= \int_{-h}^{h} G_n(t + v, t + h)dW_{hp}(v),
\end{aligned}$$

where

$$G_n(t_1, t_2) = \frac{1}{nr} \sum_{i=1}^{n} \sum_{j=1}^{r} Z_{ij} I(T_{ij} \in [t_1 \wedge t_2, t_1 \vee t_2]). \tag{8.17}$$

Define

$$V_n(t, h) = \sup_{|u| \le c} |G_n(t, t+u) - G(t, t+u)|$$

for $G(t_1, t_2) = \mathbb{E}\{G_n(t_1, t_2)\}$ and observe that

$$\begin{aligned} \sup_{t \in [0,1]} |\overline{Z}_p(t)(t) - \mathbb{E}\overline{Z}_p(t)| &\le \sup_{t \in [0,1]} V_n(t, 2h) \int_{-h}^{h} |dW_{hp}| \\ &\le \sup_{t \in [0,1]} V_n(t, 2h)\{W_{hp,1}(h) + W_{hp,2}(h)\}. \end{aligned}$$

Now, $W_{hp,1}(h) + W_{hp,2}(h) = O(h^{-1})$ by the definition of W_{hp}, and we will show below that

$$\sup_{t \in [0,1]} V_n(t, h) = O(\{\beta_n \log n/n\}^{1/2}) \quad \text{a.s.} \tag{8.18}$$

In combination, this gives us (8.16).

In proving (8.18), we can obviously treat the positive and negative parts of Z_{ij} separately and will therefore assume in the following that Z_{ij} is non-negative. Without loss of generality, assume that $1/h$ is an integer. Define an equally spaced grid $\mathcal{G} := \{v_k\}$, with $v_k = kh$ for $k = 0, \dots, 1/h$. For any $t \in [0, 1]$ and $|u| \le h$, one may now find a grid point v_k that is within h of both t and $t + u$. As

$$\begin{aligned} |G_n(t, t+u) - G(t, t+u)| \le\ & |G_n(v_k, t+u) - G(v_k, t+u)| \\ & + |G_n(v_k, t) - G(v_k, t)|, \end{aligned}$$

we can write

$$|G_n(t, t+u) - G(t, t+u)| \le 2 \sup_{t \in \mathcal{G}} V_n(t, h).$$

Thus,

$$\sup_{t \in [0,1]} V_n(t, h) \le 2 \sup_{t \in \mathcal{G}} V_n(t, h). \tag{8.19}$$

From now on, we focus on the right-hand side of (8.19). Let

$$a_n = (\beta_n \log n/n)^{1/2}, \tag{8.20}$$

$$Q_n = \beta_n / a_n \tag{8.21}$$

and define $G_n^*(t_1, t_2)$, $G^*(t_1, t_2)$, and $V_n^*(t, h)$ in the same way that we defined $G_n(t_1, t_2)$, $G(t_1, t_2)$, and $V_n(t, h)$ except with $Z_{ij}I(Z_{ij} \leq Q_n)$ replacing Z_{ij} in the expressions. Then,

$$\sup_{t \in \mathcal{G}} V_n(t, h) \leq \sup_{t \in \mathcal{G}} V_n^*(t, h) + A_{n1} + A_{n2}, \tag{8.22}$$

where

$$\begin{aligned} A_{n1} &= \sup_{t \in \mathcal{G}} \sup_{|u| \leq h} (G_n(t, t+u) - G_n^*(t, t+u)), \\ A_{n2} &= \sup_{t \in \mathcal{G}} \sup_{|u| \leq h} (G(t, t+u) - G^*(t, t+u)). \end{aligned}$$

We first consider A_{n1} and A_{n2}. For all t and u,

$$\begin{aligned} (G_n(t, t+u) - G_n^*(t, t+u)) &\leq \frac{1}{nr} \sum_{i=1}^{n} \sum_{j=1}^{r} Z_{ij}^q Z_{ij}^{1-q} I(Z_{ij} > Q_n) \\ &\leq Q_n^{1-q} \frac{1}{nr} \sum_{i=1}^{n} \sum_{j=1}^{r} Z_{ij}^q. \end{aligned}$$

It follows that

$$a_n^{-1} Q_n^{1-q} = \{\beta_n^{-1} (\log n/n)^{1-2/q}\}^{q/2} = o(1). \tag{8.23}$$

So, from (8.13) and (8.14) and (8.23), we conclude that

$$a_n^{-1} A_{n1} \overset{\text{a.s}}{\to} 0.$$

Similarly, $a_n^{-1} A_{n2} = 0$ and we have proved

$$A_{n1} + A_{n2} = o(a_n) \quad \text{a.s.} \tag{8.24}$$

To bound $V_n^*(t, h)$ for a fixed $t \in \mathcal{G}$, we perform a further partition of the interval $[t-h, t+h]$. Define $w_n = [Q_n h/a_n + 1]$ and $u_\ell = \ell h/w_n$, for $\ell = -w_n, -w_n + 1, \ldots, w_n$. Note that $w_n \to \infty$ as $a_n/(Q_n h) = h^{-1} \log n/n \leq \beta_n^{-1} \log n/n \to 0$. Now pick any u in $[-h, h]$. There is an ℓ such that $u_\ell \leq u \leq u_{\ell+1}$. Note that we have either $0 \leq u_\ell \leq u \leq u_{\ell+1}$ or $u_\ell \leq u \leq u_{\ell+1} \leq 0$. So, consider the former case as the other one can be treated similarly. Using the monotonicity of $G_n^*(t, t+u)$ in $|u|$, as $Z_{ij} \geq 0$, we obtain

$$\begin{aligned} G_n^*(t, t+u_\ell) - G^*(t, t+u_{\ell+1}) &\leq G_n^*(t, t+u) - G^*(t, t+u) \\ &\leq G_n^*(t, t+u_{\ell+1}) - G^*(t, t+u_\ell). \end{aligned}$$

The left-hand side can be written as

$$\begin{aligned}
&G_n^*(t, t+u_\ell) - G^*(t, t+u_\ell) + G^*(t, t+u_\ell) - G^*(t, t+u_{\ell+1}) \\
&\quad = G_n^*(t, t+u_\ell) - G^*(t, t+u_\ell) - G^*(t+u_\ell, t+u_{\ell+1}),
\end{aligned}$$

and, similarly, the right-hand side can be written as

$$\begin{aligned}
&G_n^*(t, t+u_{\ell+1}) - G^*(t, t+u_{\ell+1}) + G^*(t, t+u_{\ell+1}) - G^*(t, t+u_\ell) \\
&\quad = G_n^*(t, t+u_{\ell+1}) - G^*(t, t+u_{\ell+1}) + G^*(t+u_\ell, t+u_{\ell+1}).
\end{aligned}$$

From these relations, we conclude that

$$|G_n^*(t, t+u) - G^*(t, t+u)| \le \max(\xi_{n\ell}, \xi_{n,\ell+1}) + G^*(t+u_\ell, t+u_{\ell+1}),$$

where

$$\xi_{n\ell} = |G_n^*(t, t+u_\ell) - G^*(t, t+u_\ell)|.$$

Thus,

$$V_n^*(t, h) \le \max_{-w_n \le \ell \le w_n} \xi_{n\ell} + \max_{-w_n \le \ell \le w_n} G^*(t+u_\ell, t+u_{\ell+1}).$$

For all ℓ,

$$\begin{aligned}
&G^*(t+u_\ell, t+u_{\ell+1}) \\
&\quad \le Q_n \frac{1}{nr} \sum_{i=1}^{n} \sum_{j=1}^{r} \mathbb{P}(t+u_\ell \le T_{ij} \le t+u_{\ell+1}) \\
&\quad \le M_T Q_n (u_{\ell+1} - u_\ell) \le M_T a_n,
\end{aligned}$$

where M_T is the maximum of the densities of the T_{ij}'s. Therefore, for any B,

$$\mathbb{P}\{V_n^*(t, h) \ge B a_n\} \le \mathbb{P}\{\max_{-w_n \le \ell \le w_n} \xi_{n\ell} \ge (B - M_T) a_n\}. \qquad (8.25)$$

We now proceed to develop an upper bound for $\mathbb{P}\{\xi_{n\ell} \ge (B - M_T) a_n\}$. Write

$$\xi_{n\ell} = \left| \frac{1}{n} \sum_{i=1}^{n} \{Z_i - \mathbb{E}(Z_i)\} \right| \le \frac{1}{n} \sum_{i=1}^{n} |Z_i - \mathbb{E}(Z_i)|,$$

where

$$Z_i = \frac{1}{r} \sum_{j=1}^{r} Z_{ij} I(Z_{ij} \le Q_n) I(T_{ij} \in (t, t+u_\ell]).$$

It follows that

$$\begin{aligned}\mathrm{Var}(Z_i) &\le \frac{1}{r^2}\sum_{j=1}^{r}\sum_{k=1}^{r}\mathbb{E}\left(\mathbb{E}[Z_{ij}Z_{ik}|T_{ij}T_{ik}]I(T_{ij},T_{ik}\in(t,t+u_\ell])\right)\\ &\le C\frac{1}{r^2}\sum_{j=1}^{r}\sum_{k=1}^{r}\mathbb{P}(T_{ij},T_{ik}\in(t,t+u_\ell]))\end{aligned}$$

due to (8.15), where

$$\mathbb{P}(T_{ij},T_{ik}\in(t,t+u_\ell]))=\begin{cases}\mathbb{P}(T_{ij}\in(t,t+u_\ell])), & j=k,\\ \mathbb{P}(T_{ij}\in(t,t+u_\ell]))\mathbb{P}(T_{ik}\in(t,t+u_\ell])), & j\ne k.\end{cases}$$

As the densities of the T_{ij} are uniformly bounded, our derivations have established the existence of a universal constant M such that

$$\mathrm{Var}(Z_i)\le M(u_\ell^2+u_\ell/r)\le M\beta_n.$$

Bernstein's inequality, the fact that $|Z_i-\mathbb{E}(Z_i)|\le Q_n$ and (8.20) and (8.21) now imply that for $B>M_T$

$$\begin{aligned}\mathbb{P}(\xi_{nr}\ge(B-M_T)a_n) &\le \exp\left\{-\frac{(B-M_T)^2n^2a_n^2}{2\sum_{i=1}^{n}\mathrm{Var}(Z_i)+(2/3)(B-M_T)Q_nna_n}\right\}\\ &\le \exp\left\{-\frac{(B-M_T)^2n^2a_n^2}{2Mn\beta_n+(2/3)(B-M_T)n\beta_n}\right\}\\ &= \exp\left\{-B^*na_n^2/\beta_n\right\}\\ &= n^{-B^*},\end{aligned}$$

where $B^*=\frac{(B-M_T)^2}{2M+(2/3)(B-M_T)}$. Then, using (8.25) along with Boole's inequality

$$\mathbb{P}\left(\sup_{t\in\mathcal{G}}V_n^*(t,h)\ge Ba_n\right)\le h^{-1}\left(2\left[\frac{Q_nh}{a_n}+1\right]+1\right)n^{-B^*}\le C\frac{Q_n}{a_n}n^{-B^*}$$

for some finite C. Observe that $Q_n/a_n=\beta_n/a_n^2=n/\log n$. So, select B large enough that $B^*>2$ to obtain

$$\sum_{n=1}^{\infty}\mathbb{P}(\sup_{t\in\mathcal{G}}V_n^*(t,h)\ge Ba_n)<\infty.$$

We therefore conclude from the Borel–Cantelli lemma that

$$\sup_{t\in\mathcal{G}}V_n^*(t,h)=O(a_n)\quad\text{a.s.}\tag{8.26}$$

Hence, (8.18) follows from combining (8.19), (8.22), (8.24), and (8.26). □

There are two scenarios for model (8.2) that have received special attention in the literature. The so-called sparse case has r uniformly bounded with only n divergent. In contrast, for the "dense" case, both the number of sampled processes and the number of sampling points are allowed to grow large. Thus, a spectrum of rates are possible depending on the nature of r. The following corollary addresses two special instances.

To simplify subsequent presentations, we adopt the notation $a_n \lesssim b_n$ if $a_n = O(b_n)$, $a_n \gtrsim b_n$ if $b_n = O(a_n)$, and $a_n \asymp b_n$ if $a_n = O(b_n)$ and $b_n = O(a_n)$.

Corollary 8.2.3 *Under the conditions of Theorem 8.2.1,*

1. *if r is bounded*

$$\sup_{t\in[0,1]} |m_h(t) - m(t)| = O(h^2 + [(\log n/(nh)]^{1/2}) \quad a.s.$$

2. *if $r = r_n$ is such that $r_n^{-1} \lesssim h \lesssim (\log n/n)^{1/4}$,*

$$\sup_{t\in[0,1]} |m_h(t) - m(t)| = O([(\log n)/n]^{1/2}) \quad a.s.$$

Proof: For the first statement of the corollary, $(1 + (hr)^{-1}) = O(h^{-1})$ as $n \to \infty$ as r is bounded. In the second case, $(1 + (hr)^{-1}) = O(1)$ as $n \in \infty$. The condition $h \lesssim (\log n/n)^{1/4}$ then insures that the bias or h^2 term does not dominate. □

The rate in part 1 of the corollary for sparse functional data is the classical nonparametric rate for estimating a univariate function; see. e.g., Stone (1982). The second part of the result indicates that if $r \gtrsim n^{1/4}$ convergence occurs at a parametric rate.

Next, we consider estimation of the covariance function $K(s, t)$. To do that, we can proceed similarly to estimation of the mean function. First, we estimate $R(s, t) := \mathbb{E}[X(s)X(t)]$ by $R_h(s, t) = a_0$, where

$$(\hat{a}_0, \hat{a}_1, \hat{a}_2) = \arg\min_{a_0,a_1,a_2} \frac{1}{n} \sum_{i=1}^{n} \Bigg[\frac{1}{r(r-1)} \sum\sum_{\substack{1\le j,k\le r \\ j\ne k}} (y_{ij}y_{ik} - a_0 - a_1(T_{ij} - s) - a_2(T_{ik} - t))^2 W_h(T_{ij} - s)W_h(T_{ik} - t)\Bigg].$$

This produces

$$R_h(s, t) = (\mathcal{M}_1(s, t)S_{00}(s, t) - \mathcal{M}_2(s, t)S_{10}(s, t) - \mathcal{M}_3(s, t)S_{01})/\mathcal{D}(s, t),$$

where

$$\begin{aligned}
\mathscr{M}_1(s,t) &= M_{20}(s,t)M_{02}(s,t) - M_{11}^2(s,t),\\
\mathscr{M}_2(s,t) &= M_{10}(s,t)M_{02}(s,t) - M_{01}(s,t)M_{11}(s,t),\\
\mathscr{M}_3(s,t) &= M_{01}(s,t)M_{20}(s,t) - M_{10}(s,t)M_{11}(s,t),\\
\mathscr{D}(s,t) &= \mathscr{M}_1(s,t)M_{00}(s,t) - \mathscr{M}_2(s,t)M_{10}(s,t) - \mathscr{M}_3(s,t)M_{01}(s,t)
\end{aligned}$$

for

$$M_{p_1p_2}(s,t) = \frac{1}{nr(r-1)} \sum_{i=1}^{n} \sum_{\substack{1\le j,k\le r\\ j\ne k}} W_{hp_1}(T_{ij}-s)W_{hp_2}(T_{ik}-t)$$

and

$$S_{p_1p_2}(s,t) = \frac{1}{nr(r-1)} \sum_{i=1}^{n} \sum_{\substack{1\le j,k\le r\\ j\ne k}} W_{hp_1}(T_{ij}-s)W_{hp_2}(T_{ik}-t)Y_{ij}Y_{ik}.$$

We then estimate $K(s,t)$ by

$$K_h(s,t) = R_{h_R}(s,t) - m_{h_m}(s)m_{h_m}(t)$$

with $h=(h_m,h_R)$, a vector containing the two bandwidths h_m, h_R that are used for estimation of R and m, respectively.

The following result gives the convergence rates for $K_h(s,t)$.

Theorem 8.2.4 *Assume that all second-order partial derivatives of $K(s,t)$ exist and are bounded on $[0,1]^2$ and that (8.10) and (8.11) hold for some $q \in (4,\infty)$. If $h\to 0$ as $n\to\infty$ in such a way that $(h_m^2 + h_m/r)^{-1}(\log n/n)^{1-2/q} \to 0$ and $(h_R^4 + h_R^3/r + (h/r)^2)^{-1}(\log n/n)^{1-4/q} \to 0$,*

$$\sup_{s,t\in[0,1]} |K_h(s,t) - K(s,t)| = O(h_m^2 + h_R^2 + \delta_{n1}(h_m) + \delta_{n2}(h_R)) \quad a.s.$$

Proof: Define

$$\begin{aligned}
\tilde{S}_{p_1p_2}(s,t) = S_{p_1p_2}(s,t) - R(s,t)M_{p_1p_2} - h_R R^{(1,0)}(s,t)M_{(p_1+1)p_2}\\
-h_R R^{(0,1)}(s,t)M_{p_1(p_2+1)}(s,t)
\end{aligned}$$

and observe that

$$\begin{aligned}
&R_{h_R}(s,t) - R(s,t)\\
&\quad = \frac{(\mathscr{M}_1(s,t)\tilde{S}_{00}(s,t) - \mathscr{M}_2(s,t)\tilde{S}_{10}(s,t) - \mathscr{M}_3(s,t)\tilde{S}_{01}(s,t))}{\mathscr{D}(s,t)}. \qquad (8.27)
\end{aligned}$$

Note that cancellations occur that eliminate all the terms containing $R^{(1,0)}(s,t)$ and $R^{(0,1)}(s,t)$ from this last expression.

Uniformly for all s,t,

$$\mathbb{E}M_{p_1p_2}(s,t) = \{f(s)f(t) + O(h_R)\}I_{h_R,p_1,p_2}(s,t), \tag{8.28}$$

where

$$I_{h,p_1,p_2}(s,t) = \int_{\max(-1,-s/h)}^{\min(1,(1-s)/h)} \int_{\max(-1,-t/h)}^{\min(1,(1-t)/h)} u^{p_1}v^{p_2}W(u)W(v)dudv.$$

Upon applying Lemma 8.2.5 with $\kappa = q/2$ and Z_{ijk} equal to 1 and $Y_{ij}Y_{ik}$, we see that $\mathcal{M}_i(s,t), 1 \le i \le 3$ are uniformly bounded a.s. and $\mathcal{D}(s,t)$ is uniformly bounded away from 0 a.s. Thus, by (8.27), the rate for $R_{h_R}(s,t) - R(s,t)$ is determined from the rates for $\tilde{S}_{00}(s,t), \tilde{S}_{10}(s,t)$, and $\tilde{S}_{01}(s,t)$.

To analyze the behavior of the $\tilde{S}_{p_1q_2}(s,t)$, write

$$\tilde{S}_{p_1p_2}(s,t) = U_1(s,t) + U_2(s,t) + U_3(s,t) + U_4(s,t) \tag{8.29}$$

with

$$U_1(s,t) = \frac{1}{nr(r-1)} \sum_{i=1}^{n} \sum\sum_{\substack{1\le j,k\le r \\ j\ne k}} \varepsilon_{ij}\varepsilon_{ik} W_{h_Rp_1}(T_{ij}-s)W_{h_Rp_2}(T_{ik}-t),$$

$$U_2(s,t) = \frac{1}{nr(r-1)} \sum_{i=1}^{n} \sum\sum_{\substack{1\le j,k\le r \\ j\ne k}} \varepsilon_{ij}X_i(T_{ik}) W_{h_Rp_1}(T_{ij}-s)W_{h_Rp_2}(T_{ik}-t),$$

$$\begin{aligned} U_3(s,t) = \frac{1}{nr(r-1)} \sum_{i=1}^{n} \sum\sum_{\substack{1\le j,k\le r \\ j\ne k}} &[X_i(T_{ij})X_i(T_{ik}) - R(T_{ij},T_{ik})] \\ &\times W_{h_Rp_1}(T_{ij}-s)W_{h_Rp_2}(T_{ik}-t), \end{aligned}$$

$$\begin{aligned} U_4(s,t) = \frac{1}{nr(r-1)} \sum_{i=1}^{n} \sum\sum_{\substack{1\le j,k\le r \\ j\ne k}} &[R(T_{ij},T_{ik}) - R(s,t) - (T_{ij}-s)R^{(1,0)}(s,t) \\ &- (T_{ik}-t)R^{(0,1)}(s,t)] W_{h_Rp_1}(T_{ij}-s)W_{h_Rp_2}(T_{ik}-t). \end{aligned}$$

An application of Taylor's Theorem establishes that $U_4(s,t) = O(h_R^2)$ uniformly. The other U_i have zero means and Lemma 8.2.5 can be applied again to conclude the proof. □

Lemma 8.2.5 *Let Z_{ijk}, $1 \le i \le n, 1 \le j \ne k \le r$, be real-valued random variables satisfying*

$$\sup_{i,j,k} \mathbb{E}|Z_{ijk}|^{\kappa} < \infty \tag{8.30}$$

and

$$\frac{1}{nr^2} \sum_{i=1}^{n} \sum_{\substack{1\le j,k\le r \\ j\ne k}} |Z_{ijk}|^{\kappa} = O(1) \quad a.s. \tag{8.31}$$

for some $\kappa \in (2, \infty)$. Let T_{ij}, $1 \le i \le n, 1 \le j \le r$, be independent random variables with values in $[0, 1]$ with uniformly bounded probability density functions. Assume that there is a universal constant C such that

$$\sup_{i,j_1,k_1,j_2,k_2} \mathbb{E}[|Z_{ij_1k_1} Z_{ij_2k_2}||T_{ij_1}, T_{ik_1}, T_{ij_2}, T_{ik_2}] < C \quad a.s. \tag{8.32}$$

For any nonnegative integers p_1, p_2, define

$$\overline{Z}_{p_1p_2}(s,t) = \frac{1}{nr(r-1)} \sum_{i=1}^{n} \sum_{\substack{1\le j,k\le r \\ j\ne k}} Z_{ijk} W_{hp_1}(T_{ij} - s) W_{hp_2}(T_{ik} - t)$$

and let $\beta_n = h^4 + h^3/r + (h/r)^2$ be such that

$$\beta_n^{-1} (\log n/n)^{1-2/\kappa} = o(1)$$

as $h \to 0$. Then,

$$\sup_{s,t\in[0,1]} \sqrt{nh^4/(\beta_n \log n)} |\overline{Z}_{p_1p_2}(s,t) - \mathbb{E}\overline{Z}_{p_1p_2}(s,t)| = O(1) \quad a.s.$$

Proof: Write

$$\begin{aligned}
\overline{Z}_{p_1p_2}(s,t) &= \frac{1}{nr(r-1)} \sum_{i=1}^{n} \sum_{\substack{1\le j,k\le r \\ j\ne k}} Z_{ijk} I(T_{ij} \le s + h, T_{ik} \le t + h) \\
&\qquad \times W_{hp_1}(T_{ij} - s) W_{hp_2}(T_{ik} - t) \\
&= \int\int_{(u,v)\in[-h,h]^2} \frac{1}{nr(r-1)} \sum_{i=1}^{n} \sum_{\substack{1\le j,k\le r \\ j\ne k}} Z_{ijk} I(T_{ij} \in [s+u, s+h]) \\
&\qquad \times I(T_{ik} \in [t+v, t+h]) dW_{hp_1}(u) dW_{hp_2}(v) \\
&= \int\int_{(u,v)\in[-h,h]^2} G_n(s+u, t+v, s+h, t+h) dW_{hp_1}(u) dW_{hp_2}(v),
\end{aligned}$$

where

$$G_n(s_1, t_1, s_2, t_2)$$
$$= \frac{1}{nr(r-1)} \sum_{i=1}^{n} \sum \sum_{\substack{1 \le j,k \le r \\ j \ne k}} Z_{ijk} I(T_{ij} \in [s_1 \wedge s_2, s_1 \vee s_2], T_{ik} \in [t_1 \wedge t_2, t_1 \vee t_2])$$

and set $G(s_1, t_1, s_2, t_2) = \mathbb{E}\{G_n(s_1, t_1, s_2, t_2)\}$ with

$$V_n(s, t, h) = \sup_{|u_1|, |u_2| \le h} |G_n(s, t, s + u_1, t + u_2) - G(s, t, s + u_1, t + u_2)|.$$

Then,

$$\sup_{s,t \in [0,1]} |\overline{Z}_{p_1 p_2}(s, t) - \mathbb{E}\overline{Z}_{p_1 p_2}(s, t)|$$
$$\le \sup_{s,t \in [0,1]} V_n(s, t, 2h) \int\int_{(u,v) \in [-h,h]^2} |dW_{hp_1}(u)||dW_{hp_2}(v)|$$
$$= O(h^{-2}) \sup_{s,t \in [0,1]} V_n(s, t, h).$$

We will show below that

$$\sup_{s,t \in [0,1]} V_n(s, t, h) = O(\{\beta_n \log n / n\}^{1/2}) \quad \text{a.s.} \tag{8.33}$$

The proof of (8.33) is similar to that of (8.18) in Lemma 8.2.2. Accordingly, we only outline the unique aspects that arise for the covariance estimation case here.

Let a_n, Q_n be as in (8.20) and (8.21) and let $\mathscr{G}$ be the same grid defined in the proof of Lemma 8.2.2. Then we have

$$\sup_{s,t \in [0,1]} V_n(s, t, h) \le 4 \sup_{s,t \in \mathscr{G}} V_n(s, t, h). \tag{8.34}$$

Now define $G_n^*(s_1, t_1, s_2, t_2)$, $G^*(s_1, t_1, s_2, t_2)$, and $V_n^*(s, t, h)$ in the same way as we defined $G_n(s_1, t_1, s_2, t_2)$, $G(s_1, t_1, s_2, t_2)$, and $V_n(s, t, \delta)$ except now with Z_{ijk} being replaced by $Z_{ijk} I(Z_{ijk} \le Q_n)$ and observe that

$$\sup_{s,t \in \mathscr{G}} V_n(s, t, h) \le \sup_{s,t \in \mathscr{G}} V_n^*(s, t, h) + A_{n1} + A_{n2}, \tag{8.35}$$

where

$$A_{n1} = \sup_{s,t \in \mathscr{G}} \sup_{|u_1|, |u_2| \le h} |G_n(s, t, s + u_1, t + u_2) - G_n^*(s, t, s + u_1, t + u_2)|,$$
$$A_{n2} = \sup_{s,t \in \mathscr{G}} \sup_{|u_1|, |u_2| \le h} |G(s, t, s + u_1, t + u_2) - G^*(s, t, s + u_1, t + u_2)|.$$

Using arguments similar to those in the proof of Lemma 8.2.2 with relations (8.30) and (8.31), we can show that both A_{n1} and A_{n2} are $o(a_n)$ almost surely. To bound $V_n^*(s,t,h)$ for fixed (s,t), we create a further partition of the interval $[s-h, s+h] \times [t-h, t+h]$. Put $w_n = [Q_n h^2/a_n + 1]$ and $u_\ell = \ell h/w_n, \ell = -w_n, \ldots, w_n$. Clearly, $w_n \to \infty$ as

$$a_n/(Q_n h^2) = h^{-2} \log n/n \le \beta_n^{-1} \log n/n \to 0.$$

Then,

$$\begin{aligned} &V_n^*(s,t,h) \\ &\le \max_{-w_n \le \ell_1, \ell_2 \le w_n} \xi_{n\ell_1\ell_2} + \max_{-w_n \le \ell_1, \ell_2 \le w_n} \{G^*(s,t,s+u_{\ell_1+1}, t+u_{\ell_2+1}) \\ &\quad -G^*(s,t,s+u_{\ell_1}, t+u_{\ell_2})\}, \end{aligned}$$

where

$$\xi_{n\ell_1\ell_2} = |G_n^*(s,t,s+u_{\ell_1}, t+u_{\ell_2}) - G(s,t,s+u_{\ell_1}, t+u_{\ell_2})|.$$

Note that

$$\begin{aligned} \mathbb{E}\{G^*(s,t,s+u_{\ell_1+1}, t+u_{\ell_2+1}) - G^*(s,t,s+u_{\ell_1}, t+u_{\ell_2})\} &\le MQ_n h^2/w_n \\ &\le Ma_n. \end{aligned}$$

Now consider the case where u_{ℓ_1}, u_{ℓ_2} are both nonnegative with other cases being similar. Write

$$\xi_{n\ell_1\ell_2} = \left| \frac{1}{n} \sum_{i=1}^n \{Z_i - \mathbb{E}(Z_i)\} \right| \le \frac{1}{n} \sum_{i=1}^n |Z_i - \mathbb{E}(Z_i)|,$$

where

$$Z_i = \frac{1}{r(r-1)} \sum\sum_{\substack{1 \le j,k \le r \\ j \ne k}} Z_{ijk} I(Z_{ijk} \le Q_n) I(T_{ij} \in (s, s+u_{\ell_1}], T_{ik} \in (t, t+u_{\ell_2}]).$$

It follows that

$$\begin{aligned} &\mathrm{Var}(Z_i) \\ &\le \frac{C}{r^2(r-1)^2} \sum_{j_1 \ne k_1} \sum_{j_2 \ne k_2} \mathbb{P}(T_{ij_1}, T_{ij_2} \in (t, t+u_{\ell_1}], T_{ik_1}, T_{ik_2} \in (t, t+u_{\ell_2}]) \end{aligned}$$

as a result of (8.32). Considering the possible scenarios of how some of the indices j_1, j_2, k_1, k_2 are the same, we conclude that there exists a universal constant M such that

$$\mathrm{Var}(Z_i) \le M(h^4 + h^3/r + (h/r)^2) = M\beta_n.$$

The rest of the proof completely mirrors that of Lemma 8.2.2 and is omitted. □

As we did for the mean estimation problem, we present the following result that highlights the implications of Theorem 8.2.4 for the cases of sparse and dense functional data.

Corollary 8.2.6 *Under conditions of Theorem 8.2.4,*

1. *if r is bounded and $h_R^2 \lesssim h_m \lesssim h_R$,*

$$\sup_{s,t\in[0,1]} |K_h(s,t) - K(s,t)| = O(h_R^2 + \{(\log n/(nh_R^2)\}^{1/2}) \quad a.s.$$

2. *if $r = r_n$ is such that $r_n^{-1} \lesssim h_m, h_R \lesssim (\log n/n)^{1/4}$, then*

$$\sup_{s,t\in[0,1]} |K_h(s,t) - K(s,t)| = O(\{\log n/n\}^{1/2}) \quad a.s.$$

The rate in part 1 of the corollary is the classical nonparametric rate for estimating a bivariate function. In contrast, $K_h(s,t)$ has a root-n convergence rate in the dense setting.

Corollary 8.2.6 indicates that to estimate K optimally in the sparse case we should use a bandwidth $\hat{h}_R \asymp n^{-1/6}$ while Corollary 8.2.3 suggests that the optimal bandwidth for estimating $m(t)$ in this instance is $\hat{h}_m \asymp n^{-1/5}$. Thus, $\hat{h}_m$ is within the range of $[\hat{h}_R^2, \hat{h}_R]$ and Corollary 8.2.6 is applicable when optimal bandwidths are used for both estimation of m and R.

Finally, we consider the difference between the estimated and true covariance operators that correspond to the $\mathbb{L}^2[0,1]$ integral operators obtained from the true and estimated covariance kernel. For this purpose, we will study the behavior of

$$(\Delta_h e)(t) = \int_0^1 \{K_h(s,t) - K(s,t)\} e(s) ds,$$

over a class of elements e from $\mathbb{L}^2[0,1]$. One might expect the rate of convergence for Δ_h to be the same as that of $K_h(s,t) - K(s,t)$. However, our following result demonstrates that not to be the case. In Section 9.2, we will apply this work to the problem of inference for principal components.

Theorem 8.2.7 *Assume that the conditions of Theorem 8.2.4 hold. For any bounded measurable function e on $[0,1]$,*

$$\sup_{t\in[0,1]} |(\Delta_h e)(t)| = O(h_m^2 + h_R^2 + \delta_{n1}(h_m) + \delta_{n1}(h_R)) \quad a.s.$$

Proof: It follows that

$$\begin{aligned}(\Delta_h e)(t) &= \int_0^1 \{R_{h_R}(s,t) - R(s,t)\}e(s)ds \\ &\quad - \int_0^1 \{m_{h_m}(s)m_{h_m}(t) - m(s)m(t)\}e(s)ds \\ &=: A_{n1}(t) - A_{n2}(t).\end{aligned}$$

From (8.27),

$$\begin{aligned}A_{n1}(t) = \int_0^1 &\left[\{\mathscr{M}_1(s,t)\tilde{S}_{00}(s,t) - \mathscr{M}_2(s,t)\tilde{S}_{10}(s,t)\right. \\ &\left.-\mathscr{M}_3(s,t)\tilde{S}_{01}(s,t)\}/\mathscr{D}(s,t)\right] e(s)ds.\end{aligned}$$

We only consider the $\int_0^1 \left[\mathscr{M}_1(s,t)\tilde{S}_{00}(s,t)/\mathscr{D}(s,t)\right] e(s)ds$ term in this expression as the other two terms are of lower order and can be dealt with similarly.

By (8.28) and Lemma 8.2.5, $\mathscr{M}_1(s,t)/\mathscr{D}(s,t)$ behaves like a constant multiple of $1/(f(s)f(t))$. As $f(t) > 0$ for all t, it will be sufficient to focus only on $\int_0^1 \left[\tilde{S}_{00}(s,t)/f(s)\right] e(s)ds$. From (8.29),

$$\int_0^1 \left[\tilde{S}_{00}(s,t)/f(s)\right] e(s)ds = \sum_{i=1}^{3} \int_0^1 [U_i(s,t)/f(s)]e(s)ds + O(h_R^2),$$

where for each i, we can write

$$\begin{aligned}&\int_0^1 [U_i(s,t)/f(s)]e(s)ds \\ &= \frac{1}{nr(r-1)} \sum_{i=1}^{n} \sum_{\substack{1\le j,k\le r \\ j\ne k}} Z_{ijk} W_{h_R}(T_{ik} - t) \int_0^1 [W_{h_R}(T_{ij} - s)/f(s)]e(s)ds\end{aligned}$$

and the Z_{ijk} all have zero means. Express the right-hand side of this last formula as

$$\frac{1}{nr} \sum_{i=1}^{n} \sum_{k=1}^{r} Z_{ik} W_{h_R}(T_{ik} - t),$$

where

$$Z_{ik} = \frac{1}{r-1} \sum_{\substack{1\le j\le r \\ j\ne k}} Z_{ijk} \int_0^1 [W_{h_R}(T_{ij} - s)/f(s)]e(s)ds$$

Note that

$$\left| \int_0^1 \left[W_{h_R}(T_{ij} - s)/f(s) \right] e(s)ds \right| \le \sup_{s\in[0,1]} (|e(s)|/f(s)) \int_{-1}^1 W(u)du < \infty.$$

The assumptions of Lemma 8.2.2 can be easily verified for Z_{ik}, which entails that

$$\sup_{t\in[0,1]} \left| \frac{1}{nr} \sum_{i=1}^n \sum_{k=1}^r Z_{ik} W_{h_R}(T_{ik} - t) \right| = O(h_R^2 + \delta_{n1}(h_R)) \quad \text{a.s.}$$

and consequently that

$$\sup_{t\in[0,1]} \left| \int_0^1 \left[\mathscr{M}_1(s,t)\tilde{S}_{00}(s,t)/\mathscr{D}(s,t) \right] e(s)ds \right| = O(h_R^2 + \delta_{n1}(h_R)) \quad \text{a.s.}$$

Thus, we obtain the rate $\sup_{t\in[0,1]} |A_{n1}(t)| = O(h_R^2 + \delta_{n1}(h_R))$.

Finally, we write

$$\begin{aligned} A_{n2}(t) = m_{h_m}(t) &\int_0^1 \{m_{h_m}(s) - m(s)\} e(s)ds \\ +\{m_{h_m}(t) - m(t)\} &\int_0^1 m(s)e(s)ds. \end{aligned}$$

This has the uniform rate $O(h_m^2 + \delta_{n1}(h_m))$ from Theorem 8.2.1 and the proof is complete. □

8.3 Penalized least-squares estimation

Section 8.2 considered nonparametric estimation of the mean function and covariance kernel using local linear regression type estimators. In this section, we explore a slightly different development with smoothing spline variants as the estimators of choice.

The model to be considered is much the same as (8.2) with $X_1, \ldots, X_n$ iid as some second-order stochastic process X on $E = [0, 1]$. As before, let $m(t)$ and $K(s, t)$ be the mean and covariance functions of X. For simplicity, we will focus attention on the case where the T_{ij} are a random sample from the uniform distribution. Apart from that, the major difference is that we now assume that X is a random element of the Sobolev space $\mathbb{W}_q[0, 1]$ described in Section 2.8 for some $q > 1$. As $\mathbb{W}_q[0, 1]$ is an RKHS, this is the setting addressed in Section 7.5. For $f, g \in \mathbb{W}_q[0,1]$, we will use the (squared) norm and inner product

$$\begin{aligned}\|f\|^2_{\mathbb{W}_q[0,1]} &= \|f\|^2_{\mathbb{L}_2[0,1]} + \|f^{(q)}\|^2_{\mathbb{L}_2[0,1]} \\ &= \int_0^1 f^2(t)dt + \int_0^1 [f^{(q)}(t)]^2 dt \end{aligned} \tag{8.36}$$

and

$$\begin{aligned}\langle f, g\rangle_{\mathbb{W}_q[0,1]} &= \langle f, g\rangle_{\mathbb{L}_2[0,1]} + \langle f^{(q)}, g^{(q)}\rangle_{\mathbb{L}_2[0,1]} \\ &= \int_0^1 f(t)g(t)dt + \int_0^1 f^{(q)}(t)g^{(q)}(t)dt.\end{aligned}$$

Recall from Section 2.8 that a convenient basis for $\mathbb{W}_q[0,1]$ is the set of eigenfunctions $\{e_j\}$ for the differential operator $(-1)^q D^{2q}$ subject to the boundary conditions $e_j^{(k)}(0) = e_j^{(k)}(1) = 0, k = q, \ldots, 2q-1$. The e_j satisfy

$$\langle e_j, e_k\rangle_{\mathbb{L}^2[0,1]} = \int_0^1 e_j(t)e_k(t)dt = \delta_{jk}$$

and

$$\langle e_j^{(q)}, e_k^{(q)}\rangle_{\mathbb{L}^2[0,1]} = \int_0^1 e_j^{(q)}(t)e_k^{(q)}(t)dt = \gamma_j\delta_{jk}$$

for $\gamma_1 = \cdots = \gamma_q = 0$ and universal constants $C_1, C_2 \in (0,\infty)$ such that

$$C_1 j^{2q} \le \gamma_{j+q} \le C_2 j^{2q}, \quad j \ge 1.$$

Thus, any element $v \in \mathbb{W}_q[0,1]$ can be expressed as $v = \sum_{j=1}^\infty v_j e_j$ with

$$v_j = \langle v, e_j\rangle_{\mathbb{L}^2[0,1]} = \int_0^1 v(t)e_j(t)dt$$

and

$$\|v\|^2_{\mathbb{W}_q[0,1]} = \sum_{j=1}^\infty (1+\gamma_j)v_j^2. \tag{8.37}$$

In particular,

$$\|X\|^2_{\mathbb{W}_q[0,1]} = \sum_{j=1}^\infty (1+\gamma_j)\langle X, e_j\rangle^2_{\mathbb{L}_2[0,1]},$$

which is finite by the assumption that X is in $\mathbb{W}_q[0,1]$.

Let $\hat{m}(t)$ be any estimator of $m(t)$ that has been constructed from the observed data $(T_{ij}, Y_{ij}), i = 1, \ldots, n, j = 1, \ldots, r$. We will assess its departure from m via the $\mathbb{L}^2[0,1]$ norm of the difference; i.e., by

$$\|\hat{m} - m\|_{\mathbb{L}^2[0,1]} = \left\{ \int_0^1 \left(\hat{m}(t) - m(t)\right)^2 dt \right\}^{1/2}.$$

A gold standard for performance of estimators in this context has been established in Cai and Yuan (2011) that we state here as follows.

Theorem 8.3.1 *Let $\mathbb{P}(q, C_1)$ be the collection of probability measures for $\mathbb{W}_q[0,1]$ valued processes such that for any X with probability law $\mathbb{P}_X \in \mathbb{P}(q, C_1)$*

$$\mathbb{E}\|X^{(q)}\|^2_{\mathbb{L}_2[0,1]} \leq C_1$$

for some $C_1 > 0$. Then, there is a constant $C_2 > 0$ that depends only on C_1 and $\sigma^2 = \mathbb{E}[\varepsilon_{11}^2]$ such that

$$\limsup_{n\to\infty} \sup_{\mathbb{P}_X \in \mathbb{P}(q,C)} \mathbb{P}\left(\|\hat{m} - m\|^2_{\mathbb{L}^2[0,1]} > C_2 \left((nr)^{-2q/(2q+1)} + n^{-1} \right) \right) > 0.$$

Our immediate aim is to construct a rate optimal estimator in the sense of this result.

As X is second order with sample paths in $\mathbb{W}_q[0,1]$, we know from Section 7.2 that $m \in \mathbb{W}_q[0,1]$ as well. This observation leads us to consider estimation of m via the smoothing spline regression estimator discussed in Section 6.6. Specifically, we will estimate m by

$$m_\eta = \text{argmin}_{v \in \mathbb{W}_q[0,1]} f_{rn,\eta}(v)$$

for

$$f_{rn,\eta}(v) = (nr)^{-1} \sum_{i=1}^n \sum_{j=1}^r (Y_{ij} - v(T_{ij}))^2 + \eta \|v^{(q)}\|^2_{\mathbb{L}_2[0,1]}. \tag{8.38}$$

We will have shown that this choice attains the minimax lower bound of Theorem 8.3.1 upon proving

Theorem 8.3.2 *If $\eta \asymp (rn)^{-2q/(2q+1)}$, then*

$$\|m_\eta - m\|^2_{\mathbb{L}^2[0,1]} = O_p\left((nr)^{-2q/(2q+1)} + n^{-1} \right).$$

Proof: The development that follows is based on the proof of Theorem 3.2 of Cai and Yuan (2011). In keeping with their work, define

$$\begin{aligned} f_{\infty,\eta}(v) &= \mathbb{E} f_{rn,\eta}(v) \\ &= \mathbb{E}[(Y_{11} - m(T_{11}))^2] + \|m - v\|^2_{\mathbb{L}_2[0,1]} + \eta \|v^{(q)}\|^2_{\mathbb{L}^2[0,1]} \end{aligned}$$

and let

$$\overline{m}_\eta = \text{argmin}_{v \in \mathbb{W}_q[0,1]} f_{\infty,\eta}(v).$$

The Cai/Yuan proof then proceeds on the basis of the identity

$$m_\eta - m = m_\eta - \tilde{m}_\eta + \tilde{m}_\eta - \overline{m}_\eta + \overline{m}_\eta - m, \tag{8.39}$$

where

$$\tilde{m}_\eta = \overline{m}_\eta - (f''_{\infty,\eta})^{-1} f'_{rn,\eta}(\overline{m}_\eta). \tag{8.40}$$

In this last expression $f'_{rn,\eta}$ and $f''_{\infty,\eta}$ are, respectively, the first Fréchet derivative of $f_{rn,\eta}$ and the second Fréchet derivative of $f_{\infty,\eta}$ in the sense of the definitions in Section 3.6.

The validity of (8.39) is indisputable. However, the motivation behind this particular representation for $m_\eta - m$ is somewhat less obvious. The intermediate approximation $\tilde{m}_\eta$ that appears in (8.39) can be motivated by a formal Taylor type expansion wherein one writes

$$f'_{rn,\eta}(m_\eta) - f'_{rn,\eta}(\overline{m}_\eta) \approx f''_{rn,\eta}(m_\eta - \overline{m}_\eta).$$

From Theorem 3.6.3, we know that $f'_{rn,\eta}(m_\eta) = 0$ which leads to

$$(m_\eta - \overline{m}_\eta) \approx -(f''_{rn,\eta})^{-1} f'_{rn,\eta}(\overline{m}_\eta).$$

Then, a Fisher scoring type of strategy suggests that we replace $f''_{rn,\eta}$ by $f''_{\infty,\eta}$ in this last approximation to obtain

$$m_\eta \approx \overline{m}_\eta - (f''_{\infty,\eta})^{-1} f'_{rn,\eta}(\overline{m}_\eta) = \tilde{m}_\eta.$$

The task at hand is to show that each of $\|m_\eta - \tilde{m}_\eta\|^2_{\mathbb{L}^2[0,1]}$, $\|\tilde{m}_\eta - \overline{m}_\eta\|^2_{\mathbb{L}^2[0,1]}$, and $\|\overline{m}_\eta - m\|^2_{\mathbb{L}^2[0,1]}$ are $O_p((rn)^{-2q/(2q+1)} + n^{-1})$ when $\eta \asymp (rn)^{-2q/(2q+1)}$. In this regard, the most immediate result is for $\|\overline{m}_\eta - m\|^2_{\mathbb{L}^2[0,1]}$. By the definition of $\overline{m}_\eta$, we know that

$$\mathbb{E}[(Y_{11} - m(T_{11}))^2] + \|\overline{m}_\eta - m\|^2_{\mathbb{L}^2[0,1]} \leq f_{\infty,\eta}(\overline{m}_\eta) \leq f_{\infty,\eta}(v)$$

for all $v \in \mathbb{W}_q[0,1]$. In particular, if we choose $v = m$, this gives

$$\|\overline{m}_\eta - m\|^2_{\mathbb{L}^2[0,1]} \leq \eta \|m^{(q)}\|^2_{\mathbb{L}_2[0,1]} \tag{8.41}$$

with the right-hand side of this expression being of the order $(rn)^{-2q/(2q+1)}$ under the assumptions on η.

To proceed further, we must explicitly evaluate the Fréchet derivatives that appear in (8.39). □

Lemma 8.3.3 *In what follows, let h, v, v_1, v_2 be arbitrary elements of* $\mathbb{W}_q[0,1]$.

1. *The Fréchet derivative of* $f_{rn,\eta}$ *at* h *is the element* $f'_{rn,\eta}(h)$ *of* $\mathfrak{B}(\mathbb{W}_q[0,1],\mathbb{R})$ *characterized by*

$$f'_{rn,\eta}(h)v = -\frac{2}{rn}\sum_{i=1}^{n}\sum_{j=1}^{r}(Y_{ij} - h(T_{ij}))v(T_{ij}) + 2\eta\int_0^1 h^{(q)}(t)v^{(q)}(t)dt.$$

The second Fréchet derivative $f''_{rn,\eta} \in \mathfrak{B}(\mathbb{W}_q[0,1],\mathfrak{B}(\mathbb{W}_q[0,1],\mathbb{R}))$ *is characterized by*

$$f''_{rn,\eta}v_1v_2 = \frac{2}{rn}\sum_{i=1}^{n}\sum_{j=1}^{r}v_1(T_{ij})v_2(T_{ij}) + 2\eta\int_0^1 v_1^{(q)}(t)v_2^{(q)}(t)dt.$$

2. *The Fréchet derivative of* $f_{\infty,\eta}$ *at* h *is the element* $f'_{\infty,\eta}(h)$ *of* $\mathfrak{B}(\mathbb{W}_q[0,1],\mathbb{R})$ *characterized by*

$$f'_{\infty,\eta}(h)v = -2\int_0^1 (m(t) - h(t))v(t)dt + 2\eta\int_0^1 h^{(q)}(t)v^{(q)}(t)dt.$$

The second Fréchet derivative $f''_{\infty,\eta} \in \mathfrak{B}(\mathbb{W}_q[0,1],\mathfrak{B}(\mathbb{W}_q[0,1],\mathbb{R}))$ *is characterized by*

$$f''_{\infty,\eta}v_1v_2 = 2\int_0^1 v_1(t)v_2(t)dt + 2\eta\int_0^1 v_1^{(q)}(t)v_2^{(q)}(t)dt. \tag{8.42}$$

Proof: Verification of the stated results is similar in both cases. Thus, we only give the details for part 1.

In Theorem 3.6.4, take $\phi(s) = f_{rn,\eta}(h + sv), s \in \mathbb{R}$, with the consequence that

$$\phi'(0) = -\frac{2}{rn}\sum_{i=1}^{n}\sum_{j=1}^{n}(Y_{ij} - h(T_{ij}))v(T_{ij}) + 2\eta\int_0^1 h^{(q)}(t)v^{(q)}(t)dt$$

is the value of the Gâteaux derivative of $f_{rn,\eta}$ at h applied to v. This is continuous in h and therefore the Fréchet derivative and Gâteaux derivative coincide due to Theorem 3.6.2.

Now set

$$\begin{aligned}\omega(s) &= f'(h + sv_1)v_2\\ &= -\frac{2}{rn}\sum_{i=1}^{n}\sum_{j=1}^{r}(Y_{ij} - h(T_{ij}) - tv_1(T_{ij}))v_2(T_{ij})\\ &\quad +2\eta\int_0^1 [h^{(q)}(t) + sv_1^{(q)}(t)]v_2^{(q)}(t)dt\end{aligned}$$

for $s \in \mathbb{R}$. Then,

$$\omega'(0) = \frac{1}{rn} \sum_{i=1}^{n} \sum_{j=1}^{r} v_1(T_{ij}) v_2(T_{ij}) + 2\eta \int_0^1 v_1^{(q)}(t) v_2^{(q)}(t) dt.$$

This can be interpreted as follows. The second Gâteaux derivative of $f_{rn,\eta}$ at h when applied to v_1 produces a linear functional whose value at v_2 is $\omega'(0)$. As this is constant (and, hence, continuous) as a function of h the derivative is denoted by $f''_{rn,\eta}$ rather than $f''_{rn,\eta}(h)$ and must also be the second Fréchet derivative. □

The evaluation of $\tilde{m}_\eta$ involves the inverse of the operator $f''_{\infty,\eta}$ that belongs to $\mathfrak{B}(\mathbb{W}_q[0,1], \mathfrak{B}(\mathbb{W}_q[0,1], \mathbb{R}))$. It is not obvious at this point that the inverse exists and, in addition, working directly with the space $\mathfrak{B}(\mathbb{W}_q[0,1], \mathfrak{B}(\mathbb{W}_q[0,1], \mathbb{R}))$ is somewhat inconvenient. A work around for the latter problem derives from the Reisz representation theorem (Theorem 3.2.1), which tells us that there is an invertible norm-preserving mapping $\mathcal{Q}$ such that $\mathcal{Q}\mathfrak{B}(\mathbb{W}_q[0,1], \mathbb{R}) = \mathbb{W}_q[0,1]$. Thus,

$$\tilde{f}''_{\infty,\eta} := \mathcal{Q} f''_{\infty,\eta} \tag{8.43}$$

belongs to $\mathfrak{B}(\mathbb{W}_q[0,1])$ and it is invertible if and only if $f''_{\infty,\eta}$ is invertible.

We can actually obtain an explicit representation for $\tilde{f}''_{\infty,\eta}$ using the CONS for $\mathbb{W}_q[0,1]$ that was described earlier.

Lemma 8.3.4 *The operator $\tilde{f}''_{\infty,\eta}$ in (8.43) is an invertible element of $\mathfrak{B}(\mathbb{W}_q[0,1])$. For any $v = \sum_{j=1}^{\infty} v_j e_j$ in $\mathbb{W}_q[0,1]$,*

$$(\tilde{f}''_{\infty,\eta})^{-1} v = \frac{1}{2} \sum_{j=1}^{\infty} \frac{1+\gamma_j}{1+\eta\gamma_j} v_j e_j.$$

Proof: Let $v_1 \in \mathbb{W}_q[0,1]$ with the consequence that $f''_{\infty,\eta} v_1$ is in $\mathfrak{B}(\mathbb{W}_q[0,1], \mathbb{R})$ with representer $\tilde{f}''_{\infty,\eta} v_1$. Then, for any $v_2 \in \mathbb{W}_q[0,1]$, we see from (8.42) that

$$\begin{aligned} f''_{\infty,\eta} v_1 v_2 &= \langle \tilde{f}''_{\infty,\eta} v_1, v_2 \rangle_{\mathbb{W}_q[0,1]} \\ &= 2\int_0^1 v_1(t) v_2(t) dt + 2\eta \int_0^1 v_1^{(q)}(t) v_2^{(q)}(t) dt \end{aligned}$$

and, hence, that

$$\tilde{f}''_{\infty,\eta} v_1 = 2 \sum_{j=1}^{\infty} \frac{1+\eta\gamma_j}{1+\gamma_j} \langle v_1, e_j \rangle_{\mathbb{L}^2[0,1]} e_j.$$

The stated form for the inverse follows immediately from this relation. □

We now have at our disposal the requisite tools for evaluation of the norms of $\tilde{m}_\eta - m$ and $m_\eta - \tilde{m}_\eta$. For reasons that will become clear subsequently, it will be worthwhile to work with an alternative norm $\|\cdot\|_\alpha$ defined by

$$\|v\|_\alpha^2 = \sum_{j=1}^{\infty} (1+\gamma_j)^\alpha v_j^2$$

for $v = \sum_{j=1}^{\infty} v_j e_j \in \mathbb{W}_q[0,1]$ and $0 \le \alpha \le 1$. The restriction to $\alpha \in [0,1]$ means that

$$\|v\|_{\mathbb{L}^2[0,1]}^2 = \|v\|_0^2 \le \|v\|_\alpha^2 \le \|v\|_1^2 = \|v\|_{\mathbb{W}_q[0,1]}^2.$$

As $\|v\|_\alpha$ is the $\mathbb{L}^2[0,1]$ norm of $\sum_{j=1}^{\infty}(1+\gamma_j)^{\alpha/2} v_j e_j$, it is, in fact, a norm for $\mathbb{W}_q[0,1]$.

We now claim that

$$\|\tilde{m}_\eta - \overline{m}_\eta\|_\alpha^2 = \frac{1}{4}\sum_{k=1}^{\infty} \frac{(1+\gamma_k)^\alpha}{(1+\eta\gamma_k)^2}(f'_{rn,\eta}(\overline{m}_\eta)e_k)^2. \tag{8.44}$$

To see that this is so observe that, by the definitions of $\tilde{m}_\eta$ and $\|\cdot\|_\alpha$,

$$\begin{aligned}\|\tilde{m}_\eta - \overline{m}_\eta\|_\alpha^2 &= \|(f''_{\infty,\eta})^{-1} f'_{rn,\eta}(\overline{m}_\eta)\|_\alpha^2 \\ &= \sum_{k=1}^{\infty}(1+\gamma_k)^\alpha \langle (f''_{\infty,\eta})^{-1} f'_{rn,\eta}(\overline{m}_\eta), e_k\rangle_{\mathbb{L}^2[0,1]}^2.\end{aligned}$$

However, from Lemma 8.3.4,

$$\begin{aligned}&\langle (f''_{\infty,\eta})^{-1} f'_{rn,\eta}(\overline{m}_\eta), e_k\rangle_{\mathbb{L}^2[0,1]} \\ &= \frac{1}{1+\gamma_k}\langle (f''_{\infty,\eta})^{-1} f'_{rn,\eta}(\overline{m}_\eta), e_k\rangle_{\mathbb{W}_q[0,1]} \\ &= \frac{1}{1+\gamma_k}\langle \mathcal{Q} f'_{rn,\eta}(\overline{m}_\eta), (\tilde{f}''_{\infty,\eta})^{-1} e_k\rangle_{\mathbb{W}_q[0,1]} \\ &= \frac{1}{2(1+\eta\gamma_j)}\langle \mathcal{Q} f'_{rn,\eta}(\overline{m}_\eta), e_k\rangle_{\mathbb{W}_q[0,1]} \\ &= (f'_{rn,\eta}(\overline{m}_\eta)e_k)/2(1+\eta\gamma_k)\end{aligned}$$

because $\mathcal{Q} f'_{rn,\eta}(\overline{m}_\eta)$ is the representer of $f'_{rn,\eta}(\overline{m}_\eta)$ and $(\tilde{f}''_{\infty,\eta})^{-1}$ is self-adjoint.

From (8.44), we see that the probabilistic behavior of $\|\tilde{m}_\eta - \overline{m}_\eta\|_\alpha^2$ is governed by that of the random sequence $\{f'_{rn,\eta}(\overline{m}_\eta)e_k\}$. To assess the size of the sequence elements, we first observe from Theorem 3.6.3 that $f'_{\infty,\eta}(\overline{m}_\eta) = 0$.

Then, an application of Lemma 8.3.3 reveals that for every $v \in \mathbb{W}_q[0,1]$

$$\begin{aligned} f'_{rn,\eta}(\overline{m}_\eta)v &= f'_{rn,\eta}(\overline{m}_\eta)v - f'_{\infty,\eta}(\overline{m}_\eta)v \\ &= -\frac{2}{rn}\sum_{i=1}^{n}\sum_{j=1}^{r}(Y_{ij} - \overline{m}_\eta(T_{ij}))v(T_{ij}) \\ &\quad +2\int_0^1 (m(t) - \overline{m}_\eta(t))v(t)dt. \end{aligned}$$

Consequently,

$$\mathbb{E}f'_{rn,\eta}(\overline{m}_\eta)e_k = \mathbb{E}_T\mathbb{E}[f'_{rn,\eta}(\overline{m}_\eta)e_k|T] = 0$$

and

$$\begin{aligned} \mathbb{E}(f'_{rn,\eta}(\overline{m}_\eta)e_k)^2 &= \mathrm{Var}(f'_{rn,\eta}(\overline{m}_\eta)e_k) \\ &= \frac{4}{(rn)^2}\sum_{i=1}^{n}\mathrm{Var}\left(\sum_{j=1}^{r}\left(Y_{ij} - \overline{m}_\eta(T_{ij})\right)e_k(T_{ij})\right) \\ &= \frac{4}{nr^2}\mathrm{Var}\left(\sum_{j=1}^{r}\left(Y_{1j} - \overline{m}_\eta(T_{1j})\right)e_k(T_{1j})\right) \end{aligned} \tag{8.45}$$

Now, for any random variable Z, $\mathrm{Var}(Z) = \mathrm{Var}_T(\mathbb{E}[Z|T]) + \mathbb{E}_T[\mathrm{Var}(Z|T)]$. An application of this relation here produces

$$\begin{aligned} &\mathrm{Var}\left(\sum_{j=1}^{r}\left(Y_{1j} - \overline{m}_\eta(T_{1j})\right)e_k(T_{1j})\right) \\ &= \mathrm{Var}\left(\sum_{j=1}^{r}\left(m(T_{1j}) - \overline{m}_\eta(T_{1j})\right)e_k(T_{1j})\right) \\ &\quad +\mathbb{E}_T\left[\mathrm{Var}\left(\sum_{j=1}^{r}Y_{1j}e_k(T_{1j})|T\right)\right]. \end{aligned} \tag{8.46}$$

For the first term on the right hand side of this expression, we have

$$\begin{aligned} &\mathrm{Var}_T\left(\sum_{j=1}^{r}\left(m(T_{ij}) - \overline{m}_\eta(T_{ij})\right)e_k(T_{ij})\right) \\ &= r\mathrm{Var}((m(T) - \overline{m}_\eta(T))e_k(T)) \\ &\le r\int_0^1 (m(t) - \overline{m}_\eta(t))^2 e_k^2(t)dt \\ &\le r\max_{t\in[0,1]}|e_k(t)|^2\|m - \overline{m}_\eta\|^2_{\mathbb{L}^2[0,1]} \\ &= O(r\eta), \end{aligned} \tag{8.47}$$

where the bound is independent of k. This is due to (8.41) and the fact from Section 2.8 that the $|e_k|$ are uniformly bounded. Then, for the second term,

$$\mathbb{E}_T\left[\mathrm{Var}\left(\sum_{j=1}^{r} Y_{1j}e_k(T_{1j})|T\right)\right]$$
$$\leq r(r-1)\int_0^1\int_0^1 e_k(s)e_k(t)R(s,t)dsdt + r\int_0^1 e_k^2(t)R(t,t)dt. \quad (8.48)$$

Both integrals on the right are bounded as $R(s,t)$ is continuous (see the following text) and $\|e_k\|_{\mathbb{L}^2[0,1]} = 1$. In particular, $\int_0^1\int_0^1 e_k(s)e_k(t)R(s,t)dsdt$ can be expressed as $\mathbb{E}\langle e_k, X\rangle^2_{\mathbb{L}^2[0,1]}$. Combining all the bounds in (8.45)–(8.48), we obtain

$$\mathbb{E}(f'_{rn,\eta}(\overline{m}_\eta)e_k)^2 \leq \frac{4}{n}\mathbb{E}\langle e_k, X\rangle^2_{\mathbb{L}^2[0,1]} + O\left(\frac{1}{rn}\right).$$

Consequently,

$$\mathbb{E}[\|\tilde{m}_\eta - \overline{m}_\eta\|_\alpha^2] \leq \frac{4}{n}\sum_{k=1}^{\infty}\frac{(1+\gamma_k)^\alpha}{(1+\eta\gamma_k)^2}\mathbb{E}\langle e_k, X\rangle^2_{\mathbb{L}^2[0,1]}$$
$$+O\left(\frac{1}{rn}\right)\sum_{k=1}^{\infty}\frac{(1+\gamma_k)^\alpha}{(1+\eta\gamma_k)^2}.$$

However,

$$n^{-1}\sum_{k=1}^{\infty}\frac{(1+\gamma_k)^\alpha}{(1+\eta\gamma_k)^2}\mathbb{E}\langle e_k, X\rangle^2_{\mathbb{L}^2[0,1]} \leq n^{-1}\sum_{k=1}^{\infty}(1+\gamma_k)\mathbb{E}\langle e_k, X\rangle^2_{\mathbb{L}^2[0,1]}$$
$$= n^{-1}\mathbb{E}\|X\|^2_{\mathbb{W}_q[0,1]} = O(n^{-1})$$

and, from Theorem 2.8.3, there are constants $0 < C_1 \leq C_2 < \infty$ such that

$$\sum_{k=1}^{\infty}\frac{(1+\gamma_k)^\alpha}{(1+\eta\gamma_k)^2} \leq q + \sum_{k=1}^{\infty}\frac{(1+C_2k^{2q})^\alpha}{(1+C_1\eta k^{2q})^2}$$
$$\leq q + \int_0^\infty \frac{(1+C_2x^{2q})^\alpha}{(1+C_1\eta x^{2q})^2}dx$$
$$= O(\eta^{-\alpha-1/(2q)})$$

as $\eta \to 0$ if $\alpha + 1/(2q) < 2$. Thus, we are lead to the conclusion that

$$\|\tilde{m}_\eta - \overline{m}_\eta\|_\alpha^2 = O_p\left(\frac{1}{nr\eta^{\alpha+1/(2q)}} + \frac{1}{n}\right).$$

In particular, by taking $\alpha = 0$ for $\eta \asymp (rn)^{-2q/(2q+1)}$, this gives

$$\|\tilde{m}_\eta - \overline{m}_\eta\|^2_{\mathbb{L}^2[0,1]} = O_p((nr)^{-2q/(2q+1)} + n^{-1}). \tag{8.49}$$

The remaining quantity to consider is $\|m_\eta - \tilde{m}_\eta\|^2_\alpha$. The first step in this direction is to obtain a useful analytic form for $m_\eta - \tilde{m}_\eta$. In this regard, observe that

$$\begin{aligned} m_\eta - \tilde{m}_\eta &= m_\eta - \overline{m}_\eta + (f''_{\infty,\eta})^{-1} f'_{rn,\eta}(\overline{m}_\eta) \\ &= (f''_{\infty,\eta})^{-1}[f''_{\infty,\eta}(m_\eta - \overline{m}_\eta) + f'_{rn,\eta}(\overline{m}_\eta)]. \end{aligned}$$

As m_η maximizes $f_{rn,\eta}$, Theorem 3.6.3 tells us that $f'_{rn,\eta}(m_\eta) = 0$. Then, using Lemma 8.3.3, we see that for every $v \in \mathbb{W}_q[0,1]$

$$\begin{aligned} &[f''_{\infty,\eta}(m_\eta - \overline{m}_\eta) + f'_{rn,\eta}(\overline{m}_\eta)]v \\ &\quad = [f''_{\infty,\eta}(m_\eta - \overline{m}_\eta) + f'_{rn,\eta}(\overline{m}_\eta) - f'_{rn,\eta}(m_\eta)]v \\ &\quad = 2\int_0^1 (m_\eta(t) - \overline{m}_\eta(t))v(t)dt \\ &\qquad\quad - \frac{2}{rn}\sum_{i=1}^n \sum_{j=1}^r (m_\eta(T_{ij}) - \overline{m}_\eta(T_{ij}))v(T_{ij}) \\ &\quad = [f''_{\infty,0}(m_\eta - \overline{m}_\eta) - f''_{rn,0}(m_\eta - \overline{m}_\eta)]v \end{aligned}$$

or

$$m_\eta - \tilde{m}_\eta = (f''_{\infty,\eta})^{-1}[f''_{\infty,0}(m_\eta - \overline{m}_\eta) - f''_{rn,0}(m_\eta - \overline{m}_\eta)].$$

Therefore, the same argument that produced (8.44) gives us

$$\begin{aligned} &\|m_\eta - \tilde{m}_\eta\|^2_\alpha \\ &\quad = \frac{1}{4}\sum_{k=1}^\infty \frac{(1+\gamma_k)^\alpha}{(1+\eta\gamma_k)^2}([f''_{\infty,0}(m_\eta - \overline{m}_\eta) - f''_{rn,0}(m_\eta - \overline{m}_\eta)]e_k)^2 \end{aligned}$$

with

$$\begin{aligned} &[f''_{\infty,0}(m_\eta - \overline{m}_\eta) - f''_{rn,0}(m_\eta - \overline{m}_\eta)]e_k \\ &\quad = 2\int_0^1 (m_\eta(t) - \overline{m}_\eta(t))e_k(t)dt \\ &\qquad\quad - \frac{2}{rn}\sum_{i=1}^n \sum_{j=1}^r (m_\eta(T_{ij}) - \overline{m}_\eta(T_{ij}))e_k(T_{ij}). \end{aligned} \tag{8.50}$$

Now write $m_\eta - \overline{m}_\eta = \sum_{\nu=1}^{\infty} h_\nu e_\nu$ in (8.50) and apply the Cauchy–Schwarz inequality to see that for arbitrary $\theta \in (1/(2q), 1]$, we have

$$\begin{aligned}\|m_\eta - \tilde{m}_\eta\|_\alpha^2 &= \sum_{k=1}^{\infty} \frac{(1+\gamma_k)^\alpha}{(1+\eta\gamma_k)^2}\left[\sum_{\nu=1}^{\infty} h_\nu \left\{\delta_{k\nu} - \frac{1}{rn}\sum_{i=1}^{n}\sum_{j=1}^{r} e_k(T_{ij})e_\nu(T_{ij})\right\}\right]^2 \\ &\le \|m_\eta - \overline{m}_\eta\|_\theta^2 \sum_{k=1}^{\infty} \frac{(1+\gamma_k)^\alpha}{(1+\eta\gamma_k)^2} \sum_{\nu=1}^{\infty} (1+\gamma_\nu)^{-\theta} V_{k\nu}^2\end{aligned}$$

with

$$V_{k\nu} = \delta_{k\nu} - \frac{1}{rn}\sum_{i=1}^{n}\sum_{j=1}^{r} e_k(T_{ij})e_\nu(T_{ij}).$$

We have $\mathbb{E}V_{k\nu} = 0$ so that

$$\begin{aligned}\mathbb{E}V_{k\nu}^2 &= \mathrm{Var}(V_{k\nu}) = \frac{1}{rn}\mathrm{Var}(e_k(T)e_\nu(T)) \\ &\le \frac{1}{rn}\mathbb{E}[e_k^2(T)e_\nu^2(T)] = O\left(\frac{1}{rn}\right)\end{aligned}$$

due to the uniform boundedness of the basis functions. As $\sum_{\nu=1}^{\infty}(1+\gamma_\nu)^{-\theta}$ is finite for $\theta > 1/(2q)$, it follows that

$$\|m_\eta - \tilde{m}_\eta\|_\alpha^2 = O_p\left(\frac{1}{rn\eta^{\alpha+1/(2q)}}\right)\|m_\eta - \overline{m}_\eta\|_\theta^2.$$

In particular, if $\alpha > 1/(2q)$

$$\|m_\eta - \tilde{m}_\eta\|_\alpha^2 = O_p\left(\frac{1}{rn\eta^{\alpha+1/(2q)}}\right)\|m_\eta - \overline{m}_\eta\|_\alpha^2. \tag{8.51}$$

From (8.51), we can conclude that $\|m_\eta - \tilde{m}_\eta\|_\alpha^2 = o_p(\|m_\eta - \overline{m}_\eta\|_\alpha^2)$. So,

$$\begin{aligned}\|\tilde{m}_\eta - \overline{m}_\eta\|_\alpha &\ge \|m_\eta - \overline{m}_\eta\|_\alpha - \|m_\eta - \tilde{m}_\eta\|_\alpha \\ &= (1 - o_p(1))\|m_\eta - \overline{m}_\eta\|_\alpha;\end{aligned}$$

i.e., $\|m_\eta - \overline{m}_\eta\|_\alpha^2 = O_p(\|\tilde{m}_\eta - \overline{m}_\eta\|_\alpha^2)$. Using this fact in combination with (8.51) and (8.49) reveals that for $\alpha > 1/(2q)$

$$\begin{aligned}\|m_\eta - \tilde{m}_\eta\|_{\mathbb{L}^2[0,1]}^2 &\le \|m_\eta - \tilde{m}_\eta\|_\alpha^2 \\ &= O_p\left(\frac{1}{rn\eta^{\alpha+1/(2q)}}\right) O_p(\|\tilde{m}_\eta - \overline{m}_\eta\|_\alpha^2) \\ &= O_p\left(\frac{1}{rn\eta^{\alpha+1/(2q)}}\right) O_p\left(\frac{1}{nr\eta^{\alpha+1/(2q)}} + \frac{1}{n}\right) \\ &= o_p(n^{-1} + (nr\eta^{1/(2q)})^{-1})\end{aligned}$$

provided that $\eta \to 0$ while $rn \to \infty$ in such a way that $rn\eta^{\alpha+1/(2q)} \to \infty$. This latter condition holds for $\eta \asymp (rn)^{-2q/(2q+1)}$ if $\alpha < 1/2$. Our results were predicated on having $\alpha > 1/(2q)$. Beyond that, the choice for α was arbitrary and both bounds can be satisfied whenever $q > 1$.

We now turn our attention to the problem of estimating the covariance kernel for the X process. To simplify matters, we assume that $\mathbb{E}[X(t)] = 0$ for all $t \in [0, 1]$; i.e., the mean function is either known or identically zero. For estimation in the general case, one may replace the response data by $Y_{ij} - \hat{m}(T_{ij})$ with $\hat{m}$ some suitable estimator of m. Provided this estimator converges sufficiently fast the results that follow will not be affected by this substitution. In particular, it follows from Theorem 4 of Cai and Yuan (2010) that the smoothing spline estimator in Theorem 8.3.2 will work in this capacity.

As before, the sample paths of the X process are presumed to lie in $\mathbb{W}_q[0, 1]$. Theorem 7.5.2, therefore has the consequence that the process covariance kernel is in the direct product Hilbert space

$$\mathbb{H} = \mathbb{W}_q[0, 1] \otimes \mathbb{W}_q[0, 1].$$

Similar to our scheme for estimation of the mean function, we estimate K by

$$K_\eta = \operatorname{argmin}_{v\in\mathbb{H}} f_{rn,\eta}(v)$$

with $f_{rn,\eta}$ now defined by

$$f_{rn,\eta}(v) = (r(r-1)n)^{-1} \sum_{i=1}^{n} \sum_{1\le j<k\le r} (Y_{ij}Y_{ik} - v(T_{ij}, T_{ik}))^2 + \eta\|v\|^2_{\mathbb{H}}. \tag{8.52}$$

Theorem 6.2.1 may be applied here with $\mathscr{W} = \eta I$ and

$$\mathscr{T}v = (v(T_{11}, T_{12}), \ldots, v(T_{n(r-1)}, T_{nr}))^T$$

to establish the existence and uniqueness of the estimator as well as characterize its form. However, a somewhat more direct path toward the latter goal can be obtained using Theorem 2.7.13. An application of this result indicates that if R_q is the rk for $\mathbb{W}_q[0, 1]$ under the norm determined by (8.36) then $\mathbb{H}$ is an RKHS with rk given by

$$R(s_1, t_1, s_2, t_2) = R_q(s_1, t_1)R_q(s_2, t_2)$$

for any $s_1, s_2, t_1, t_2 \in [0, 1]$. One can then use the reproducing property of R in combination with an argument like the one employed in the proof of

Theorem 6.4.1 to see that

$$K_\eta(s,t) = \sum_{i=1}^{n} \sum_{1\le j<k\le r} a_{ijk} R(s, T_{ij}, t, T_{ik})$$
$$= \sum_{i=1}^{n} \sum_{1\le j<k\le r} a_{ijk} R_q(s, T_{ij}) R_q(t, T_{ik})$$

for some set of coefficients a_{ijk}. Equivalently, we can write

$$K_\eta(s,t) = \sum_{i=1}^{n} R_i(s)^T A_i R_i(t) \tag{8.53}$$

for symmetric matrices $A_i, i = 1, \dots, n$ with zero diagonal entries and

$$R_i(s) = (R_q(s, T_{i1}), \dots, R_q(s, T_{ir}))^T. \tag{8.54}$$

By substituting such explicit forms into the criterion function $f_{rn,\eta}$, one finds that the coefficients may be computed directly via, e.g., quadratic programming methodology.

The performance of K_η will be measured by

$$\|K_\eta - K\|^2_{\mathbb{L}^2([0,1]\times[0,1])} = \int_0^1 \int_0^1 (K_\eta(s,t) - K(s,t))^2 ds dt.$$

In this regard, we will show the following.

Theorem 8.3.5 *Assume that* $\mathbb{E}\varepsilon_{11}^4 < \infty$ *and that there is a* $C > 0$ *such*

$$\mathbb{E}X^4(t) \le C[\mathbb{E}X^2(t)]^2 \tag{8.55}$$

for all $t \in [0,1]$ *and for all* $k \ge 1$

$$\mathbb{E}\left(\int_0^1 X(t)e_k(t)dt\right)^4 \le C\left[\mathbb{E}\left(\int_0^1 X(t)e_k(t)dt\right)^2\right]^2. \tag{8.56}$$

Then, if $\eta \asymp (\log n/rn)^{2q/(2q+1)}$

$$\|K_\eta - K\|^2_{\mathbb{L}^2([0,1]\times[0,1])} = O_p\left(\left(\frac{\log n}{rn}\right)^{2q/(2q+1)} + n^{-1}\right). \tag{8.57}$$

Note that $X \in \mathbb{W}_q[0,1]$ entails that

$$\mathbb{E}X^4(t) = \mathbb{E}\langle X, R_q(\cdot,t)\rangle^4_{\mathbb{W}_q[0,1]} \le R_q^2(t,t)\mathbb{E}\|X\|^4_{\mathbb{W}_q[0,1]}.$$

So, condition (8.55) is satisfied if $\|X\|^4_{\mathbb{W}_q[0,1]}$ has finite expectation. Condition (8.56) holds for normal processes with $C = 3$.

In the special case where $X = Ae$ for a random coefficient A and some function $e \in \mathbb{W}_q[0,1]$ of unit norm, Cai and Yuan (2010) show that a minimax lower bound for estimation of K corresponds to squared $\mathbb{L}^2([0,1]\times[0,1])$ norm convergence at the rate $(rn)^{-2q/(2q+1)} + n^{-1}$. Thus, K_η can be viewed as near optimal from this perspective.

Proof: The proof is quite similar to that of Theorem 8.3.2. So, we merely point out the major differences here.

In keeping with the notation that was used in the previous proof, we first define

$$f_{rn}(v) = \frac{1}{r(r-1)n}\sum_{i=1}^{n}\sum_{1\le j<k\le r}(Y_{ij}Y_{ik} - v(T_{ij},T_{ik}))^2$$

and take

$$\begin{aligned} f_\infty(v) &= \mathbb{E}f_{rn}(v) \\ &= \mathrm{Var}(Y_{11}Y_{12}) + \int_0^1\int_0^1 (v(s,t) - K(s,t))^2 ds dt. \end{aligned}$$

Then, $f_{rn,\eta}(v) = f_{rn}(v) + \eta\|v\|^2_{\mathbb{H}}$ and

$$\begin{aligned} f_{\infty,\eta}(v) &= \mathbb{E}f_{rn,\eta}(v) \\ &= \mathrm{Var}(Y_{11}Y_{12}) + \int_0^1\int_0^1 (v(s,t) - K(s,t))^2 ds dt + \eta\|v\|^2_{\mathbb{H}} \end{aligned}$$

with

$$\overline{K}_\eta = \mathrm{argmin}_{v\in\mathbb{H}} f_{\infty,\eta}(v).$$

The proof now proceeds exactly as for the mean function estimation case using identity (8.39) except that now $f_{rn,\eta}$ refers to criterion (8.52). Thus, we seek bounds for the (squared) $\mathbb{L}^2([0,1]\times[0,1])$ norms of $K_\eta - \tilde{K}_\eta$, $\tilde{K}_\eta - \overline{K}_\eta$, and $\overline{K}_\eta - K$ with

$$\tilde{K}_\eta = \overline{K}_\eta - (f''_{\infty,\eta})^{-1} f'_{rn,\eta}(\overline{K}_\eta).$$

In this regard, we can immediately conclude from the definition of $\overline{K}_\eta$ that

$$\int_0^1\int_0^1 (\overline{K}_\eta(s,t) - K(s,t))^2 ds dt \le \eta\|K\|^2_{\mathbb{H}}.$$

Let $\mathcal{Q}$ be the congruence mapping from $\mathfrak{B}(\mathbb{H}, \mathbb{R})$ onto $\mathbb{H}$ from Theorem 3.2.1. Then, one finds that $\mathcal{Q}f''_{\infty\eta}$ is an invertible element of $\mathfrak{B}(\mathbb{H})$ with inverse determined by

$$(\mathcal{Q}f''_{\infty,\eta})^{-1}v = \frac{1}{2}\sum_{\nu=1}^{\infty}\sum_{\tau=1}^{\infty}\frac{(1+\gamma_\nu)(1+\gamma_\tau)}{1+\eta(1+\gamma_\nu)(1+\gamma_\tau)}v_{\nu\tau}e_\nu e_\tau$$

for any $v = \sum_{\nu=1}^{\infty}\sum_{\tau=1}^{\infty} v_{\nu\tau}e_\nu e_\tau \in \mathbb{H}$. In addition, because $f'_{\infty,\eta}(\overline{K}_\eta) = 0$,

$$\begin{aligned}
f'_{rn,\eta}(\overline{K}_\eta)v &= [f'_{rn}(\overline{K}_\eta) - f'_{\infty}(\overline{K}_\eta)]v \\
&= -\frac{2}{r(r-1)n}\sum_{i=1}^{n}\sum_{1\le j<k\le r}(Y_{ij}Y_{ik} - \overline{K}_\eta(T_{ij}, T_{ik}))v(T_{ij}, T_{ik}) \\
&\quad + 2\int_0^1\int_0^1 (K(s,t) - \overline{K}_\eta(s,t))v(s,t)dsdt.
\end{aligned}$$

Thus,

$$\begin{aligned}
\|\tilde{K}_\eta - \overline{K}_\eta\|^2_{\mathbb{L}^2([0,1]\times[0,1])} &= \sum_{\nu=1}^{\infty}\sum_{\tau=1}^{\infty}\langle (f''_{\infty,\eta})^{-1}f'_{rn,\eta}(\overline{K}_\eta), e_\nu e_\tau\rangle^2_{\mathbb{L}([0,1]\times[0,1])} \\
&= \sum_{\nu=1}^{\infty}\sum_{\tau=1}^{\infty}\frac{\langle (f''_{\infty,\eta})^{-1}f'_{rn,\eta}(\overline{K}_\eta), e_\nu e_\tau\rangle^2_{\mathbb{H}}}{(1+\gamma_\nu)(1+\gamma_\tau)} \\
&= \sum_{\nu=1}^{\infty}\sum_{\tau=1}^{\infty}\frac{V^2_{\nu\tau}}{(1+\eta(1+\gamma_\nu)(1+\gamma_\tau))^2},
\end{aligned}$$

where

$$\begin{aligned}
V_{\nu\tau} &= -\frac{1}{r(r-1)n}\sum_{i=1}^{n}\sum_{1\le j<k\le r}(Y_{ij}Y_{ik} - \overline{K}_\eta(T_{ij}, T_{ik}))e_\nu(T_{ij})e_\tau(T_{ij}) \\
&\quad + \int_0^1\int_0^1 (K(s,t) - \overline{K}_\eta(s,t))e_\nu(s)e_\tau(t)dsdt.
\end{aligned}$$

Somewhat more generally, we may work with

$$\|v\|^2_\alpha = \sum_{\nu=1}^{\infty}\sum_{\tau=1}^{\infty}(1+\gamma_\nu)^\alpha(1+\gamma_\tau)^\alpha v^2_{\nu\tau}$$

for $v = \sum_{\nu=1}^{\infty}\sum_{\tau=1}^{\infty} v_{\nu\tau}e_\nu e_\tau \in \mathbb{H}$ and $0 \le \alpha \le 1$. The previous calculation then leads to

$$\|\tilde{K}_\eta - \overline{K}_\eta\|^2_\alpha = \sum_{\nu=1}^{\infty}\sum_{\tau=1}^{\infty}\frac{(1+\gamma_\nu)^\alpha(1+\gamma_\tau)^\alpha}{(1+\eta(1+\gamma_\nu)(1+\gamma_\tau))^2}V^2_{\nu\tau}.$$

Now $\mathbb{E}V_{\nu\tau} = 0$ and, hence,

$$\begin{aligned}
&\mathbb{E}V_{\nu\tau}^2 \\
&= \frac{1}{nr^2(r-1)^2}\mathrm{Var}\left(\sum_{1\le j<k\le r}\left(K(T_{1j},T_{1k}) - \overline{K}_\eta(T_{1j},T_{1k})\right)e_\nu(T_{1j})e_\tau(T_{1k})\right) \\
&+\frac{1}{nr^2(r-1)^2}\mathbb{E}_T\left[\mathrm{Var}\left(\sum_{1\le j<k\le r} Y_{1j}Y_{1k}e_\nu(T_{1j})e_\tau(T_{1k})T\right|\right)\right].
\end{aligned}$$

For the first term

$$\begin{aligned}
&\mathrm{Var}\left(\sum_{1\le j<k\le r}\left(K(T_{1j},T_{1k}) - \overline{K}_\eta(T_{1j},T_{1k})\right)e_\nu(T_{1j})e_\tau(T_{1k})\right) \\
&\le \mathbb{E}\left(\sum_{1\le j<k\le r}\left(K(T_{1j},T_{1k}) - \overline{K}_\eta(T_{1j},T_{1k})\right)e_\nu(T_{1j})e_\tau(T_{1k})\right)^2 \\
&\le r^4\|K-\overline{K}_\eta\|^2_{\mathbb{L}^2([0,1]\times[0,1])} + O(r^3) \\
&= O(r^4\eta) + O(r^3).
\end{aligned}$$

For the second term, write

$$\begin{aligned}
\sum_{1\le j<k\le r} Y_{1j}Y_{1k}e_\nu(T_{1j})e_\tau(T_{1k}) = &\sum_{1\le j<k\le r} X(T_{1j})X(T_{1k})e_\nu(T_{1j})e_\tau(T_{1k}) \\
&+\sum_{1\le j<k\le r} \varepsilon_{1j}X(T_{1k})e_\nu(T_{1j})e_\tau(T_{1k}) \\
&+\sum_{1\le j<k\le r} X(T_{1j})\varepsilon_{1k}e_\nu(T_{1j})e_\tau(T_{1k}) \\
&+\sum_{1\le j<k\le r} \varepsilon_{1j}\varepsilon_{1k}e_\nu(T_{1j})e_\tau(T_{1k}).
\end{aligned}$$

Denote the four terms in this last expression by U_1, U_2, U_3, and U_4 with indices corresponding to their location in the sum. Then,

$$\mathbb{E}_T\left[\mathrm{Var}\left(\sum_{i=1}^4 U_i T\right|\right)\right] \le 4\sum_{i=1}^4 \mathbb{E}[U_i^2]. \tag{8.58}$$

Each of the $\mathbb{E}[U_i^2]$ values in (8.58) can now be treated in turn under the conditions stated in the theorem to derive the bound we need. We will illustrate how this works for U_1 as the other three are somewhat simpler to handle and of order at most r^3.

First note that

$$
\begin{aligned}
\mathbb{E}U_1^2 &= r(r-1)(r-2)(r-3)\mathbb{E}[A_{1\nu}^2 A_{1\tau}^2] \\
&\quad + r(r-1)(r-2)\mathbb{E}\left[A_{1\nu}A_{1\tau}\left(\int_0^1 X^2(t)e_\nu(t)e_\tau(t)dt\right)\right] \\
&\quad + r(r-1)(r-2)\mathbb{E}[A_{1\nu}^2 A_{2\tau} + A_{2\nu}A_{1\tau}^2] + r(r-1)\mathbb{E}[A_{2\nu}A_{2\tau}],
\end{aligned}
$$

where

$$
\begin{aligned}
A_{1\nu} &= \int_0^1 X(t)e_\nu(t)dt, \\
A_{2\nu} &= \int_0^1 X^2(t)e_\nu^2(t)dt.
\end{aligned}
$$

The first term that involves $A_{1\nu}^2 A_{1\tau}^2$ requires the most delicate treatment because it will eventually provide the coefficient for the order n^{-1} part of our approximation to $\|\tilde{K}_\eta - \overline{K}_\eta\|_\alpha^2$. It needs to be summable in a sense that will be made specific momentarily. In contrast, the other terms can simply be bounded by constants and we will give an illustration of how this can be accomplished for a typical case.

From the Cauchy–Schwarz inequality and assumption (8.56),

$$
\begin{aligned}
\mathbb{E}[A_{1\nu}^2 A_{1\tau}^2] &\le C\mathbb{E}[A_{1\nu}^2]\mathbb{E}[A_{1\tau}^2] \\
&= Ca_{\nu\nu}a_{\tau\tau}
\end{aligned}
$$

with

$$
a_{\nu\nu} = \int_0^1\int_0^1 K(s,t)e_\nu(s)e_\nu(t)dsdt = \mathbb{E}[A_{1\nu}^2].
$$

Then, one finds that

$$
\begin{aligned}
\sum_{\nu=1}^{\infty}(1+\gamma_\nu)a_{\nu\nu} &= \mathbb{E}\left[\sum_{\nu=1}^{\infty}(1+\gamma_\nu)A_{1\nu}^2\right] \\
&= \mathbb{E}\|X\|_{\mathbb{W}_2[0,1]}^2.
\end{aligned}
$$

We will have use for this feature subsequently.

Next observe that

$$\mathbb{E}\left[A_{1\nu}A_{1\tau}\left(\int_0^1 X^2(t)e_\nu(t)e_\tau(t)dt\right)\right]$$
$$\leq (\mathbb{E}[A_{1\nu}^2A_{1\tau}^2])^{1/2}\left(\mathbb{E}\left[\left(\int_0^1 X^2(t)e_\nu(t)e_\tau(t)dt\right)^2\right]\right)^{1/2}.$$

Then, assumption (8.56) gives us

$$\mathbb{E}\left[\left(\int_0^1 X^2(t)e_\nu(t)e_\tau(t)dt\right)^2\right]$$
$$= \int_0^1\int_0^1 \mathbb{E}[X^2(s)X^2(t)]e_\nu(s)e_\tau(s)e_\nu(t)e_\tau(t)dsdt$$
$$\leq \int_0^1\int_0^1 (\mathbb{E}[X^4(s)])^{1/2}(\mathbb{E}[X^4(t)])^{1/2}|e_\nu(s)e_\tau(s)e_\nu(t)e_\tau(t)|dsdt$$
$$\leq C\left(\int_0^1 \mathbb{E}\left[X^2(t)\right]|e_\nu(s)e_\tau(s)|ds\right)^2$$
$$= C\left(\int_0^1 K(s,s)|e_\nu(s)e_\tau(s)|ds\right)^2.$$

The $|e_\nu|$ are uniformly bounded and as $K \in \mathbb{H}$

$$K(s,s) = \langle K, R(\cdot, s, \star, s)\rangle_{\mathbb{H}} \leq \|K\|_{\mathbb{H}}R_q(s,s).$$

We already know that $\mathbb{E}[A_{1\nu}^2A_{1\tau}^2] \leq Ca_{\nu\nu}a_{\tau\tau}$ and these nonnegative quantities are uniformly bounded by, e.g., their sum.

The final result of all the requisite approximations is that

$$\mathbb{E}V_{\nu\tau}^2 = O(n^{-1}[a_{\nu\nu}a_{\nu\nu} + \eta + r^{-1}]).$$

As,

$$\sum_{\nu=1}^{\infty}\sum_{\tau=1}^{\infty}\frac{(1+\gamma_\nu)^\alpha(1+\gamma_\tau)^\alpha}{(1+\eta(1+\gamma_\nu)(1+\gamma_\tau))^2}a_{\nu\nu}a_{\tau\tau} \leq \left(\sum_{\nu=1}^{\infty}(1+\gamma_\nu)a_{\nu\nu}\right)^2$$
$$= (\mathbb{E}\|X\|_{\mathbb{W}_2[0,1]}^2)^2$$

and, from Lemma 2.3 of Lin (2000),

$$\sum_{\nu=1}^{\infty}\sum_{\tau=1}^{\infty}\frac{(1+\gamma_\nu)^\alpha(1+\gamma_\tau)^\alpha}{(1+\eta(1+\gamma_\nu)(1+\gamma_\tau))^2} = O(\eta^{-(\alpha+1/(2q))}\log(1/\eta))$$

it follows that

$$\|\overline{K}_\eta - \tilde{K}_\eta\|_\alpha^2$$
$$= O_p(n^{-1} + (nr)^{-1}\eta^{-(\alpha+1/(2q))}\log(1/\eta) + n^{-1}\eta^{1-(\alpha+1/(2q))}\log(1/\eta)).$$

The final step is to deal with $K_\eta - \tilde{K}_\eta$. As in the mean estimation case, we find that

$$\tilde{K}_\eta - K_\eta = (f''_{\infty,\eta})^{-1}[f''_\infty(K_\eta - \overline{K}_\eta) - f''_{rn}(K_\eta - \overline{K}_\eta)]$$

with

$$[f''_\infty - f''_{rn}]v = \frac{1}{r(r-1)n}\sum_{i=1}^{n}\sum_{1\le j<k\le r} v(T_{ij}, T_{ik}) - \int_0^1\int_0^1 v(s,t)dsdt.$$

Thus,

$$\|K_\eta - \tilde{K}_\eta\|_\alpha^2 = \sum_{\nu=1}^{\infty}\sum_{\tau=1}^{\infty}\frac{(1+\gamma_\nu)^\alpha(1+\gamma_\tau)^\alpha}{(1+\eta(1+\gamma_\nu)(1+\gamma_\tau))^2}V_{\nu\tau}^2$$

with $V_{\nu\tau}$ now defined by

$$V_{\nu\tau} = \frac{1}{r(r-1)n}\sum_{i=1}^{n}\sum_{1\le j<k\le r}(K_\eta - \overline{K}_\eta)(T_{ij}, T_{ik})$$
$$- \int_0^1\int_0^1 (K_\eta - \overline{K}_\eta)(s,t)dsdt.$$

At this point, we take $K_\eta - \overline{K}_\eta = \sum_{\nu=1}^{\infty}\sum_{\tau=1}^{\infty} h_{\nu\tau}e_\nu e_\tau$ and proceed exactly as in the mean estimation case to complete the proof. □

We conclude this chapter with some remarks about how the estimation methods in Sections 8.2 and 8.3 compare to one another. First observe that if $q = 2$, the convergence rates for the local linear smoothers and penalized least-squares estimators roughly coincide. By assuming more derivatives in the local smoothing approach and using a higher (than linear) order polynomial in the fit, this rate similarity would presumably extend to $q > 2$. It is, however, noteworthy that the rates for the penalized least-squares approach are obtained under stronger assumptions in that X must have differentiable sample paths. On the other hand, the penalized least-squares approach is to be preferred from the perspective of computational efficiency.

9

Principal components analysis

In Section 1.1, we considered a p-dimensional random vector X with $\mathbb{E}X = m$ and covariance matrix $\mathscr{K}$. If

$$\mathscr{K} = \sum_{j=1}^{p} \lambda_j e_j e_j^T$$

is the eigenvalue–eigenvector decomposition for the covariance matrix, we saw that X could be decomposed as

$$X = m + \sum_{j=1}^{p} Z_j e_j$$

for $Z_j = e_j^T(X - m)$ zero mean, uncorrelated random variables having $\mathrm{Var}(Z_j) = \lambda_j$. The Z_j provide a new set of variables that are called the principal components of X. One may, for example, opt to use the collection $Z_1, \ldots, Z_r$ for some $r < p$ as a surrogate for X, thereby achieving a form of dimension reduction that can have both conceptual and practical utility when p is large. The process of obtaining the Z_j is generally called principal components analysis or just pca subsequently.

As one would anticipate, dimension reduction in the infinite dimensional environments of fda and related fields can be particularly worthwhile. One common way to accomplish this is through the functional analog of the principal components concept. This chapter is devoted to that topic.

Suppose now that χ is a random element of a Hilbert space $\mathbb{H}$ with $\mathbb{E}\|\chi\|^2 < \infty$. Then, we know from Theorem 7.2.6 that χ has a covariance operator $\mathscr{K}$ that admits the spectral decomposition

$$\mathscr{K} = \sum_{j=1}^{\infty} \lambda_j e_j \otimes e_j \tag{9.1}$$

Theoretical Foundations of Functional Data Analysis, with an Introduction to Linear Operators, First Edition. Tailen Hsing and Randall Eubank.

and, from Theorem 7.2.7, that with probability one,

$$\chi = m + \sum_{j=1}^{\infty} Z_j e_j \tag{9.2}$$

for $m = \mathbb{E}[\chi]$ and $Z_j = \langle (\chi - m), e_j \rangle$ zero mean, uncorrelated random variables with $\mathrm{Var}(Z_j) = \lambda_j$. The mva version of pca is now seen to be just a special case of (9.1) and (9.2) that applies when $\mathbb{H} = \mathbb{R}^p$. Accordingly, we will continue to refer to the Z_j as principal components and term the process of evaluating (9.1) and (9.2) pca.

The fda version of pca arises when we have a stochastic process $X = \{X(t), t \in E\}$ that is also a random element in a Hilbert space $\mathbb{H}$. We discussed two such settings in Sections 7.4 and 7.5. In this chapter, we focus on the former where $\mathbb{H} = \mathbb{L}^2(E, \mathscr{B}(E), \mu)$. Theorem 7.4.3 then has the consequence that knowledge of the covariance operator $\mathscr{K}$ is equivalent to knowing the process covariance kernel

$$K(s, t) = \mathrm{Cov}(X(s), X(t)).$$

Mercer's Theorem (Theorem 4.6.5) tells us that if, e.g., X is mean-square continuous,

$$K(s, t) = \sum_{j=1}^{\infty} \lambda_j e_j(s) e_j(t) \tag{9.3}$$

with $\{(\lambda_j, e_j)\}_{j=1}^{\infty}$, the eigen sequence for the $\mathbb{L}^2(E, \mathscr{B}(E), \mu)$ integral operator corresponding to K and that the sequence converges absolutely and uniformly in s and t. The analog of the principal component decomposition for a Hilbert space random element in this instance derives from the Karhunen–Lòeve expansion (Theorem 7.3.5) that allows us to write

$$X(t) = m(t) + \sum_{j=1}^{\infty} Z_j e_j(t)$$

with $m(t) = \mathbb{E}[X(t)]$ and

$$Z_j = \int_E (X(t) - m(t))\, e_j(t) d\mu(t)$$

whose convergence properties were the subject of Theorem 7.3.5. The potential benefits of dimension reduction are decidedly more profound here which is why estimating principal components is at least as important for fda as it is for mva. Somewhat more generally, we can view functional pca with data as a means of carrying out inference about the properties of the process covariance structure with dimension reduction being just a useful by-product. It is also a useful theoretical tool with, e.g., sample estimators of eigenfunctions providing a convenient set of basis functions for use in large sample work.

In subsequent sections, we will investigate the large sample properties of three schemes for empirical pca. Our treatment of the topic runs parallel to developments in Chapter 8. We first deal with the case of completely observed sample paths using the sample covariance operator. Then, we turn to the case of discretely observed noisy data using the covariance kernel estimators from Sections 8.2 and 8.3.

9.1 Estimation via the sample covariance operator

We begin by examining the asymptotic properties of the eigenvalues and eigenvectors of the sample covariance operator. For that purpose, let χ be a second-order random element of a Hilbert space $\mathbb{H}$ with associated covariance operator $\mathscr{K}$. Denote the nonrepeated eigenvalues for $\mathscr{K}$ by $\lambda_1 > \lambda_2 > \cdots$ where, for each j, λ_j has multiplicity $r_j \geq 1$. The projection operator for the r_j dimensional eigen-space corresponding to λ_j will be indicated by $\mathscr{P}_j$.

Now assume that $\chi_1, \ldots, \chi_n$ are independent copies of χ with sample covariance operator

$$\mathscr{K}_n = \frac{1}{n-1} \sum_{i=1}^{n} (\chi_i - m_n) \otimes (\chi_i - m_n)$$

for $m_n = n^{-1} \sum_{i=1}^{n} \chi_i$. Let $\{\lambda_{jn}\}_{j=1}^{n-1}$ be the (possibly) repeated eigenvalues for $\mathscr{K}_n$. For any fixed j and n sufficiently large, we can then define associated projection operators $\mathscr{P}_{jn}$ as in Theorem 5.1.4. Specifically, if e_{jn} is the eigenvector that corresponds to λ_{jn}, we take

$$\mathscr{P}_{jn} = \sum_{k=N_{j-1}+1}^{N_j} e_{kn} \otimes e_{kn}$$

for $N_j = \sum_{i=1}^{j} r_i$.

Concerning the eigen-projection operators, we will show

Theorem 9.1.1 *If* $\mathbb{E}(\|\chi\|^4) < \infty$,

$$n^{1/2}(\mathscr{P}_{jn} - \mathscr{P}_j) \xrightarrow{d} \mathcal{S}_j \mathfrak{Z} \mathscr{P}_j + \mathscr{P}_j \mathfrak{Z} \mathcal{S}_j,$$

where $\mathfrak{Z}$ *is the distributional limit of* $n^{1/2}(\mathscr{K}_n - \mathscr{K})$ *from Theorem 8.1.2 and*

$$\mathcal{S}_j = \sum_{\lambda_k \neq \lambda_j} \frac{1}{\lambda_j - \lambda_k} \mathscr{P}_k.$$

Proof: Let $\Delta_n = \mathscr{K}_n - \mathscr{K}$ with $\mathscr{R}$ and $\mathscr{R}_n$ the resolvents of $\mathscr{K}$ and $\mathscr{K}_n$, respectively. Define $\eta_j = (1/2)\min_{k\neq j}|\lambda_k - \lambda_j|$ and denote by Γ the circle in the complex plane centered at λ_j with radius η_j. By Theorem 8.1.2, $\Delta_n \xrightarrow{a.s.} 0$ and, hence,

$$\mathbb{P}(\|\Delta_n\| < \eta_j \text{ eventually}) = 1.$$

Then, Theorem 5.1.4 implies that

$$\mathbb{P}\left(\mathscr{P}_{jn} - \mathscr{P}_j = \mathcal{S}_j\Delta_n\mathscr{P}_j + \mathscr{P}_j\Delta_n\mathcal{S}_j + \frac{1}{2\pi i}\oint_\Gamma \mathscr{M}_n(z)dz \text{ eventually}\right) = 1,$$

where

$$\mathscr{M}_n(z) = \mathscr{R}(z)\sum_{k=2}^{\infty}\{-\Delta_n\mathscr{R}(z)\}^k.$$

Corollary 5.1.5 and (5.17) show that

$$\mathscr{P}_{jn} - \mathscr{P}_j = \mathcal{S}_j\Delta_n\mathscr{P}_j + \mathscr{P}_j\Delta_n\mathcal{S}_j + O_p(\|\Delta_n\|^2),$$

where $\|\Delta_n\|^2 = O_p(n^{-1})$ and the proof is complete. □

The following result describes the asymptotic distribution of the eigenvalues for $\mathscr{K}_n$.

Theorem 9.1.2 *Assume that $\mathbb{E}(\|\chi\|^4) < \infty$. Then, for any fixed j,*

$$n^{1/2}(\lambda_{kn} - \lambda_j)_{k=N_{j-1}+1,\ldots,N_j} \xrightarrow{d} \lambda(\mathscr{P}_j\mathfrak{Z}\mathscr{P}_j),$$

where $\mathscr{P}_j$ is the projection operator for the eigen-space of λ_j, $\mathfrak{Z}$ is the distributional limit of $n^{1/2}(\mathscr{K}_n - \mathscr{K})$ and $\lambda(\mathscr{P}_j\mathfrak{Z}\mathscr{P}_j)$ are the eigenvalues of $\mathscr{P}_j\mathfrak{Z}\mathscr{P}_j$.

Proof: It is easy to see that $n^{1/2}(\lambda_{kn} - \lambda_j)_{k=N_{j-1}+1,\ldots,N_j}$ are the eigenvalues of $n^{1/2}(\mathscr{P}_{jn}\mathscr{K}_n\mathscr{P}_{jn} - \lambda_j\mathscr{P}_{jn})$. As in the proof of Theorem 5.1.6, write

$$\begin{aligned}
&\mathscr{P}_{jn}\mathscr{K}_n\mathscr{P}_{jn} - \lambda_j\mathscr{P}_{jn}\\
&\quad = \mathscr{P}_{jn}\Delta_n\mathscr{P}_{jn} + (\mathscr{P}_{jn} - \mathscr{P}_j)(\mathscr{K} - \lambda_j I)(\mathscr{P}_{jn} - \mathscr{P}_j)\\
&\quad = \mathscr{P}_j\Delta_n\mathscr{P}_j + (\mathscr{P}_{jn} - \mathscr{P}_j)\Delta_n\mathscr{P}_{jn} + \mathscr{P}_j\Delta_n(\mathscr{P}_{jn} - \mathscr{P}_j)\\
&\qquad + (\mathscr{P}_{jn} - \mathscr{P}_j)(\mathscr{K} - \lambda_j I)(\mathscr{P}_{jn} - \mathscr{P}_j).
\end{aligned}$$

As all three terms $(\mathscr{P}_{jn} - \mathscr{P}_j)\Delta_n\mathscr{P}_{jn}$, $\mathscr{P}_j\Delta_n(\mathscr{P}_{jn} - \mathscr{P}_j)$, and $(\mathscr{P}_{jn} - \mathscr{P}_j)(\mathscr{K} - \lambda_j I)(\mathscr{P}_{jn} - \mathscr{P}_j)$ have rates $O_p(n^{-1})$ by Theorem 9.1.1, we conclude that $n^{1/2}(\lambda_{kn} - \lambda_j), k = N_{j-1} + 1, \ldots, N_j$ are asymptotically equal to the eigenvalues of $\mathscr{P}_j(n^{1/2}\Delta_n)\mathscr{P}_j$. □

Theorems 9.1.1 and 9.1.2 are of particular interest when $r_j = 1$ for some j.

Theorem 9.1.3 *Assume that* $\mathbb{E}(\|\chi\|^4) < \infty$ *and that, for some fixed* j, λ_j *is an eigenvalue with unit multiplicity that has* e_j *as its corresponding eigenfunction. If* λ_{jn}, e_{jn} *are the matching eigenvalue and eigenfunction for* $\mathscr{K}_n$,

$$n^{1/2}(\lambda_{jn} - \lambda_j) \xrightarrow{d} \langle \mathfrak{Z} e_j, e_j \rangle$$

and

$$n^{1/2}(e_{jn} - e_j) \xrightarrow{d} \sum_{k \neq j} (\lambda_j - \lambda_k)^{-1} \mathscr{P}_k \mathfrak{Z} e_j$$

with $\mathfrak{Z}$ *the distributional limit of* $n^{1/2}(\mathscr{K}_n - \mathscr{K})$.

Proof: The first conclusion follows from Theorem 9.1.2, whereas the second is a consequence of Theorems 5.1.8 and 8.1.2. □

9.2 Estimation via local linear smoothing

In the following two sections, we focus on the fda setting where we have a second-order stochastic process X that is a random element of $\mathbb{L}^2[0,1]$. The process sample paths are then observed discretely with additive random noise and realized in terms of response values

$$Y_{ij} = X_i(T_{ij}) + \varepsilon_{ij}, j = 1, \ldots, r, i = 1, \ldots, n,$$

with the X_i being independent replicas of X and the T_{ij} and ε_{ij} are iid sampling points and random errors. In Section 8.2, we used the Y data to construct an estimator of the X covariance kernel that had the form

$$K_h(s,t) = R_{h_R}(s,t) - m_{h_m}(s) m_{h_m}(t),$$

where $m_{h_m}(t)$ and $R_{h_R}(s,t)$ are estimators of the mean function $m(t) = \mathbb{E}[X(t)]$ and $R(s,t) = \mathbb{E}[X(s)X(t)]$ obtained via local linear smoothing of the responses Y_{ij} and response products $Y_{ij}Y_{ik}, j \neq k$. The vector $h = (h_m, h_R)$ contains the two bandwidths h_m and h_R that are used for estimation of m and R.

Rates of convergence for K_h as an estimator of the X process covariance kernel K were provided in Theorem 8.2.4. Here, our goal is somewhat different. Write $K_h(s,t)$ as

$$K_h(s,t) = \sum_{j=1}^{\infty} \hat{\lambda}_j \hat{e}_j(s) \hat{e}_j(t),$$

with $\{(\hat{\lambda}_j, \hat{e}_j)\}_{j=1}^{\infty}$ the eigen sequence for the $\mathbb{L}^2[0,1]$ integral operator that can be obtained from K_h. Our aim is to assess how well these entities perform as estimators of the eigenvalues and eigenfunctions $\{(\lambda_j, e_j)\}_{j=1}^{\infty}$ that appear in the analogous decomposition for K in (9.3). To simplify matters, we assume in what follows that the λ_j are all distinct.

Notice that the eigenfunctions e_j and $\hat{e}_j$ are uniquely identifiable only up to a sign change. This causes no problem in practice but represents a technicality that must be addressed in order to discuss the convergence rate of $\hat{e}_j$. Our resolution is to simply let e_j take an arbitrary sign while choosing $\hat{e}_j$ to minimize $\|\hat{e}_j - e_j\|$ over the two possible signs; i.e., $\hat{e}_j$ is chosen so that $\langle \hat{e}_j, e_j \rangle \geq 0$.

Let Δ_h be the integral operator with kernel $K_h(s,t) - K(s,t)$. Then, Theorem 8.2.7 provides the rate of convergence for $\|\Delta_h\|$ in our estimation setting. If we now take $\hat{\mathscr{P}}_j = \hat{e}_j \otimes \hat{e}_j$ and $\mathscr{P}_j = e_j \otimes e_j$, the perturbation result we obtained in (5.17) implies that $\|\hat{\mathscr{P}}_j - \mathscr{P}_j\| = O_p(\|\Delta_h\|/\gamma_j)$ for all j such that $\|\Delta_h\|/\gamma_j \to 0$, where $\gamma_j = (1/2)\min_{\lambda_k \neq \lambda_j} |\lambda_k - \lambda_j|$. Thus, Theorem 8.2.7 provides the rate of convergence of $\hat{\mathscr{P}}_j$ to all such $\mathscr{P}_j$ simultaneously. In the following, we address the rates of convergence for estimation of λ_j and e_j.

For a given bandwidth h, we will again need the notation

$$\delta_{n1}(h) = \left(\{1 + (hr)^{-1}\}\log n/n\right)^{1/2},$$
$$\delta_{n2}(h) = \left(\{1 + (hr)^{-1} + (hr)^{-2}\}\log n/n\right)^{1/2}$$

that was employed in Section 8.2. Using these two definitions, we can state our first result concerning eigenvalue estimation as follows.

Theorem 9.2.1 *Assume that the conditions of Theorem 8.2.4 hold. Then, for any j with $\lambda_j > 0$,*

$$\hat{\lambda}_j - \lambda_j = O((\log n/n)^{1/2} + h_m^2 + h_R^2 + \delta_{n1}^2(h_m) + \delta_{n2}^2(h_R)) \quad a.s.$$

Proof: By (5.17) and (5.21),

$$\begin{aligned}
\hat{\lambda}_j - \lambda_j &= \int_0^1 \int_0^1 (K_h - K)(s,t)e_j(s)e_j(t)dsdt + O(\|\Delta_h\|^2) \\
&= \int_0^1 \int_0^1 (R_{h_R} - R)(s,t)e_j(s)e_j(t)dsdt \\
&\quad - \int_0^1 \int_0^1 \left(m_{h_m}(s)m_{h_m}(t) - m(s)m(t)\right) e_j(s)e_j(t)dsdt \\
&\quad + O(\|\Delta_h\|^2) \\
&= A_{n1} - A_{n2} + O(\|\Delta_h\|^2)
\end{aligned} \tag{9.4}$$

with

$$A_{n1} = \int_0^1 \int_0^1 (R_{h_R} - R)(s,t)e_j(s)e_j(t)dsdt$$

and

$$\begin{aligned} A_{n2} &= \int_0^1 \int_0^1 \left(m_{h_m}(s)m_{h_m}(t) - m(s)m(t)\right) e_j(s)e_j(t)dsdt \\ &= \int_0^1 \left(m_{h_m}(s) - m(s)\right) e_j(s)ds \int_0^1 m_{h_m}(t)e_j(t)dt \\ &\quad + \int_0^1 m(s)e_j(s)ds \int_0^1 \left(m_{h_m}(t) - m(t)\right) e_j(t)dt. \end{aligned}$$

An application of Theorem 8.2.1 using this last relation shows that

$$A_{n2} = O((\log n/n)^{1/2} + h_m^2) \quad \text{a.s.}$$

and we now only need to deal with A_{n1}.

Using the notation of Section 8.2, we can write

$$\begin{aligned} A_{n1} = \int_0^1 \int_0^1 & \left[\left(\mathcal{M}_1(s,t)\tilde{S}_{00}(s,t) - \mathcal{M}_2(s,t)\tilde{S}_{10}(s,t)\right.\right. \\ & \left.\left. -\mathcal{M}_3(s,t)\tilde{S}_{01}(s,t)\right) / \mathcal{D}(s,t)\right] e_j(s)e_j(t)dsdt. \end{aligned}$$

From (8.28) and Lemma 8.2.5, $\mathcal{M}_1(s,t)/\mathcal{D}(s,t)$ behaves like a constant multiple of $1/(f(s)f(t))$ with f the common density for the T_{ij}. So, using (8.29), it suffices to work with

$$\begin{aligned} \int_0^1 \left[\frac{\tilde{S}_{00}(s,t)}{f(s)f(t)}\right] e_j(s)e_j(t)dsdt = \sum_{i=1}^{3} \int_0^1 \int_0^1 \frac{U_i(s,t)}{f(s)f(t)} e_j(s)e_j(t)dsdt \\ +O(h_R^2). \qquad (9.5) \end{aligned}$$

For each i

$$\int_0^1 \int_0^1 \left[U_i(s,t)/(f(s)f(t))\right] e_j(s)e_j(t)dsdt = \frac{1}{n}\sum_{i=1}^{n} Z_{ni},$$

where

$$\begin{aligned} Z_{ni} = \frac{1}{r(r-1)} \sum\sum_{\substack{1\le j,k \le r \\ j \ne k}} Z_{ijk} \int_0^1 \left[W_{h_R}(T_{ij} - s)/f(s)\right] e_j(s)ds \\ \times \int_0^1 \left[W_{h_R}(T_{ik} - t)/f(t)\right] e_j(t)dt \end{aligned}$$

and Z_{ijk} is either $\varepsilon_{ij}\varepsilon_{ik}$, $\varepsilon_{ij}X_i(T_{ik})$ or $X_i(T_{ij})X_i(T_{ik}) - R(T_{ij}, T_{ik})$. For example, when $Z_{ijk} = \varepsilon_{ij}\varepsilon_{ik}$,

$$|Z_{ni}| \le \frac{Cr}{r-1}\left(\frac{1}{r}\sum_{j=1}^{r}|\varepsilon_{ij}|\right)^2,$$

where C is a constant such that

$$\left|\int_0^1 \left[W_{h_R}(T_{ij} - s)/f(s)\right] e_j(s)ds \int_0^1 \left[W_{h_R}(T_{ik} - t)/f(t)\right] e_j(t)dt\right| < C$$

uniformly. Then, from Jensen's inequality,

$$|Z_{ni}|^\delta \le \left(\frac{Cr}{r-1}\right)^\delta \frac{1}{r}\sum_{j=1}^{r}|\varepsilon_{ij}|^{2\delta}.$$

Thus, the conditions of Lemma 9.2.2 are met and we are led to the conclusion that $n^{-1}\sum_{i=1}^n Z_{ni} = O((\log n/n)^{1/2})$ a.s. The other two terms in A_{n1} can be dealt with similarly to see that

$$A_{n1} = O((\log n/n)^{1/2} + h_R^2).$$

Now use our approximations for A_{n1}, A_{n2} in (9.4) and apply Theorem 8.2.4 to complete the proof. □

The following is the technical result we employed in the previous proof.

Lemma 9.2.2 *Assume that Z_{ni}, $1 \le i \le n$, are independent random variables with mean zero satisfying*

$$\sup_{i,n} \mathbb{E}|Z_{ni}|^\delta < \infty$$

and

$$\frac{1}{n}\sum_{i=1}^{n}|Z_{ni}|^\delta = O(1) \quad a.s.$$

for some $\delta \in (2, \infty)$. Then,

$$\frac{1}{n}\sum_{i=1}^{n} Z_{ni} = O((\log n/n)^{1/2}) \quad a.s.$$

Proof: Let $a_n = (\log n/n)^{1/2}$ and write

$$Z_{ni} = Z_{ni>} + Z_{ni<} := Z_{ni}I(|Z_{ni}| > a_n^{-1}) + Z_{ni}I(|Z_{ni}| \le a_n^{-1}).$$

Then,

$$\left|\frac{1}{a_n n}\sum_{i=1}^{n} Z_{ni>}\right| \leq \frac{1}{a_n n}\sum_{i=1}^{n} |Z_{ni>}|^{\delta}|Z_{ni>}|^{1-\delta} \leq a_n^{\delta-2}\frac{1}{n}\sum_{i=1}^{n} |Z_{ni}|^{\delta} \to 0 \quad \text{a.s.}$$

The mean of the left-hand side is also tending to zero by the same argument. Thus, $n^{-1}\sum_{i=1}^{n}(Z_{ni>} - \mathbb{E}\{Z_{ni>}\}) = o(a_n)$ a.s. Next, by Bernstein's inequality,

$$\mathbb{P}\left(\frac{1}{n}\sum_{i=1}^{n}(Z_{ni<} - \mathbb{E}\{Z_{ni<}\}) > Ba_n\right) \leq \exp\left\{-\frac{B^2n^2a_n^2}{2n\sigma^2 + (2/3)Bn}\right\}$$
$$= \exp\left\{-\frac{B^2\log n}{2\sigma^2 + (2/3)B}\right\},$$

which is summable for large enough B. The result now follows from the Borel–Cantelli Lemma. □

The following theorem deals with estimation of eigenfunctions.

Theorem 9.2.3 *Assume that the conditions of Theorem 8.2.4 hold. Then, for any j such that $\lambda_j > 0$,*

$$\|\hat{e}_j - e_j\| = O(h_m^2 + h_R^2 + \delta_{n1}(h_m) + \delta_{n1}(h_R)) \quad a.s. \tag{9.6}$$

and

$$\sup_t |\hat{e}_j(t) - e_j(t)| = O(h_m^2 + h_R^2 + \delta_{n1}(h_m) + \delta_{n1}(h_R) + \delta_{n2}^2(h_R)) \quad a.s. \tag{9.7}$$

Proof: Theorem 5.1.8 gives the expansion:

$$\hat{e}_j - e_j = \sum_{k\neq j}(\lambda_j - \lambda_k)^{-1}\langle\Delta_h e_j, e_k\rangle e_k + O(\|\Delta_h\|^2).$$

By Bessel's inequality, this leads to

$$\|\hat{e}_j - e_j\| \leq C(\|\Delta_h e_j\| + \|\Delta_h\|^2)$$

for some finite $C > 0$. Then, from Theorems 8.2.4 and 8.2.7,

$$\|\Delta_h\|^2 = O(h_m^4 + h_R^4 + \delta_{n1}^2(h_m) + \delta_{n2}^2(h_R)) \quad \text{a.s.}$$

and

$$\|\Delta_h e_j\| = O(h_m^2 + h_R^2 + \delta_{n1}(h_m) + \delta_{n1}(h_R)) \quad \text{a.s.}$$

Thus,

$$\|\hat{e}_j - e_j\| = O(h_m^2 + h_R^2 + \delta_{n1}(h_m) + \delta_{n1}(h_R)) \quad \text{a.s.,}$$

thereby proving the first assertion of the theorem.

To establish (9.7), let t be any value in $[0, 1]$ and observe that

$$\begin{aligned}
&\hat{\lambda}_j\hat{e}_j(t) - \lambda_j e_j(t) \\
&\quad = \int_0^1 K_h(s,t)\hat{e}_j(s)ds - \int_0^1 K(s,t)e_j(s)ds \\
&\quad = \int_0^1 (K_h(s,t) - K(s,t))\, e_j(s)ds + \int_0^1 K_h(s,t)\left(\hat{e}_j(s) - e_j(s)\right) ds.
\end{aligned}$$

The first term can be bounded using Theorem 8.2.7 and the Cauchy–Schwarz inequality implies that uniformly for all $t \in [0, 1]$

$$\begin{aligned}
\left|\int_0^1 K_h(s,t)\left(\hat{e}_j(s) - e_j(s)\right) ds\right| &\le \left\{\int_0^1 K_h^2(s,t)ds\right\}^{1/2} \|\hat{e}_j - e_j\| \\
&\le \sup_{s,t} |K_h(s,t)|\,\|\hat{e}_j - e_j\| \\
&= O(\|\hat{e}_j - e_j\|) \quad \text{a.s.}
\end{aligned}$$

Thus, using (9.6), we see that

$$\hat{\lambda}_j\hat{e}_j(t) - \lambda_j e_j(t) = O(h_m^2 + h_R^2 + \delta_{n1}(h_m) + \delta_{n1}(h_R)) \quad \text{a.s.}$$

Then, the triangle inequality in combination with Theorem 9.2.1 gives

$$\begin{aligned}
&\lambda_j|\hat{e}_j(t) - e_j(t)| \\
&\quad = |\hat{\lambda}_j\hat{e}_j(t) - \lambda_j e_j(t) - (\hat{\lambda}_j - \lambda_j)\hat{e}_j(t)| \\
&\quad \le |\hat{\lambda}_j\hat{e}_j(t) - \lambda_j e_j(t)| + |\hat{\lambda}_j - \lambda_j| \sup_t |\hat{e}_j(t)| \\
&\quad = O((\log n/n)^{1/2} + h_m^2 + h_R^2 + \delta_{n1}(h_m) + \delta_{n1}(h_R) + \delta_{n2}^2(h_R)) \quad \text{a.s.}
\end{aligned}$$

As $(\log n/n)^{1/2} = o(\delta_{n1}(h_m))$, the proof is complete. □

We conclude the section with an eigenvalue/eigenfunction analog of Corollaries 8.2.3 and 8.2.6 that dealt with the cases of sparse and dense functional data.

Corollary 9.2.4 *Assume that the conditions of Theorem 8.2.4 hold and that $\lambda_j > 0$. If $r \le M$ for some fixed M,*

1. *if $(\log n/n)^{1/2} \lesssim h_m \lesssim (\log n/n)^{1/4}$ and $h_R \asymp (\log n/n)^{1/4}$,*

$$\hat{\lambda}_j - \lambda_j = O(\{\log n/n\}^{1/2}) \quad a.s.$$

and

2. *if* $h_R \asymp h_m$ *and* $h_R \gtrsim (\log n/n)^{1/3}$, *both* $\|\hat{e}_j - e_j\|$ *and* $\sup_t |\hat{e}_j(t) - e_j(t)|$ *have the rate* $O(h_R^2 + \{\log n/(nh_R)\}^{1/2})$ *a.s.*

If $r = r_n \to \infty$ *in such a way that* $r_n^{-1} \lesssim h_m, h_R \lesssim (\log n/n)^{1/4}$, *then* $\hat{\lambda}_j - \lambda_j$, $\|\hat{e}_j - e_j\|$ *and* $\sup_t |\hat{e}_j(t) - e_j(t)|$ *have the rate* $O(\{\log n/n\}^{1/2})$ *a.s.*

Corollary 9.2.4 shows that in both the dense and sparse functional data settings, the λ_j can be estimated at a root-n rate, while in the sparse data case, the eigenfunctions can be estimated at the optimal nonparametric rate for one-dimensional functions.

9.3 Estimation via penalized least squares

Section 9.2 used the local linear regression estimator of the covariance kernel from Section 8.2 for estimation of the eigenfunctions of K. Under this same basic model formulation, we can also employ the eigenfunctions of the smoothing spline type estimator from Section 8.3 for that purpose. In this section, we briefly explore what can be said about that option.

Recall from (8.53) and (8.54) that our penalized least-squares estimator of K had the form

$$K_\eta(s,t) = \sum_{i=1}^{n} R_i(s)^T \mathscr{A}_i R_i(t) \tag{9.8}$$

for symmetric matrices $\mathscr{A}_i, i = 1, \ldots, n$ with zero diagonal entries and

$$R_i(s) = \left(R_q(s, T_{i1}), \ldots, R_q(s, T_{ir})\right)^T$$

with R_q the rk for $\mathbb{W}_q[0,1]$. As a result of (9.8), any eigenfunction of K_η must be expressible as $\hat{e}(\cdot) = v^T R(\cdot)$ for some vector v and

$$R(\cdot) = \left(R_1(\cdot)^T, \ldots, R_n(\cdot)^T\right)^T. \tag{9.9}$$

We now claim that all such vectors have the form $v = \mathscr{G}^{-1/2}u$, where u is an eigenvector of the matrix

$$\mathscr{B} = \mathscr{G}^{1/2} \mathscr{A} \mathscr{G}^{1/2} \tag{9.10}$$

with

$$\mathscr{A} = \operatorname{diag}(\mathscr{A}_1, \ldots, \mathscr{A}_n)$$

and

$$\mathscr{G} = \left\{\mathscr{G}_{ij}\right\}_{i,j=1:n} \tag{9.11}$$

for

$$\mathcal{G}_{ij} = \left\{ \int_0^1 R_q(s, T_{iv}) R_q(s, T_{j\tau}) ds \right\}_{v,\tau=1:r}.$$

To see why this is so let $v_i = \mathcal{G}^{-1/2} u_i$ and $v_j = \mathcal{G}^{-1/2} u_j$ correspond to any two eigenvectors u_i, u_j of $\mathcal{B}$ in (9.10). Then, take $\hat{e}_i(\cdot) = v_i^T R(\cdot)$ and $\hat{e}_j(\cdot) = v_j^T R(\cdot)$ as our candidates for the corresponding eigenvectors of K_η. These choices give us orthonormal element of $\mathbb{L}^2[0, 1]$ because

$$\int_0^1 \hat{e}_i(s)\hat{e}_j(s)ds = v_i^T \mathcal{G} v_j = u_i^T u_j = \delta_{ij}$$

and, if $\hat{\lambda}_j$ is the eigenvalue of $\mathcal{B}$ that corresponds to u_j,

$$\begin{aligned}
\sum_{j=1}^{rn} \hat{\lambda}_j \hat{e}_j(s)\hat{e}_j(t) &= R(s)^T \left(\sum_{j=1}^{rn} \hat{\lambda}_j v_j v_j^T \right) R(t) \\
&= R(s)^T \mathcal{G}^{-1/2} \left(\sum_{j=1}^{rn} \hat{\lambda}_j u_j u_j^T \right) \mathcal{G}^{-1/2} R(t) \\
&= R(s)^T \mathcal{A} R(t) = K_\eta(s, t).
\end{aligned}$$

Let λ_j be the jth largest eigenvalue of the $\mathbb{L}^2[0, 1]$ integral operator generated by the X process covariance kernel K with associated eigenfunction e_j. If $\hat{\lambda}_j$ is the jth largest eigenvalue of the matrix $\mathcal{B}$ in (9.10) with corresponding eigenvector u_j, we estimate λ_j by $\hat{\lambda}_j$ and then estimate $e_j(\cdot)$ by $\hat{e}_j(\cdot) = u_j^T R(\cdot)$ with R defined in (9.9). As in Theorem 9.2.3, we can without loss assume that u_j is such that $\langle e_j, \hat{e}_j \rangle \geq 0$. The asymptotic behavior of these estimators is described in the following theorem.

Theorem 9.3.1 *Assume that λ_j has unit multiplicity and that the smoothing parameter for K_η satisfies $\eta \asymp (\log n/(rn)))^{2q/(2q+1)}$. Under the conditions of Theorem 8.3.5, $\hat{\lambda}_j - \lambda_j = O_p(\delta_n)$ for*

$$\delta_n = \left((\log n/(rn))^{2q/(2q+1)} + n^{-1} \right)^{1/2}$$

and

$$\|\hat{e}_j - e_j\|^2 = O_p\left(\delta_n^2\right),$$

where $\|\cdot\|$ is the $\mathbb{L}^2[0, 1]$ norm.

Proof: Let Δ_η be the $\mathbb{L}^2[0,1]$ operator with kernel $K_\eta - K$. Then, from Theorem 8.3.5, $\|\Delta_\eta\|^2 = O_p(\delta_n^2)$ and the projection operators corresponding to λ_j and $\hat{\lambda}_j$ satisfy

$$\begin{aligned}\|\hat{\mathscr{P}}_j - \mathscr{P}_j\| &= \|\hat{e}_j \otimes \hat{e}_j - e_j \otimes e_j\| \\ &= O_p(\delta_n)\end{aligned}$$

due to (5.17). Using this fact in part 2 of Lemma 5.1.7 gives

$$\begin{aligned}\|\hat{e}_j - e_j\|^2 &= 2\left(1 - \left(1 - \|\hat{\mathscr{P}}_j - \mathscr{P}_j\|^2\right)^{1/2}\right) \\ &= O_p(\delta_n^2)\end{aligned}$$

because $(1-x)^{1/2} = 1 - x/2 + o(x)$ as $x \to 0$. The eigenvalue result can now be obtained by an application of (5.19). □

10

Canonical correlation analysis

In this chapter, we examine the concept of canonical correlation. The idea was introduced in Chapter 1 as a problem of finding maximally correlated linear combinations of two random vectors. Our goal here is to extend this notion to a sufficiently general setting where it becomes applicable to fda and other related abstract data analysis problems.

From our work in Chapter 7, we know that functional data can be viewed from two perspectives: namely, as realizations of Hilbert space valued random elements or of second-order, continuous time, stochastic processes. These two views overlap; but, they are not, in general, equivalent. In particular, we saw how, depending on which perspective one employs, slightly different definitions are obtained for the covariance operator (Sections 7.2 and 7.3) and linear span (Section 7.6) while different considerations also arise for estimating the mean and covariance functions (Chapter 8). In this chapter, we, for the most part, adhere to the random element viewpoint; but, we also mention the key differences that result when data is collected from a second-order process. Thus, unless otherwise stated, we consider random elements χ_1, χ_2 of some separable Hilbert space $\mathbb{H}$ defined on a common probability space $(\Omega, \mathscr{F}, \mathbb{P})$. Both χ_1 and χ_2 are presumed to be of second order in the sense that $\mathbb{E}\|\chi_i\|^2 < \infty, i = 1, 2$. Note that the notation $\|\cdot\|$ and $\langle\cdot,\cdot\rangle$ are reserved for norm and inner product of $\mathbb{H}$ throughout this chapter.

Because the mean element plays no real role in our initial development, we can without loss take $\mathbb{E}[\chi_1] = \mathbb{E}[\chi_2] = 0$ for the time being. We should also mention that the assumption of a common space $\mathbb{H}$ for χ_1, χ_2 is for notational convenience. All the results we develop below can be readily extended to situations where χ_1 and χ_2 take values in different spaces.

Our problem of finding maximally correlated "linear combinations" of χ_1 and χ_2 can now be loosely formulated as finding vectors f_1, f_2 that maximize

Theoretical Foundations of Functional Data Analysis, with an Introduction to Linear Operators, First Edition. Tailen Hsing and Randall Eubank.

an extended version of (1.10) such as

$$\rho^2(f_1, f_2) = \frac{\mathrm{Cov}^2\left(\langle \chi_1, f_1\rangle, \langle \chi_2, f_2\rangle\right)}{\mathrm{Var}\left(\langle \chi_1, f_1\rangle\right)\mathrm{Var}\left(\langle \chi_2, f_2\rangle\right)}.$$

In order to make this rigorous, one of our first tasks will be to find appropriate spaces for f_1 and f_2 so that this optimization problem has a solution.

We know from Section 7.2 that there are covariance operators $\mathscr{K}_1$ and $\mathscr{K}_2$ corresponding to each of the random elements such that

$$\mathbb{E}[\langle \chi_i, f_i\rangle\langle \chi_i, f_i'\rangle] = \langle f_i, \mathscr{K}_i f_i'\rangle$$

with $f_i, f_i' \in \mathbb{H}, i = 1, 2$. Similarly, the cross-covariance operator $\mathscr{K}_{12}$ from (7.8) satisfies

$$\mathbb{E}\left[\langle \chi_1, f_1\rangle\langle \chi_2, f_2\rangle\right] = \langle f_1, \mathscr{K}_{12} f_2\rangle$$

with $\mathscr{K}_{21} = \mathscr{K}_{12}^*$. Using these operators, the squared correlation can be rewritten as

$$\rho^2(f_1, f_2) = \frac{\langle f_1, \mathscr{K}_{12} f_2\rangle^2}{\langle f_1, \mathscr{K}_1 f_1\rangle\langle f_2, \mathscr{K}_2 f_2\rangle}$$

provide that $f_1, f_2 \in \mathbb{H}$. Our work in Sections 1.1 and 4.3 would then lead us to anticipate that the optimal correlations and associated canonical variables would be obtained from a singular value expansion of $\mathscr{R}_{12} = \mathscr{K}_1^{-1/2}\mathscr{K}_{12}\mathscr{K}_2^{-1/2}$. However, as observed in Section 7.2, the inverses in this expression are not well defined for anything but the finite-dimensional, multivariate analysis, setting. From Theorem 7.2.10, we know that there is a bounded extension of $\mathscr{R}_{12}$ to all of $\mathbb{H}$. However, it still true that this operator is only precisely defined when calculations are restricted to elements of $\mathbb{H}$ that lie in the range of $\mathscr{K}_2^{1/2}$. So, we must proceed cautiously and lay a sound mathematical foundation for our optimization problem that allows us to deal with infinite dimensions while, at the same time, including the multivariate analysis version of canonical correlation as a special case.

Section 10.1 provides a detailed treatment of our abstract cca problem. Sections 10.3–10.5 then explore the implications, this has for problems of prediction, regression, factor analysis, analysis of variance, and discriminant analysis. These developments serve to show that in each of these settings there are well-defined population entities with direct ties to cca that represent the corresponding natural inferential targets. While the best inferential techniques for such disparate contexts are likely to be problem specific, it is nonetheless important to establish the existence of, e.g., consistent point estimators in some general context if for no other purpose than conceptual validation. This is accomplished in Section 10.2 where we use the perturbation tools from Chapter 5 to analyze the large sample properties of estimators for canonical correlations that derive from a type of nonparametric series estimator.

10.1 CCA for random elements of a Hilbert space

Let us first deal with the problem of finding an appropriate domain for the cca optimization problem. To do so, take $\{(\lambda_{ij}, e_{ij})\}_{j=1}^{\infty}$ to be the eigenvalue–eigenvector sequence corresponding to the two trace class covariance operators $\mathscr{K}_i, i = 1, 2$ for χ_1, χ_2, and let $\mathbb{G}_i := \mathbb{G}(\mathscr{K}_i)$ be the corresponding Hilbert spaces constructed as in (7.43) and (7.44), namely,

$$\mathbb{G}_i = \left\{ \sum_{j=1}^{\infty} \lambda_{ij} a_{ij} e_{ij} : \sum_{j=1}^{\infty} \lambda_{ij} a_{ij}^2 < \infty \right\} \tag{10.1}$$

with inner product

$$\left\langle \sum_{j=1}^{\infty} \lambda_{ij} a_{ij} e_{ij}, \sum_{j=1}^{\infty} \lambda_{ij} a_{ij} e_{ij} \right\rangle_{\mathbb{G}_i} = \sum_{j=1}^{\infty} \lambda_j a_{ij} b_{ij}. \tag{10.2}$$

To simplify notation, we will write

$$\|\cdot\|_i := \|\cdot\|_{\mathbb{G}_i} \text{ and } \langle\cdot,\cdot\rangle_i := \langle\cdot,\cdot\rangle_{\mathbb{G}_i}$$

for $i = 1, 2$, subsequently. As in (7.45), for $f_i = \sum_{j=1}^{\infty} \lambda_{ij} a_{ij} e_{ij} \in \mathbb{G}_i$, let

$$Z_i(f_i) = \sum_{j=1}^{\infty} a_{ij} \langle \chi_i, e_{ij} \rangle, \tag{10.3}$$

where, as noted in the preamble, $\langle\cdot,\cdot\rangle$ is the inner product for $\mathbb{H}$. From Section 7.6, we know that the $Z_i(\cdot)$ in (10.3) generate Hilbert spaces $\mathbb{L}^2(\chi_i)$ defined as in (7.42) that are congruent to the $\mathbb{G}_i$.

The spaces $\mathbb{L}^2(\chi_i)$ consist of linear combinations of the random variables $\langle \chi_i, e_{ij} \rangle$ that arise in the Karhunen–Lòeve type decompositions of the random elements χ_1, χ_2 provided by Theorem 7.2.7. The coefficients in the linear combinations are precisely the ones that endow them with finite variances and thereby provide the natural candidates for consideration as possible canonical variables. With this in mind, we now consider the problem of finding $f_i \in \mathbb{G}_i$ to maximize

$$\rho^2(f_1, f_2) = \frac{\text{Cov}^2\,(Z_1(f_1), Z_2(f_2))}{\text{Var}\,(Z_1(f_1))\,\text{Var}\,(Z_2(f_2))}.$$

Specifically, if it exists, we take the first squared canonical correlation to be

$$\rho_1^2 = \sup_{f_i \in \mathbb{G}_i, i=1,2} \rho^2(f_1, f_2). \tag{10.4}$$

We now proceed to obtain conditions under which (10.4) is well defined and characterize the canonical variables that attain the maximum correlation.The first step in that direction is

Theorem 10.1.1 *There are operators* $\mathscr{C}_{12} \in \mathfrak{B}(\mathbb{G}_2, \mathbb{G}_1)$ *and* $\mathscr{C}_{21} \in \mathfrak{B}(\mathbb{G}_1, \mathbb{G}_2)$ *with* $\mathscr{C}_{12}^* = \mathscr{C}_{21}$ *such that*

$$\begin{aligned}\mathrm{Cov}\,(\mathrm{Z}_1(f_1), \mathrm{Z}_2(f_2)) &= \langle f_1, \mathscr{C}_{12} f_2 \rangle_1 \\ &= \langle \mathscr{C}_{21} f_1, f_2 \rangle_2\end{aligned}$$

and $\|\mathscr{C}_{12}\| = \|\mathscr{C}_{21}\| \leq 1$.

Proof: Define the functional

$$\ell_{f_2}(f_1) = \mathrm{Cov}(\mathrm{Z}_1(f_1), \mathrm{Z}_2(f_2))$$

on $\mathbb{G}_1$. As covariance is bilinear and $\mathrm{Z}_1(\alpha f_1 + \alpha' f_1') = \alpha \mathrm{Z}_1(f_1) + \alpha' \mathrm{Z}_1(f_1')$ for any scalars α, α' and any $f_1, f_1' \in \mathbb{G}_1$, ℓ_{f_2} is linear in its f_1 argument. In addition, by the Cauchy–Schwarz inequality,

$$|\ell_{f_2}(f_1)| \leq \sqrt{\mathrm{Var}(\mathrm{Z}_1(f_1))\mathrm{Var}(\mathrm{Z}_2(f_2))} = \|f_1\|_1 \|f_2\|_2.$$

Thus, ℓ_{f_2} is a bounded linear functional on $\mathbb{G}_1$ and the Riesz representation theorem (Theorem 3.2.1) tells us that there is an element $\mathscr{C}_{12} f_2 \in \mathbb{G}_1$ such that

$$\ell_{f_2}(f_1) = \mathrm{Cov}(\mathrm{Z}_1(f_1), \mathrm{Z}_2(f_2)) = \langle f_1, \mathscr{C}_{12} f_2 \rangle_1. \tag{10.5}$$

We may use the bilinear nature of covariance again here to see that the mapping that takes f_2 to $\mathscr{C}_{12} f_2$ is linear. Thus, we need only verify the bound for the norm.

Observe that

$$\|\mathscr{C}_{12} f_2\|_1 = \left\langle \mathscr{C}_{12} f_2, \frac{\mathscr{C}_{12} f_2}{\|\mathscr{C}_{12} f_2\|_1} \right\rangle_1.$$

Thus,

$$\begin{aligned}\|\mathscr{C}_{12} f_2\|_1 &\leq \sup_{\|f_1\|_1 = 1} |\langle \mathscr{C}_{12} f_2, f_1 \rangle_1| \\ &\leq \|f_2\|_2.\end{aligned}$$

The remainder of the proof follows by interchanging the roles of f_1 and f_2 in the previous arguments and using part 2 of Theorem 3.3.7. □

The solution to the canonical correlation problem is now rather elementary provided that the operator $\mathscr{C}_{12}$ in the previous theorem is compact.

Theorem 10.1.2 *Assume that $\mathscr{C}_{12}$ is a compact element of $\mathfrak{B}(\mathbb{G}_2, \mathbb{G}_1)$ with associated singular system $\{(\rho_j, f_{1j}, f_{2j})\}_{j=1}^{\infty}$ and define*

$$U_{ij} = Z_i(f_{ij})$$

for $i = 1, 2$ and $j = 1, \ldots$ Then,

$$\rho_1^2 = \max_{f_1 \in \mathbb{G}_1, f_2 \in \mathbb{G}_2} \mathrm{Corr}^2\left(Z_1(f_1), Z_2(f_2)\right) = \mathrm{Corr}^2(U_{11}, U_{21})$$

and, for $k > 1$,

$$\rho_k^2 = \max_{\substack{f_1 \in \mathbb{G}_1 : \mathrm{Cov}(Z_1(f_1), Z_1(f_{1i}))=0, i=1,\ldots,k-1 \\ f_2 \in \mathbb{G}_2 : \mathrm{Cov}(Z_2(f_2), Z_2(f_{2i}))=0, i=1,\ldots,k-1}} \mathrm{Corr}^2\left(Z_1(f_1), Z_2(f_2)\right) = \mathrm{Corr}^2(U_{1k}, U_{2k}).$$

Proof: The Cauchy–Schwarz inequality and Parseval's relation give

$$\begin{aligned} |\mathrm{Cov}(Z_1(f_1), Z_2(f_2))| &= |\langle f_1, \mathscr{C}_{12} f_2 \rangle_1| \\ &= \left| \sum_{j=1}^{\infty} \rho_j \langle f_1, f_{1j} \rangle_1 \langle f_2, f_{2j} \rangle_2 \right| \\ &\le \rho_1 \|f_1\|_1 \|f_2\|_2 \end{aligned}$$

with equality when $f_1 = f_{11}, f_2 = f_{21}$. The general result follows similarly. □

Analogous to the finite-dimensional case, Theorem 10.1.2 provides us with a sequence of canonical variable pairs $\{(U_{1j}, U_{2j})\}_{j=1}^{\infty}$. Different elements of the sequence are uncorrelated while any particular pair has the maximum possible correlation subject to being uncorrelated with its sequence predecessors. In particular, $U_{11} = Z_1(f_{11})$ and $U_{21} = Z_2(f_{21})$ are the maximally correlated linear functionals of the random elements χ_1 and χ_2 that were the object of our initial inquiry.

The operator $\mathscr{C}_{12}$ in Theorem 10.1.1 and the operator $\mathscr{R}_{12}$ from Theorem 7.2.10 are, of course, related. To see this, first observe that $\{\tilde{e}_{1j}\}_{j=1}^{\infty}$ and $\{\tilde{e}_{2j}\}_{j=1}^{\infty}$ with

$$\tilde{e}_{ij} = \sqrt{\lambda_{ij}} e_{ij}$$

for $i = 1, 2$, provide CONSs for $\mathbb{G}_1$ and $\mathbb{G}_2$. Thus, $\mathscr{C}_{12}$ can be represented as

$$\mathscr{C}_{12} = \sum_{i=1}^{\infty} \sum_{j=1}^{\infty} \langle \tilde{e}_{1i}, \mathscr{C}_{12} \tilde{e}_{2j} \rangle_1 \tilde{e}_{2j} \otimes_2 \tilde{e}_{1i}.$$

The connection between $\mathscr{C}_{12}$ and $\mathscr{R}_{12}$ is then explained by the following theorem.

Theorem 10.1.3 *For all* $i, j > 1$, $\langle \tilde{e}_{1i}, \mathscr{C}_{12}\tilde{e}_{2j}\rangle_1 = \langle e_{1i}, \mathscr{R}_{12}e_{2j}\rangle$.

Proof: The result is immediate from

$$\begin{aligned}\langle \tilde{e}_{1i}, \mathscr{C}_{12}\tilde{e}_{2j}\rangle_1 &= \mathrm{Cov}\left(Z_1(\tilde{e}_{1i}), Z_2(\tilde{e}_{2j})\right) \\ &= \frac{\mathrm{Cov}\left(\langle \chi_1, e_{1i}\rangle, \langle \chi_2, e_{2j}\rangle\right)}{\sqrt{\lambda_{1i}}\sqrt{\lambda_{2j}}} \\ &= \frac{\langle e_{1i}, \mathscr{K}_{12}e_{2j}\rangle}{\sqrt{\lambda_{1i}}\sqrt{\lambda_{2j}}} = \langle e_{1i}, \mathscr{R}_{12}e_{2j}\rangle.\end{aligned}$$

□

Example 10.1.4 *Suppose that we are in the setting of Examples 2.7.8, 7.6.2, and 7.6.5. The two random elements now have representations as zero mean random vectors* $X_1 = \left(X_1(t_{11}), \ldots, X_1(t_{1p})\right)^T$ *and* $X_2 = \left(X_2(t_{21}), \ldots, X_2(t_{2q})\right)^T$. *In keeping with our previous treatments for this situation, we will use* $X_1(\cdot), X_2(\cdot)$ *and* $f_1(\cdot), f_2(\cdot)$ *to represent elements of* $\mathbb{L}^2(X_1), \mathbb{L}^2(X_2)$, *and* $\mathbb{G}_1, \mathbb{G}_2$ *while indicating their corresponding vector representations by* X_1, X_2, *and* f_1, f_2. *The covariance operators in this instance are equivalent to finite-dimensional, nonnegative matrices* $\mathscr{K}_1$ *and* $\mathscr{K}_2$ *that, for simplicity, we take to be of full-rank. The cross-covariance operator is just the matrix*

$$\begin{aligned}\mathscr{K}_{12} &= \{\mathrm{Cov}\left(X_1(t_{1i}), X_2(t_{2j})\right)\} \\ &= \mathbb{E}\left[X_1X_2^T\right].\end{aligned}$$

We now wish to apply the results from this section. However, there is a minor technical issue that arises when doing so; unless $p = q$, *the random vectors* X_1 *and* X_2 *take values in different Hilbert spaces. As noted in the chapter introduction, one may check that all the previous arguments remain valid for random elements on two different Hilbert spaces with only some additional notational baggage. Thus, a suitably extended version of Theorem 10.1.3 allows us to write* $\mathscr{C}_{12}$ *as*

$$\mathscr{C}_{12} = \sum_{i=1}^{p}\sum_{j=1}^{q}\langle e_{1i}, \mathscr{R}_{12}e_{2j}\rangle \tilde{e}_{2j} \otimes_2 \tilde{e}_{1i},$$

for $\langle\cdot,\cdot\rangle$ *now an ordinary Euclidean inner product and* $\tilde{e}_{ij} = \sqrt{\lambda_{ij}}e_{ij}$ *with* e_{ij} *the eigenvector for* $\mathscr{K}_i$ *that corresponds to its jth largest eigenvalue* λ_{ij} *for* $i = 1, 2$. *It is clear from this expression that the singular values for* $\mathscr{C}_{12}$ *are the* *same as those for the matrix $\mathscr{R}_{12}$ in (1.11) which, in turn, agree with the original canonical correlations in Hotelling (1936) as demonstrated in Kshirsagar (1972). To see that the scores also coincide, let $\tilde{a}_2$ be a right singular vector for $\mathscr{R}_{12}$ and define*

$$\begin{aligned} f_2(\cdot) &= \sum_{j=1}^{q} \langle \tilde{a}_2, e_{2j} \rangle \tilde{e}_{2j}(\cdot) \\ &= \sum_{j=1}^{q} \lambda_{2j} \frac{\langle \tilde{a}_2, e_{2j} \rangle}{\sqrt{\lambda_{2j}}} e_{2j}(\cdot). \end{aligned}$$

Then, $f_2(\cdot)$ is a right singular function for $\mathscr{C}_{12}$ and the corresponding canonical variable is

$$\begin{aligned} U_{2j} &= Z_2(f_2(\cdot)) \\ &= \sum_{j=1}^{q} \frac{\langle \tilde{a}_2, e_{2j} \rangle}{\sqrt{\lambda_{2j}}} \langle X_2, e_{2j} \rangle = \sum_{j=1}^{q} \langle \tilde{a}_2, \mathscr{K}_2^{-1/2} e_{2j} \rangle \langle X_2, e_{2j} \rangle \\ &= \sum_{j=1}^{q} \langle \mathscr{K}_2^{-1/2} \tilde{a}_2, e_{2j} \rangle \langle X_2, e_{2j} \rangle = \langle \mathscr{K}_2^{-1/2} \tilde{a}_2, X_2 \rangle \end{aligned}$$

exactly as in multivariate analysis case.

In the remainder of this section, we consider the formulation of cca for mean-square continuous processes. The development largely parallels that for the random element setting and, consequently, this requires only a brief digression from the main theme of our narrative.

Let $\{X_i(t) : t \in E\}, i = 1, 2$, be mean-square continuous processes defined on a common probability space $(\Omega, \mathscr{F}, \mathbb{P})$, where E is a compact metric space. We take the parameter space E to be the same for X_1 and X_2 purely for notational convenience and to be consistent with our assumption in the random element setting. In addition, for convenience, assume that the processes have mean zero for all t.

We begin by summarizing some of the differences between how the covariance operators and linear spans are defined for Hilbert space random elements and mean-square continuous processes. These are described in more detail in Chapter 7.

The covariance operators in the process setting are integral operators with covariance functions as their associated kernels. Specifically, the auto and cross-covariance functions

$$\begin{aligned} K_i(t_1, t_2) &= \mathrm{Cov}(X_i(t_1), X_i(t_2)), \quad i = 1, 2, \\ K_{12}(t_1, t_2) &= \mathrm{Cov}(X_1(t_1), X_2(t_2)), \end{aligned} \tag{10.6}$$

for $t_1, t_2 \in E$ produce the auto and cross-covariance operators

$$(\mathscr{K}_i f)(t) = \int_E K_i(s,t) f(s) d\mu(s),$$

$$(\mathscr{K}_{12} f)(t) = \int_E K_{12}(t,s) f(s) d\mu(s),$$

$$(\mathscr{K}_{21} f)(t) = \int_E K_{12}(s,t) f(s) d\mu(s)$$

for $f \in \mathbb{L}^2(E, \mathscr{B}(E), \mu)$, where μ is a finite measure.

Let $\{(\lambda_{ij}, e_{ij})\}_{j=1}^{\infty}$ denote the eigenvalue–eigenfunction sequence corresponding to $\mathscr{K}_i, i = 1, 2$. With this system, define $\mathbb{G}_i$ as in (10.1) and (10.2) for the random element case and, for $f_i = \sum_{j=1}^{\infty} \lambda_{ij} a_{ij} e_{ij} \in \mathscr{G}_i$, let

$$Z_i(f_i) = \sum_{j=1}^{\infty} a_{ij} I_{X_i}(e_{ij}), i = 1, 2,$$

where $I_{X_i}(e_{ij})$ is the score of X_i that corresponds to e_{ij}. The closure of the set of all $Z_i(f_i)$ for $f_i \in \mathbb{G}_i$ constitutes $\mathbb{L}^2(X_i)$, which is congruent to $\mathbb{G}_i$ through the mapping $Z_i(\cdot)$.

We know from Theorem 7.6.4 that $\mathbb{G}_i = \mathbb{H}(K_i)$ and hence $K_i(t, \cdot) \in \mathbb{G}_i$ for any t. The following result shows that membership in $\mathbb{G}_i$ can also be established for the cross-covariance function.

Theorem 10.1.5 *For fixed $t \in E$, $K_{12}(t, \cdot) \in \mathbb{G}_2$ and $K_{12}(\cdot, t) \in \mathbb{G}_1$.*

Proof: If, for example, $\mathscr{P}_2$ is the $\mathbb{L}^2(\Omega, \mathscr{B}, \mathbb{P})$ projection for $\mathbb{L}^2(X_2)$,

$$\begin{aligned} K_{12}(t, \cdot) &= \mathbb{E}[X_1(t) X_2(\cdot)] \\ &= \mathbb{E}[\mathscr{P}_2 X_1(t) X_2(\cdot)] \end{aligned}$$

for fixed $t \in E$. This means that $K_{12}(t, \cdot) \in \mathbb{G}_2$ due to Theorem 7.6.3. □

From this theorem, we conclude that if $f_2 = \sum_{j=1}^{\infty} \lambda_{2j} a_{2j} e_{2j} \in \mathbb{G}_2$,

$$\begin{aligned} (\mathscr{C}_{12} f_2)(t) &:= \langle K_{12}(t, \cdot), f_2 \rangle_2 \\ &= \sum_{j=1}^{\infty} \lambda_{2j} a_{2j} \langle K_{12}(t, \cdot), e_{2j} \rangle_2 \\ &= \sum_{j=1}^{\infty} a_{2j} \langle K_{12}(t, \cdot), e_{2j} \rangle \\ &= \sum_{j=1}^{\infty} a_{2j} \int_E K_{12}(t,s) e_{2j}(s) d\mu(s) \end{aligned} \tag{10.7}$$

defines a bounded operator from $\mathbb{G}_2$ into $\mathbb{G}_1$. By Theorem 7.3.6 and derivations similar to those of Theorem 7.3.3, we now obtain

$$\begin{aligned}
\mathrm{Cov}\,(Z_1(f_1), Z_2(f_2)) &= \sum_{i=1}^{\infty}\sum_{j=1}^{\infty} a_{1i}a_{2j}\mathrm{Cov}\left(I_{X_1}(e_{1i}), I_{X_2}(e_{2j})\right) \\
&= \sum_{i=1}^{\infty}\sum_{j=1}^{\infty} a_{1i}a_{2j}\int_E\int_E K_{12}(t,s)e_{1i}(s)e_{2j}(t)d\mu(t)d\mu(s) \\
&= \sum_{i=1}^{\infty}\sum_{j=1}^{\infty} \lambda_{2j}a_{1i}a_{2j}\int_E \langle K_{12}(t,\cdot), e_{2j}(\cdot)\rangle_2 e_{1i}(t)d\mu(t) \\
&= \sum_{i=1}^{\infty}\sum_{j=1}^{\infty} \lambda_{1i}\lambda_{2j}a_{1i}a_{2j}\langle e_{1i}(\star), \langle K_{12}(\star,\cdot), e_{2j}(\cdot)\rangle_2\rangle_1 \\
&= \langle f_1, \mathscr{C}_{12}f_2\rangle_1.
\end{aligned}$$

This parallels Theorem 10.1.1 for the random element setting and therefore we can proceed to define canonical correlations in exactly the same way as we did in that setting. In particular, if X_1, X_2 are also random elements of $\mathbb{H} = \mathbb{L}^2(E, \mathscr{B}(E), \mu)$ then the two definitions of canonical correlations are identical. However, we can prove the following additional result in the process setting that shows the singular system of $\mathscr{C}_{12}$ translates into a canonical representation for $\mathscr{K}_{12}$.

Theorem 10.1.6 *If $\mathscr{C}_{12}$ is compact with singular system $\{(\rho_j, f_{1j}, f_{2j})\}_{j=1}^{\infty}$*

$$K_{12}(t_1,t_2) = \sum_{j=1}^{\infty}\rho_j f_{1j}(t_1)f_{2j}(t_2)$$

with the series converging uniformly in $t_1, t_2 \in E$.

Proof: Using the reproducing property and Theorem 10.1.5,

$$\begin{aligned}
K_{12}(t_1,t_2) &= \langle K_{12}(t_1,\cdot), K_2(t_2,\cdot)\rangle_2 \\
&= \left(\mathscr{C}_{12}\,(K_2(t_2,\cdot)\right)(t_1) \\
&= \sum_{j=1}^{\infty}\rho_j\left(\left(f_{2j}\otimes_2 f_{1j}\right)K_2\,(t_2,\cdot)\right)(t_1) \\
&= \sum_{j=1}^{\infty}\rho_j\langle f_{2j}, K_2(t_2,\cdot)\rangle_2 f_{1j}(t_1) \\
&= \sum_{j=1}^{\infty}\rho_j f_{2j}(t_2)f_{1j}(t_1).
\end{aligned}$$

Now let $\mathscr{C}_{12}^{(n)} = \sum_{j=1}^{n} \rho_j f_{2j} \otimes_2 f_{1j}$ with $K_{12}^{(n)}$ defined similarly. Then,

$$\begin{aligned}|K_{12}(t_1,t_2) - K_{12}^{(n)}(t_1,t_2)| &= |\langle K_1(t_1,\star), \left[\mathscr{C}_{12} - \mathscr{C}_{12}^{(n)}\right](K_2(t_2,\cdot))(\star)\rangle_1| \\ &\leq \|\mathscr{C}_{12} - \mathscr{C}_{12}^{(n)}\| \|K_1(t_1,\cdot)\|_1 \|K_2(t_2,\cdot)\|_2 \\ &= \|\mathscr{C}_{12} - \mathscr{C}_{12}^{(n)}\| \sqrt{K_1(t_1,t_1)K_2(t_2,t_2)}\end{aligned}$$

from which the uniformity of convergence follow. □

In the fda literature, it is commonly assumed that X_1, X_2 are in the intersection of both of the perspectives, we have discussed with $\mathbb{H} = \mathbb{L}(E, \mathscr{B}(E), \mu) = \mathbb{L}^2[0,1]$. In that event, this particular spin on cca has appeared a number of times in the literature. It has usually (e.g., He et al., 2003) been formulated as an optimization problem over $\mathbb{L}^2[0,1]$ rather than on $\mathbb{G}_1, \mathbb{G}_2$. This has the unfortunate consequence of requiring the imposition of additional (unnecessary) conditions to insure the existence of a solution. This approach avoids this problem and coincides with the treatment in Eubank and Hsing (2007).

10.2 Estimation

We now take a brief foray into the area of point estimation of the canonical correlations and canonical scores. To be somewhat more precise, let $\mathscr{C} := \mathscr{C}_{21}$ be the operator from Theorem 10.1.1 with singular system $\{(\rho_j, f_{1j}, f_{2j})\}_{j=1}^{\infty}$. We will then develop consistent estimators/predictors for the ρ_j and their associated scores.

Suppose now that we have two zero mean, second-order random elements χ_1, χ_2 taking values in a Hilbert space $\mathbb{H}$. A random sample has then provided us with independent random element pairs $(\chi_{1r}, \chi_{2r}), r = 1, \ldots, n$, that are all identically distributed as (χ_1, χ_2). We wish to use these observations for statistical inference concerning $\mathscr{C}$.

As in Section 10.1, let $\{(\lambda_{1j}, e_{1j})\}_{j=1}^{\infty}$ and $\{(\lambda_{2j}, e_{2j})\}_{j=1}^{\infty}$ be the eigen systems that correspond to the covariance operators $\mathscr{K}_1$ and $\mathscr{K}_2$. Our interest is directed toward the infinite-dimensional case here as the finite-dimensional case has already been extensively investigated; see, e.g., Muirhead and Waternaux (1980). If we define $\tilde{e}_{ij} = \sqrt{\lambda_{ij}} e_{ij}$, then $\{\tilde{e}_{1j}\}$ and $\{\tilde{e}_{2j}\}$ are CONSs for $\mathbb{G}_1$ and $\mathbb{G}_2$. With this latter property in mind, one approach to estimation can now be formulated along the lines of Example 5.2.3.

First observe that

$$\mathscr{C} = \sum_{i=1}^{\infty} \sum_{j=1}^{\infty} \langle \tilde{e}_{2i}, \mathscr{C} \tilde{e}_{1j} \rangle_2 \tilde{e}_{1j} \otimes_1 \tilde{e}_{2i}$$

so that if, for some integer $k \geq 1$, we define

$$\mathscr{C}^{(k)} = \sum_{i=1}^{k} \sum_{j=1}^{k} \langle \tilde{e}_{2i}, \mathscr{C} \tilde{e}_{1j} \rangle_2 \tilde{e}_{1j} \otimes_1 \tilde{e}_{2i}$$

then

$$\begin{aligned} \|\mathscr{C} - \mathscr{C}^{(k)}\| &\leq \|\mathscr{C} - \mathscr{C}^{(k)}\|_{HS} \\ &=: \delta_k, \end{aligned} \tag{10.8}$$

which decays to zero as k diverges provided that $\mathscr{C}$ is an HS operator.

Now recall that

$$\begin{aligned} \langle \tilde{e}_{2i}, \mathscr{C} \tilde{e}_{1j} \rangle_2 &= \mathrm{Cov}(Z_1(\tilde{e}_{1j}), Z_2(\tilde{e}_{2i})) \\ &= \mathrm{Cov}\left(\left\langle \chi_1, \frac{e_{1j}}{\sqrt{\lambda_{1j}}} \right\rangle, \left\langle \chi_2, \frac{e_{2i}}{\sqrt{\lambda_{2i}}} \right\rangle \right) \\ &= \mathbb{E} \left\langle \chi_1, \frac{e_{1j}}{\sqrt{\lambda_{1j}}} \right\rangle \left\langle \chi_2, \frac{e_{2i}}{\sqrt{\lambda_{2i}}} \right\rangle \\ &=: a_{ij} \end{aligned}$$

for which a natural estimator is

$$a_{ijn} := n^{-1} \sum_{r=1}^{n} \left\langle \chi_{1r}, \frac{e_{1jn}}{\sqrt{\lambda_{1jn}}} \right\rangle \left\langle \chi_{2r}, \frac{e_{2in}}{\sqrt{\lambda_{2in}}} \right\rangle,$$

where $\{(\lambda_{1jn}, e_{1jn})\}_{j=1}^{\infty}$ and $\{(\lambda_{2jn}, e_{2jn})\}_{j=1}^{\infty}$ are the eigen systems associated with some generic estimators $\mathscr{K}_{1n}$ and $\mathscr{K}_{2n}$ of $\mathscr{K}_1$ and $\mathscr{K}_2$ such as ones that might derive from the developments in Chapter 8. In particular, we impose the natural requirement that $\mathscr{K}_{1n}$ and $\mathscr{K}_{2n}$ are random elements of $\mathfrak{B}(\mathbb{H})$ and are also covariance operators, i.e., operators that are nonnegative definite, self-adjoint, and trace class.

Thus, we are led to consideration of the estimator

$$\mathscr{C}_n^{(k)} := \sum_{i=1}^{k} \sum_{j=1}^{k} a_{ijn} \tilde{e}_{1jn} \otimes_{1n} \tilde{e}_{2in},$$

where $\tilde{e}_{kin} = \sqrt{\lambda_{kin}} e_{kin}$ for $k = 1, 2$, and $\otimes_{1n}$ denotes the tensor product in the space $\mathbb{G}(\mathscr{K}_{1n})$. The fact that $\mathscr{C}_n^{(k)}$ operates on a different space (Theorem 7.5.4) than $\mathscr{C}$ and $\mathscr{C}^{(k)}$ creates difficulties in comparing the estimated and true singular vectors and motivates the approach that is taken in the following.

As in Example 5.2.3, the singular values and vectors for $\mathscr{C}^{(k)}$ and $\mathscr{C}_n^{(k)}$ can be obtained directly through singular value decompositions of $\mathscr{A} = \{a_{ij}\}_{i,j=1:k}$ and $\mathscr{A}_n = \{a_{ijn}\}_{i,j=1:k}$, respectively, in $\mathbb{R}^k$. If ρ_{jn} is the jth singular value and $u_{jn} = (u_{1jn}, \dots, u_{kjn})^T$ and $v_{jn} = (v_{1jn}, \dots, v_{kjn})^T$ are the left and right singular vectors for $\mathscr{A}_n$, then the jth singular value of $\mathscr{C}_n^{(k)}$ is ρ_{jn} and the corresponding left and right singular vectors are

$$f_{1jn} = \sum_{i=1}^{k} u_{ijn}\tilde{e}_{1in}$$

and

$$f_{2jn} = \sum_{i=1}^{k} v_{ijn}\tilde{e}_{2in}.$$

As mentioned earlier, these estimated singular vectors cannot be directly compared with the true ones as they may not be elements of common Hilbert spaces. However, the estimated singular vectors lead to sample canonical variables or scores: namely,

$$U_{1jn} := Z_{1n}(f_{1jn}) = \sum_{i=1}^{k} \frac{u_{ijn}}{\sqrt{\lambda_{1in}}}\langle \chi_1, e_{1in}\rangle \tag{10.9}$$

and

$$U_{2jn} := Z_{2n}(f_{2jn}) = \sum_{i=1}^{k} \frac{v_{ijn}}{\sqrt{\lambda_{2in}}}\langle \chi_2, e_{2in}\rangle, \tag{10.10}$$

where Z_{1n} and Z_{2n} are the isometries defined in (7.40) based on $\mathscr{K} = \mathscr{K}_{1n}$ and $\mathscr{K}_{2n}$, respectively. Comparison of the true and estimated scores becomes feasible because they live in the same spaces; i.e., they are both elements of $\mathbb{L}^2(\chi_i)$.

Our goal in the following is to show that the estimated canonical correlations and variables are close to the true canonical correlations and variables under rather general conditions if both n and k are large with k selected from the data. With that in mind, observe from Theorem 10.1.3 that the svds of $\mathscr{A}_n$ and $\mathscr{A}$ in $\mathbb{R}^k$ are equivalent to those of the operators

$$\mathscr{R}^{(k)} = (\mathscr{K}_1^{(k)})^{-1/2}\mathscr{K}_{12}(\mathscr{K}_2^{(k)})^{-1/2}$$

and

$$\mathscr{R}_n^{(k)} = (\mathscr{K}_{1n}^{(k)})^{-1/2}\mathscr{K}_{12n}(\mathscr{K}_{2n}^{(k)})^{-1/2}$$

in $\mathbb{H}$, where $\mathscr{K}_i^{(k)}$ and $\mathscr{K}_{in}^{(k)}$ are truncated versions of $\mathscr{K}_i$ and $\mathscr{K}_{in}$ using only the first k terms in their respective eigen expansions in $\mathbb{H}$.

To assess the behavior of $\|\mathscr{R}^{(k)} - \mathscr{R}_n^{(k)}\|$, we first prove a general lemma.

Lemma 10.2.1 *Let $\mathscr{K}$ be an infinite-dimensional covariance operator in $\mathfrak{B}(\mathbb{H})$ with positive eigenvalues $\lambda_1 > \lambda_2 > \cdots > 0$ and $\mathscr{K}_n$ an estimator of $\mathscr{K}$ with eigenvalues $\lambda_{1n} > \lambda_{2n} > \cdots$. Assume that $\epsilon_n := \|\mathscr{K}_n - \mathscr{K}\| \xrightarrow{p} 0$ as $n \to \infty$ and let $\{b_n\}$ be a sequence of constants tending to ∞ and satisfying $b_n = O_p(\epsilon_n^{-2/3})$. Define $k = k_n$ to be the largest positive integer k such that*

$$k(b_n\epsilon_n)^{1/2} \xrightarrow{p} 0, \quad \frac{\epsilon_n^{1/2}}{\min_{j\le k}\eta_{jn}} \xrightarrow{p} 0 \quad \textit{and} \quad (\lambda_{kn}b_n)^{-1} \xrightarrow{p} 0 \qquad (10.11)$$

for $\eta_{jn} = (1/2)\inf_{s\neq j}|\lambda_{jn} - \lambda_{sn}|$. Then,

1. *k diverges in probability,*
2. $\max_{j\le k}|\lambda_{jn}^{-1/2} - \lambda_j^{-1/2}| = o_p\left(b_n^{3/2}\epsilon_n\right)$, *and*
3. $\|(\mathscr{K}_n^{(k)})^{-1/2} - (\mathscr{K}^{(k)})^{-1/2}\| = o_p\left(b_n^{3/2}\epsilon_n\right) + O_p\left(k(b_n\epsilon_n)^{1/2}\right)$.

Proof: Theorem 4.2.8 has the implication that

$$\sup_j |\lambda_{jn} - \lambda_j| \le \epsilon_n$$

and, therefore, for any fixed J

$$\lambda_{Jn} \ge \lambda_J - \epsilon_n$$

and

$$\min_{j\le J} \eta_{jn} \ge \min_{j\le J} \eta_j - \epsilon_n$$

with $\eta_j = (1/2)\inf_{s\neq j}|\lambda_j - \lambda_s|$. Part 1 of the lemma is a straightforward consequence of these two relations.

A Taylor expansion reveals that

$$x^{-1/2} - x_0^{-1/2} = -\frac{1}{2}x_0^{-3/2}(x - x_0) + \frac{3}{8}\hat{x}^{-5/2}(x - x_0)^2,$$

where $\hat{x}$ is some value between x and x_0. An application of this fact then produces

$$|\lambda_{jn}^{-1/2} - \lambda_j^{-1/2}| = \lambda_j^{-3/2}O_p\left(|\lambda_{jn} - \lambda_j|\right)$$
$$+\left(\lambda_j - O_p(\epsilon_n)\right)^{-5/2}O_p\left(|\lambda_{jn} - \lambda_j|^2\right)$$

for $j \le k$. Thus,

$$\max_{j\le k} |\lambda_{jn}^{-1/2} - \lambda_j^{-1/2}| = o_p\left(b_n^{3/2}\epsilon_n\right) + o_p\left(b_n^{5/2}\epsilon_n^2\right) = o_p\left(b_n^{3/2}\epsilon_n\right)$$

and part 2 has been proved.

To show part 3, write

$$\begin{aligned}
&(\mathscr{K}_n^{(k)})^{-1/2} - (\mathscr{K}^{(k)})^{-1/2} \\
&\quad = \sum_{j=1}^k \lambda_{jn}^{-1/2}\mathscr{P}_{jn} - \sum_{j=1}^k \lambda_j^{-1/2}\mathscr{P}_j \\
&\quad = \sum_{j=1}^k \lambda_{jn}^{-1/2}(\mathscr{P}_{jn} - \mathscr{P}_j) + \sum_{j=1}^k (\lambda_{jn}^{-1/2} - \lambda_j^{-1/2})\mathscr{P}_j. \qquad (10.12)
\end{aligned}$$

Now apply part 2 of the lemma to see that

$$\left\| \sum_{j=1}^k (\lambda_{jn}^{-1/2} - \lambda_j^{-1/2})\mathscr{P}_{jn} \right\| = o_p\left(b_n^{3/2}\epsilon_n\right). \qquad (10.13)$$

Using (5.17) with $\mathscr{P}_j$ and $\tilde{\mathscr{P}}_j$ there chosen to be $\mathscr{P}_{jn}$ and $\mathscr{P}_j$ in this context gives

$$\|\mathscr{P}_{jn} - \mathscr{P}_j\| \le \frac{\delta_{jn}}{1-\delta_{jn}}$$

for $\delta_{jn} = \|\mathscr{K}_n - \mathscr{K}\|/\eta_{jn} = O(\epsilon_n^{1/2})$ uniformly for all $j \le k$. Thus,

$$\begin{aligned}
\left\| \sum_{j=1}^k \lambda_{jn}^{-1/2}(\mathscr{P}_{jn} - \mathscr{P}_j) \right\| &\le \sum_{j=1}^k \lambda_{jn}^{-1/2}\|\mathscr{P}_{jn} - \mathscr{P}_j\| \\
&= O_p\left(\epsilon_n^{1/2}\right)\sum_{j=1}^k \lambda_{jn}^{-1/2} \\
&= O_p\left(\epsilon_n^{1/2} k b_n^{1/2}\right) \qquad (10.14)
\end{aligned}$$

and part 3 follows from (10.12)–(10.14). □

We now apply Lemma 10.2.1 to our cca problem. As discussed in Chapter 9, estimation of $\mathscr{K}_i$ in fda can be achieved with a root-n rate for dense data while a slower, nonparametric rate is the best that can be expected for the sparse data scenario.

Theorem 10.2.2 *For $i = 1, 2$, let $\mathscr{K}_i$ be infinite-dimensional with distinct eigenvalues $\lambda_{i1} > \lambda_{i2} > \cdots > 0$ and let $\mathscr{K}_{in}$ be an estimator of $\mathscr{K}_i$ with eigenvalues $\lambda_{i1n} > \lambda_{i2n} > \cdots$. In addition, let $\mathscr{K}_{12n}$ be an estimator of $\mathscr{K}_{12}$. Assume that $\epsilon_n := \max(\|\mathscr{K}_{1n} - \mathscr{K}_1\|, \|\mathscr{K}_{2n} - \mathscr{K}_2\|, \|\mathscr{K}_{12n} - \mathscr{K}_{12}\|) = O_p(1)$ and let $b_n = O_p(\epsilon_n^{-2/3})$ with $k = k_n$ the largest integer such that (10.11) holds for both $\mathscr{K}_{1n}$ and $\mathscr{K}_{2n}$. Then*

$$\|\mathscr{R}_n^{(k)} - \mathscr{R}^{(k)}\| = o_p\left(b_n^{3/2}\epsilon_n\right) + O_p\left(k(b_n\epsilon_n)^{1/2}\right). \tag{10.15}$$

Proof: Write

$$\begin{aligned}\mathscr{R}_n^{(k)} - \mathscr{R}^{(k)} &= \left((\mathscr{K}_{1n}^{(k)})^{-1/2} - (\mathscr{K}_1^{(k)})^{-1/2}\right)\mathscr{K}_{12n}(\mathscr{K}_{2n}^{(k)})^{-1/2} \\ &\quad + (\mathscr{K}_1^{(k)})^{-1/2}(\mathscr{K}_{12n} - \mathscr{K}_{12})(\mathscr{K}_{2n}^{(k)})^{-1/2} \\ &\quad + (\mathscr{K}_1^{(k)})^{-1/2}\mathscr{K}_{12}\left((\mathscr{K}_{2n}^{(k)})^{-1/2} - (\mathscr{K}_2^{(k)})^{-1/2}\right).\end{aligned} \tag{10.16}$$

The first term is

$$\begin{aligned}&\left((\mathscr{K}_{1n}^{(k)})^{-1/2} - (\mathscr{K}_1^{(k)})^{-1/2}\right)\mathscr{K}_{12n}(\mathscr{K}_{2n}^{(k)})^{-1/2} \\ &\quad = \left((\mathscr{K}_{1n}^{(k)})^{-1/2} - (\mathscr{K}_1^{(k)})^{-1/2}\right)(\mathscr{K}_{1n}^{(k)})^{1/2}\left\{(\mathscr{K}_{1n}^{(k)})^{-1/2}\mathscr{K}_{12n}(\mathscr{K}_{2n}^{(k)})^{-1/2}\right\},\end{aligned}$$

which has the rate of the theorem as a result of Lemma 10.2.1 and the fact that $(\mathscr{K}_{1n}^{(k)})^{1/2}$ and $(\mathscr{K}_{1n}^{(k)})^{-1/2}\mathscr{K}_{12n}(\mathscr{K}_{2n}^{(k)})^{-1/2}$ are bounded in probability. The third term in (10.16) can be dealt with in the same manner. Finally, write the second term of (10.16) as

$$\begin{aligned}&(\mathscr{K}_1^{(k)})^{-1/2}(\mathscr{K}_{12n} - \mathscr{K}_{12})\left((\mathscr{K}_{2n}^{(k)})^{-1/2} - (\mathscr{K}_2^{(k)})^{-1/2}\right) \\ &\quad + (\mathscr{K}_1^{(k)})^{-1/2}(\mathscr{K}_{12n} - \mathscr{K}_{12})(\mathscr{K}_2^{(k)})^{-1/2},\end{aligned}$$

where both terms in the sum can be shown to be dominated by the rate in (10.15) due to Lemma 10.2.1. □

Theorem 10.2.3 *Assume that $\mathscr{C}$ is HS with singular values $\rho_1 \geq \rho_2 \geq \cdots \geq 0$. In addition, let the assumptions of Theorem 10.2.2 hold and let $\rho_{1n} \geq \rho_{2n} \geq \cdots \geq 0$ be the singular values of $\mathscr{R}_n^{(k)}$. Then, as $n \to \infty$,*

$$\sup_j |\rho_{jn}^2 - \rho_j^2| = o_p\left(b_n^{3/2}\epsilon_n\right) + O_p\left(\delta_k + k(b_n\epsilon_n)^{1/2}\right), \tag{10.17}$$

where δ_k is defined as in (10.8). In addition, if the nonzero singular values of $\mathscr{C}, \mathscr{C}^{(k)}$ and $\mathscr{C}_n^{(k)}$ are all distinct, for any fixed $\rho_j \neq 0$ and $i = 1, 2$,

$$\mathbb{E}^{1/2}\left\{\left(U_{ij} - U_{ijn}\right)^2 \mid (\chi_{1r}, \chi_{2r}), r = 1, \ldots, n\right\}$$
$$= o_p\left(b_n^{3/2}\epsilon_n\right) + O_p\left(\delta_k + k(b_n\epsilon_n)^{1/2}\right), \tag{10.18}$$

where (χ_1, χ_2) is assumed to be independent of $(\chi_{1r}, \chi_{2r}), r = 1, \ldots, n$, in this calculation.

Proof: The fact that (10.17) holds is a consequence of (10.8), Theorem 10.2.2, and Theorem 4.2.8. We will show (10.18) in two steps. The first of these is to establish that

$$\|U_{ij}^{(k)} - U_{ij}\|_{\mathbb{L}(\chi_i)} = O_p(\delta_k) \tag{10.19}$$

for $i = 1, 2$, where $U_{1j}^{(k)}, U_{2j}^{(k)}$ denote the canonical variables computed using $\mathscr{C}^{(k)}$. Recall that $\|U_{ij}^{(k)} - U_{ij}\|_{\mathbb{L}(\chi_i)}^2 = \mathbb{E}\left(U_{ij}^{(k)} - U_{ij}\right)^2$. If we take

$$\Delta = \mathscr{C}^{(k)} - \mathscr{C},$$

then Theorem 5.2.2 provides the justification for (10.19) once we observe that for any $\mathscr{T} \in \mathfrak{B}(\mathbb{G}_1, \mathbb{G}_2)$

$$\left\|\sum_{l \neq j} \frac{\rho_l\langle f_{2l}, \mathscr{T} f_{1j}\rangle_2 + \rho_j\langle f_{2j}, \mathscr{T} f_{1l}\rangle_2}{\rho_l^2 - \rho_j^2} f_{1l}\right\|_1^2$$
$$\leq \frac{2}{\min_{j \neq l}\left(\rho_j^2 - \rho_l\right)^2}\left[\sum_{j \neq l} \rho_l^2\langle f_{2l}, \mathscr{T} f_{1j}\rangle_2^2 + \rho_j^2 \sum_{j \neq l}\langle f_{2j}, \mathscr{T} f_{1l}\rangle_2^2\right]$$
$$\leq \frac{2\rho_1^2}{\min_{j \neq l}\left(\rho_j^2 - \rho_l\right)^2}\left[\|\mathscr{T} f_{1j}\|_2^2 + \|\mathscr{T}^* f_{2j}\|_1^2\right]$$
$$= O\left(\|\mathscr{T}\|^2\right).$$

Thus, the congruence between $\mathbb{G}_i$ and $\mathbb{L}(\chi_i)$ gives (10.19).

The next step is to show that

$$\mathbb{E}^{1/2}\left\{\left(U_{ijn}^{(k)} - U_{ij}^{(k)}\right)^2 \mid (\chi_{1r}, \chi_{2r}), r = 1, \ldots, n\right\}$$
$$= o_p\left(b_n^{3/2}\epsilon_n\right) + O_p\left(k(b_n\epsilon_n)^{1/2}\right)$$

for $i = 1, 2$. The proof of this result is analogous to that of (10.19). As $\mathscr{R}_n^{(k)}$ and $\mathscr{R}^{(k)}$ are finite dimensional, we let

$$\Delta = \mathscr{R}_n^{(k)} - \mathscr{R}^{(k)}$$

in $\mathfrak{B}(\mathbb{H})$ and apply the above-mentioned argument together with Theorem 10.2.2. □

10.3 Prediction and regression

We know from Section 1.1 that cca provides the basis for other mva techniques such as MANOVA and discriminant analysis. This remains true more generally and, in particular, in the context of fda. For the following several sections, we will expand on this comment by working exclusively with situations where the two random elements correspond to zero mean stochastic processes X_1, X_2 that are jointly measurable in t, ω and can also be viewed as random elements of $\mathbb{L}^2 := \mathbb{L}^2(E, \mathscr{B}(E), \mu)$. Recall from Section 10.1 that canonical correlations defined from the process and random element perspective are identical in this instance. We aim to develop parallels of many of the ideas in Section 1.1 in this infinite-dimensional environment.

Our first such foray proceeds in the direction of optimal prediction. Suppose that our interest is in the X_1 process but only the X_2 process will actually be observed. In that case, it may be of interest to assess the value of $X_1(t)$ for $t \in E$ using a best linear predictor based on the X_2 process. By this, we mean that we want to find $\overline{X}_1(t) \in \mathbb{L}^2(X_2)$ such that

$$\mathbb{E}|X_1(t) - \overline{X}_1(t)|^2 = \inf_{Y \in \mathbb{L}^2(X_2)} \mathbb{E}|X_1(t) - Y|^2. \tag{10.20}$$

In this regard, we have the following.

Theorem 10.3.1 *The best linear predictor of $X_1(t)$ is $\overline{X}_1(t) = Z_2(K_{12}(t, \cdot))$.*

Proof: From Theorem 10.1.5, we know that $K_{12}(t_1, \cdot) \in \mathbb{G}_2$ so that $\overline{X}_1(t_1)$ is well defined. The result will follow once we have shown that $\overline{X}_1(t_1)$ is the orthogonal projection of $X_1(t)$ onto the Hilbert space $\mathbb{L}^2(X_2)$.

Take $Y \in \mathbb{L}^2(X_2)$ and select a sequence of random variables of the form

$$Y_n = \sum_{j=1}^{n} a_{jn} X_2(t_{jn})$$

for real numbers a_{jn} and points $t_{jn} \in E$ such that $\mathbb{E}|Y - Y_n|^2 \to 0$ as $n \to \infty$. Then,

$$\begin{aligned}
\mathbb{E}\left[\left(X_1(t_1) - \overline{X}_1(t_1)\right) Y_n\right] &= \sum_{j=1}^{n} a_{jn}\mathbb{E}[X_1(t_1)X_2(t_{jn})] \\
&\quad - \sum_{j=1}^{n} a_{jn}\mathbb{E}[X_2(t_{jn})Z_2(K_{12}(t_1,\cdot))] \\
&= \sum_{j=1}^{n} a_{jn}K_{12}(t_1,t_{jn}) \\
&\quad - \sum_{j=1}^{n} a_{jn}\langle K_{12}(t_1,\cdot), K_2(t_{jn},\cdot)\rangle_2 \\
&= 0.
\end{aligned}$$

The continuity of the inner product along with (2.14) implies the result. □

An application of Theorem 10.3.1 in combination with Theorem 10.1.6 produces an fda version of identity (1.16) that connects the prediction problem with cca.

Corollary 10.3.2 *If $\mathscr{C}_{12}$ is compact with singular system $\{(\rho_j, f_{1j}, f_{2j})\}$, $\overline{X}_1(\cdot) = \sum_{j=1}^{\infty} \rho_j Z_2(f_{2j}) f_{1j}(\cdot)$.*

Example 10.3.3 *Let us consider the application of Corollary 10.3.2 in the finite-dimensional case of Example 10.1.4. In that instance, the vector representations of the singular functions $f_{1j}(\cdot), f_{2j}(\cdot)$ have the form $f_{1j} = \mathscr{K}_1 a_{1j}, f_{2j} = \mathscr{K}_2 a_{2j}$ and*

$$Z_2(f_{2j}) = a_{2j}^T X_2 =: U_{2j}$$

with $\mathscr{K}_1^{1/2} a_{1j}, \mathscr{K}_2^{1/2} a_{2j}$ the singular vectors that correspond to the jth singular value ρ_j of the matrix $\mathscr{R}_{12}$ in (1.11). Thus,

$$\begin{aligned}
\overline{X}_1 &= \sum_{j=1}^{\infty} \rho_j Z_2(f_{2j}) f_{1j} \\
&= \sum_{j=1}^{\infty} \rho_j U_{2j} \mathscr{K}_1 a_{1j},
\end{aligned}$$

which agrees with (1.16).

An alternative take on the prediction problem derives from imposition of a linear model such as in (1.17). An analog of that finite-dimensional relation that could be used here is

$$X_1(t) = \int_E \beta(t,s)X_2(s)d\mu(s) + \varepsilon(t) \tag{10.21}$$

with $\beta(t,\cdot) \in \mathbb{L}^2$ and $\varepsilon(\cdot)$ a zero mean process that is uncorrelated with $X_2(\cdot)$. However, a bit of thought is needed to make such a formulation rigorous.

In we view $\int_E \beta(t,s)X_2(s)d\mu(s)$ as a limit of weighted sums of finitely many values from the X_2 process, this leads us to the assumption that it should be an element of $\mathbb{L}_2(X_2)$. However, if that is true then it must be the image of some element from $\mathbb{G}_2$ under the isometric mapping that connects the two spaces. With that in mind, we can now advance a tenable form for β.

Let $\mathscr{T}$ be an element of $\mathfrak{B}(\mathbb{G}_2, \mathbb{G}_1)$ with associated kernel

$$R(\cdot,t) = \mathscr{T}^* K_1(\cdot,t) \tag{10.22}$$

from Section 4.7. We will refer to R as a *regression kernel* for reasons that will become clear shortly. If, for example, $\mathscr{T}$ is HS, Theorem 4.7.2 tells us that we can write

$$R(s,t) = \sum_{i=1}^{\infty}\sum_{j=1}^{\infty} \lambda_{1i}\lambda_{2j}b_{ij}e_{1i}(t)e_{2j}(s)$$

with

$$\sum_{i=1}^{\infty}\sum_{j=1}^{\infty} \lambda_{1i}\lambda_{2j}b_{ij}^2 < \infty.$$

An updated version of the regression model (10.21) now appears as

$$X_1(t) = Z_2(R(\cdot,t)) + \varepsilon(t) \tag{10.23}$$

with R as defined in (10.22) and $\varepsilon(\cdot)$ as before. Now

$$Z_2(R(\cdot,t)) = \sum_{i=1}^{\infty}\sum_{j=1}^{\infty} \lambda_{1i}b_{ij}e_{1i}(t)\langle X_2, e_{2j}\rangle. \tag{10.24}$$

If $\sum_{i=1}^{\infty}\sum_{j=1}^{\infty} \lambda_{1i}^2 b_{ij}^2 < \infty$, (10.23) and (10.21) coincide with

$$\beta(s,t) = \sum_{i=1}^{\infty}\sum_{j=1}^{\infty} \lambda_{1i}b_{ij}e_{1i}(t)e_{2j}(s).$$

The fact that $X_2(\cdot)$ and $\varepsilon(\cdot)$ are uncorrelated means that $\varepsilon(\cdot)$ is uncorrelated with $\langle X_2, e_{2j}\rangle$ for all j due to Theorem 3.1.7. The (X_1, X_2) cross-covariance kernel is therefore seen to be

$$\begin{aligned} K_{12}(t_1, t_2) &= \mathbb{E}\left[Z_2(R(\cdot, t_1))X_2(t_2)\right] \\ &= \langle R(\cdot, t_1), K_2(\cdot, t_2)\rangle_2 \\ &= R(t_2, t_1). \end{aligned}$$

Canonical correlation now gives us a canonical form for the regression kernel as a result of Theorem 10.1.6. In addition, our operator $\mathscr{T}$ coincides with the operator $\mathscr{C}_{12}$ in (10.7) because

$$\begin{aligned} (\mathscr{C}_{12}f)(t) &= \langle R(\cdot, t), f(\cdot)\rangle_2 = \langle K_1(\cdot, t), (\mathscr{T}f)(\cdot)\rangle_1 \\ &= (\mathscr{T}f)(t). \end{aligned}$$

Among other things, we can conclude here that, just as in the mva setting, a linear regression relationship can exist between X_1 and X_2 only when at least one of the canonical correlations is nonzero.

10.4 Factor analysis

In this section, we explore the problem of factor analysis under the same conditions as Section 10.3. For that purpose, we consider the signal-plus-noise model given by

$$X_1(t) = X_2(t) + \varepsilon(t) \tag{10.25}$$

for $t \in E$. Here X_1, X_2, and ε are all zero mean, $\mathbb{L}^2(E, \mathscr{B}(E), \mu)$ valued processes with covariance kernels K_1, K_2, and K_ε. The signal process, X_2, is assumed to be uncorrelated with the noise process ε. With this latter specification model (10.25) can be viewed as the natural extension of the mva factor model (1.21) to continuous time.

Following along the lines of our factor analysis development in Section 1.1, our first step should be canonical analysis of the X_1, X_2 processes. In this instance, we find that the cross-covariance kernel (10.6) is $K_{12} = K_2$ and the covariance kernel for X_1 is $K_1 = K_2 + K_\varepsilon$. Theorem 2.7.10 tells us that $\mathbb{G}_1$ consists of sums of functions from the RKHSs $\mathbb{G}_2$ and $\mathbb{H}(K_\varepsilon) = \mathbb{G}(\mathscr{K}_\varepsilon)$ and, in particular, any $f_2 \in \mathbb{G}_2$ is also an element of $\mathbb{G}_1$. Thus, by (10.7),

$$\begin{aligned} (\mathscr{C}_{12}f_2)(t) &= \langle K_{12}(t, \cdot), f_2\rangle_2 \\ &= \langle K_2(t, \cdot), f_2\rangle_2 \\ &= f_2(t) \end{aligned}$$

and

$$(\mathscr{C}_{12}^{*}\mathscr{C}_{12}f_2)(t) = \langle K_{21}(t,\cdot), f_2\rangle_1 = \langle K_2(t,\cdot), f_2\rangle_1.$$

A squared canonical correlations ρ^2 and its associated singular function f_2 for the X_2 space must therefore satisfy

$$\begin{aligned}\langle K_2(t,\cdot), f_2\rangle_1 &= \langle K_1(t,\cdot) - K_\varepsilon(t,\cdot), f_2\rangle_1\\ &= f_2(t) - \langle K_\varepsilon(t,\cdot), f_2\rangle_1\\ &= \rho^2 f_2(t)\end{aligned}$$

for all $t \in E$. This gives us

$$\langle K_\varepsilon(t,\cdot), f_2\rangle_1 = (1-\rho^2)f_2(t).$$

However, when we use Theorem 4.3.1 here we see that the singular functions for the X_1 and X_2 spaces are related by $f_1 = f_2/\rho$. Hence, we have the identity

$$\langle K_\varepsilon(t,\cdot), f_1\rangle_1 = (1-\rho^2)f_1(t) \tag{10.26}$$

that characterizes the singular functions for the X_1 space.

Example 10.4.1 *Consider the mva setting of Example 10.1.4 with $\mathscr{K}_1, \mathscr{K}_2$, and $\mathscr{K}_\varepsilon$ now representing the variance–covariance matrices for the random p-vectors X_1, X_2, and ε. In this case, we know that $f_1(\cdot) \in \mathbb{G}_1$ means that $f_1 = \mathscr{K}_1 a_1$ for some $a_1 \in \mathbb{R}^p$ and (10.26) becomes*

$$\begin{aligned}\mathscr{K}_\varepsilon \mathscr{K}_1^{-1}\mathscr{K}_1 a_1 &= \mathscr{K}_\varepsilon a_1\\ &= (1-\rho^2)\mathscr{K}_1 a_1,\end{aligned}$$

thereby returning us to relation (1.22) that was used to develop factor analysis in the finite-dimensional case.

As in the mva setting, to complete the factor model, we need to add structure to the X_2 process. Specifically, we will assume that X_2 admits the representation

$$X_2(t) = \sum_{j=1}^{\infty} Z_j \phi_j(t),$$

where the Z_j are zero mean, uncorrelated random variables with unit variance and $\{\phi_j\}$ is an orthogonal sequence of functions in $\mathbb{H}(K_\varepsilon)$ with $\|\phi_j\|^2_{\mathbb{H}(K_\varepsilon)} = \gamma_j, j = 1, \ldots$ for positive values γ_j that satisfy $\sum_{j=1}^{\infty}\gamma_j < \infty$. In the language of Section 1.1, the Z_j occupy the role of the factors while the ϕ_j play the part of factor loadings.

The covariance kernel for the X_2 process and, hence, the cross-covariance kernel is now seen to have the form

$$\begin{aligned} K_{12}(s,t) &= K_2(s,t) \\ &= \sum_{j=1}^{\infty} \gamma_j \tilde{\phi}_i(s)\tilde{\phi}_i(t) \end{aligned} \tag{10.27}$$

with

$$\tilde{\phi}_j = \phi_j / \sqrt{\gamma_j}$$

being an orthonormal sequence in $\mathbb{H}(K_\varepsilon)$. The function $K_2(\cdot, t)$ is a well-defined element of $\mathbb{H}(K_\varepsilon)$, which has the consequence that the elements of $\mathbb{H}(K_1)$ must be the same as those in $\mathbb{H}(K_\varepsilon)$.

The kernels K_1 and K_ε can be directly related using the operator $\tilde{\Phi} \in \mathfrak{B}(\mathbb{H}(K_\varepsilon))$ defined by

$$\tilde{\Phi} = \sum_{j=1}^{\infty} \gamma_j \tilde{\phi}_j \otimes_{\mathbb{H}(K_\varepsilon)} \tilde{\phi}_j. \tag{10.28}$$

This is a nonnegative operator whose eigenvector $\tilde{\phi}_j$ corresponds to the eigenvalue γ_j with the consequence that $(I + \tilde{\Phi})$ is invertible. Keeping that in mind, we claim that

$$\langle f, g \rangle_1 = \langle (I + \tilde{\Phi})^{-1} f, g \rangle_{\mathbb{H}(K_\varepsilon)}. \tag{10.29}$$

To establish (10.29) observe that for $f \in \mathbb{H}(K_1)$

$$\begin{aligned} \langle (I + \tilde{\Phi})f, K_1(\cdot, t)\rangle_1 &= \langle f, K_1(\cdot, t)\rangle_1 + \langle \tilde{\Phi} f, K_1(\cdot, t)\rangle_1 \\ &= f(t) + (\tilde{\Phi} f)(t) \\ &= \langle f, K_\varepsilon(\cdot, t)\rangle_{\mathbb{H}(K_\varepsilon)} + \langle \tilde{\Phi} f, K_\varepsilon(\cdot, t)\rangle_{\mathbb{H}(K_\varepsilon)} \\ &= \langle f, K_\varepsilon + K_2\rangle_{\mathbb{H}(K_\varepsilon)} \\ &= \langle f, K_1\rangle_{\mathbb{H}(K_\varepsilon)} \end{aligned}$$

because

$$\begin{aligned} \langle \tilde{\Phi} f, K_\varepsilon(\cdot, t)\rangle_{\mathbb{H}(K_\varepsilon)} &= \sum_{j=1}^{\infty} \gamma_j \langle \tilde{\phi}_j, f\rangle_{\mathbb{H}(\mathscr{K}_\varepsilon)} \tilde{\phi}_j(t) \\ &= \langle K_2(\cdot, t), f\rangle_{\mathbb{H}(K_\varepsilon)}. \end{aligned}$$

As linear combinations of K_1 are dense in $\mathbb{G}_1$, we conclude that $\langle (I + \tilde{\Phi})f, g\rangle_1 = \langle f, g\rangle_{\mathbb{H}(\mathscr{K}_\varepsilon)}$ for every $f, g \in \mathbb{G}_1$ and the claim has been verified.

As a result of (10.29), we see that for all $f \in \mathbb{G}_1$

$$f(t) = \langle K_1(\cdot, t), f\rangle_1 = \langle (I + \tilde{\Phi})^{-1} K_1(\cdot, t), f\rangle_{\mathbb{H}(K_\varepsilon)}$$

which has the implication that $(I + \tilde{\Phi})^{-1} K_1(\cdot, t) = K_\varepsilon(\cdot, t)$. Using this in (10.26) produces

$$\begin{aligned}(1 - \rho^2) f_1(t) &= \langle K_\varepsilon(t, \cdot), f_1\rangle_1 \\ &= \langle (I + \tilde{\Phi})^{-1} K_1(\cdot, t), f_1\rangle_1 \\ &= \langle K_1(\cdot, t), (I + \tilde{\Phi})^{-1} f_1\rangle_1 \\ &= \left((I + \tilde{\Phi})^{-1} f\right)(t)\end{aligned}$$

or

$$\tilde{\Phi} f_1 = \frac{\rho^2}{1 - \rho^2} f_1.$$

However, $\tilde{\Phi}\tilde{\phi}_j = \gamma_j \tilde{\phi}_j$ and it follows that $\tilde{\phi}_j$ is, in fact, the singular function for the X_1 space with associated canonical correlation ρ_j. The value of γ_j can be recovered from the relation $\gamma_j = \rho_j^2/(1 - \rho_j^2)$.

The idealized prescription for factor analysis would now proceed something like this. If we knew both K_1 and K_ε, we could determine the X_1 singular vectors and canonical correlations using (10.26) and thereby create the cross-covariance kernel $K_{12} = K_2$ in (10.27). Theorem 10.3.1 now gives us the best linear predictor of $X_2(t)$ as $Z_1(K_2(\cdot, t))$. However, the aim is to predict the factors $Z_1, Z_2, \ldots$. That problem can be addressed similarly using the cross-covariance kernel between the Z_i and X_1: namely,

$$\mathrm{Cov}\,(Z_i, X_1(t)) = \tilde{\phi}(t).$$

Another application of Theorem 10.3.1 reveals that $Z_1(\tilde{\phi}_i)$ is the best linear predictor of the ith factor.

Example 10.4.2 *Continuing with Example 10.4.1, the vector representation of the $X_2(\cdot)$ process under the factor model appears as $X_2 = \Phi Z$ with $\Phi = \{\phi_j(t_i)\}$ a $p \times r$ matrix of coefficients and Z a random r-vector with mean zero and an identity for its variance–covariance matrix. The operator in (10.28) becomes*

$$\tilde{\Phi} = \sum_{j=1}^{r} \gamma_j \tilde{\phi}_j \tilde{\phi}_j^T$$

with $\tilde{\phi}_j := \mathscr{K}_\varepsilon^{-1/2} \phi_j$ for ϕ_j the jth column of the matrix Φ. Thus, under the present formulation, the factor analysis constraint from Section 1.1 that $\phi_i^T \mathscr{K}_\varepsilon^{-1} \phi_j = 0$ for $i \neq j$ can be appreciated as an orthogonality condition for the loading vectors in the Hilbert space $\mathbb{G}(\mathscr{K}_\varepsilon)$ that represents their home.

10.5 MANOVA and discriminant analysis

One formulation of functional analysis of variance (ANOVA) and discriminant analysis can be developed by assuming that we are observing a stochastic process X_1 whose sample paths can lie in one of J possible populations. If π_j is the probability that the reading comes from population j, we can view the problem as seeing the process pair (X_1, X_2), where X_1 takes values in $\mathbb{L}^2(E, \mathscr{B}(E), \mu)$ and $X_2 = \{X_2(j) : j = 1, \ldots, J\}$ for dichotomous random variables $X_2(j)$ that take the values 1 or 0 with probability π_j and $1 - \pi_j$, respectively, depending on whether or not X_1 derives from the jth population.

Let

$$m_j(\cdot) = \mathbb{E}[X_1(\cdot)|X_2 = j]$$

for $j = 1, \ldots, J$ be the conditional mean functions for X_1 that correspond to the J different populations. The overall or grand mean for X_1 is then

$$m(\cdot) = \mathbb{E}[X_1(\cdot)] = \sum_{j=1}^{J} \pi_j m_j(\cdot).$$

Unlike the developments in previous sections, we can no longer assume that m or the m_j vanish. To account for this, we now need to work with the X_1 covariance kernel

$$R_1(s,t) = \mathbb{E}[(X_1(s) - m(s))(X_1(t) - m(t))]$$

and cross-covariance kernel

$$\begin{aligned} R_{12}(s,j) &= \mathbb{E}[(X_1(s) - m(s))(X_2(j) - \pi_j)] \\ &= \pi_j \left(m_j(s) - m(s)\right) \end{aligned}$$

The kernel R_1 generates an RKHS $\mathbb{G}_1$ with norm and inner product we will continue to denote by $\langle \cdot, \cdot \rangle_1$ and $\| \cdot \|_1$. Movement between the spaces $\mathbb{G}_1$ and $\mathbb{L}^2(X_1)$ is then carried out via the mapping Z_1 as before.

For our more immediate purposes, it will suffice to work with the centered process $\tilde{X}_2 = \{\tilde{X}_2(j) : j = 1, \ldots, J\}$ with $\tilde{X}_2(j) = X_2(j) - \pi_j, j = 1, \ldots, J$. Its covariance kernel is

$$\begin{aligned} K_2(i,j) &= \mathbb{E}[\tilde{X}_2(i)\tilde{X}_2(j)] = \mathbb{E}[(X_2(i) - \pi_i)(X_2(j) - \pi_j)] \\ &= \delta_{ij}\pi_j - \pi_i\pi_j \end{aligned}$$

and, of course,

$$R_{12}(s,j) = \mathbb{E}[(X_1(s) - m(s))\tilde{X}_2(j)].$$

The K_2 kernel has a matrix representation as

$$\mathscr{K}_2 = \text{diag}(\pi_1, \ldots, \pi_J) - \pi\pi^T$$

with $\pi = (\pi_1, \ldots \pi_J)^T$ and $\sum_{j=1}^J \pi_j = 1$. Thus, from Example 2.7.8, we know that $\mathbb{G}_2 := \mathbb{H}(K_2)$ consists of functions on $\{1, \ldots, J\}$ of the form

$$f_2(\cdot) = \sum_{j=1}^J a_{2j} K_2(\cdot, j)$$

for $a_2 = (a_{21}, \ldots a_{2J})^T$ an element of the orthogonal complement of the null space of the matrix $\mathscr{K}_2$: i.e., $\sum_{j=1}^J a_{2j} = 0$.

A convenient choice for a generalized inverse of $\mathscr{K}_2$ is

$$\mathscr{K}_2^- = \text{diag}(\pi_1^{-1}, \ldots, \pi_J^{-1})$$

and using this option in (2.34) reveals that the inner product of two functions f_2, f_2' in $\mathbb{G}_2$ is

$$\langle f_2, f_2' \rangle_2 = \sum_{j=1}^J \frac{f_2(j) f_2'(j)}{\pi_j}.$$

The isometric mapping that connects $\mathbb{L}^2(\tilde{X}_2)$ and $\mathbb{G}_2$ is found to be

$$\begin{aligned} Z_2(f_2) &= \sum_{j=1}^J \frac{f_2(j)[X_2(j) - \pi_j]}{\pi_j} \\ &= \sum_{j=1}^J \frac{f_2(j) X_2(j)}{\pi_j} \end{aligned}$$

for $f_2 \in \mathbb{G}_2$.

As in (10.7), cca now revolves around the singular value expansion of the operator $\mathscr{C}_{12}$ defined by

$$\begin{aligned} (\mathscr{C}_{12} f_2)(\cdot) &= \sum_{j=1}^J \frac{R_{12}(\cdot, j) f_2(j)}{\pi_j} \\ &= \sum_{j=1}^J f_2(j) m_j(t). \end{aligned}$$

Thus, $\mathscr{C}_{12}$ maps elements of $\mathbb{G}_2$ to contrasts among the population mean functions with the consequence that all the mean functions coincide if and only if the $\mathscr{C}_{12}$ singular values $\rho_1, \ldots, \rho_{J-1}$ vanish. Testing the functional ANOVA hypothesis that $m_1(\cdot) = \cdots = m_J(\cdot)$ is therefore equivalent to the hypothesis that all the canonical correlations are zero.

If $\{(\rho_k, f_{1k}, f_{2k})\}_{k=1}^{J-1}$ is the singular system for $\mathscr{C}_{12}$, the canonical variables of the X_1 and $\tilde{X}_2$ spaces are $U_{1k} := Z_1(f_{1k})$ and

$$\begin{aligned} U_{2k} &= Z_2(f_{2k}) \\ &= \sum_{j=1}^{J} \frac{f_{2k}(j) X_2(j)}{\pi_j}. \end{aligned}$$

These variables may be used for classification purposes. For example, suppose we define the centered X_1 canonical variables

$$\begin{aligned} U_{1j}^c &= U_{1j} - \mathbb{E}[U_{1j}] \\ &= U_{1j} - \langle f_{1j}, m \rangle_1. \end{aligned}$$

Then, $\rho_j U_{2j}$ provides a linear predictor of U_{1j}^c and, in particular, $U_{2j} = f_{2j}(k)/\pi_k$ when X_1 comes from the kth population. Thus, a regression argument suggests classification via minimization with respect to k of a distance measure such as

$$\sum_{j=1}^{r} \frac{1}{1-\rho_j^2} (U_{1j}^c - \rho_j \pi_k^{-1} f_{2j}(k))^2 \tag{10.30}$$

for some $r \le J-1$.

Fisher's approach to discriminant analysis described in Section 1.1 has also been generalized to the stochastic processes setting in work by Shin (2008). She defines the discriminant functions to be random variables $\ell \in \mathbb{L}^2(X_1)$ that maximize

$$\frac{\mathrm{Var}(\mathbb{E}[\ell | X_2])}{\mathbb{E}[\mathrm{Var}(\ell | X_2)]}.$$

Thus, the first linear discriminant function ℓ_1 satisfies

$$\mathrm{Var}(\mathbb{E}[\ell_1 | X_2]) = \sup_{\ell \in \mathbb{L}^2(X_1)} \mathrm{Var}(\mathbb{E}[\ell | X_2]),$$

where ℓ is subject to $\mathbb{E}[\mathrm{Var}(\ell | X_2)] = 1$. The ith linear discriminant function for $i > 1$ is defined similarly subject to the additional restriction $\mathbb{E}[\mathrm{Cov}(\ell_i, \ell_k | X_2)] = 0, k < i$.

Assume that the within group covariance kernel R_W is the same across the J populations: i.e.,

$$R_W(s,t) = \mathbb{E}[(X_1(s) - m_j(s))(X_1(t) - m_j(t)) | X_2 = j]$$

for all $j = 1, \ldots, J$. Denote the RKHS generated by R_W as $\mathbb{G}_W$ with associated norm and inner product indicated by $\|\cdot\|_W$ and $\langle \cdot, \cdot \rangle_W$. In addition, define the kernel function

$$R_B(s,t) = \sum_{j=1}^{J} \pi_j (m_j(s) - m(s))(m_j(t) - m(t))$$

for $s, t \in E$.

Arguments in Shin (2008) establish that if $m_j \in \mathbb{G}_W$ for all $j = 1, \ldots, J$, there exists a one-to-one linear mapping Z_W from $\mathbb{G}_W$ onto $\mathbb{L}^2(X_1)$ defined by

$$Z_W : R_W(\cdot, t) \mapsto X_1(t)$$

for every $t \in E$ with the properties that

1. $\mathbb{E}[Z_W(h)] = \langle h, m \rangle_W$,
2. $\mathbb{E}[Z_W(h)|X_2 = j] = \langle h, m_j \rangle_W$, and
3. $\mathbb{E}[\mathrm{Var}(Z_W(h)|X_2)] = \|h\|_W^2$ for $h \in \mathbb{G}_W$.

Using Z_W, we see that

$$\mathrm{Var}(\mathbb{E}[Z_W(h)|X_2]) = \langle h, \mathscr{C}_B h \rangle_W$$

with

$$(\mathscr{C}_B h)(t) := \langle R_B(t, \cdot), h(\cdot) \rangle_W \tag{10.31}$$

for $h \in \mathbb{G}_W$.

The operator $\mathscr{C}_B$ in (10.31) has the spectral decomposition

$$\mathscr{C}_B = \sum_{j=1}^{J-1} \gamma_j h_j \otimes_{\mathbb{G}_W} h_j,$$

with $\gamma_1 \geq \cdots \geq \gamma_{J-1} \geq 0$ the eigenvalues and $h_j, j = 1, \ldots, J-1$, the associated eigenfunctions for the operator. The linear discriminant functions are then seen to be

$$\ell_j = Z_W(h_j)$$

for $j = 1, \ldots, J-1$ with classification obtained from squared Mahalanobis distance based on the first $r \leq J-1$ linear discriminant functions. That is, for the kth population, we employ the distance measure

$$\sum_{j=1}^{r} \left(Z_W(h_j) - \langle h_j, m_k \rangle_W \right)^2 \tag{10.32}$$

that compares the process "score" vector $(Z_W(h_1), \ldots ,Z_W(h_r))^T$ to its conditional mean $(\langle h_1, m_k\rangle_W, \ldots , \langle h_r, m_k\rangle_W)^T$ for the kth population.

In Section 1.1, we saw that cca and Fisher's method were equivalent in the case of finite-dimensional mva. We now show that this equivalence continues to hold for fda situations.

Theorem 10.5.1 *Assume that $m_j \in \mathbb{G}_W, j = 1, \ldots , J$. Then,*

$$\gamma_j = \frac{\rho_j^2}{1-\rho_j^2},$$
$$h_j = (1-\rho_j^2)^{1/2} f_{1j}$$
$$\ell_j = \frac{U_{1j}}{\sqrt{1-\rho_j^2}}$$

for $j = 1, \ldots , J-1$.

Proof: For the theorem to be true, it must be that both $\mathscr{C}_{12}$ and $\mathscr{C}_B$ are defined on the same space: i.e., $\mathbb{G}_W$ and $\mathbb{G}_1$ must consist of the same elements. To see that, this is so first use the identity

$$\mathrm{Cov}(Y, Z) = \mathrm{Cov}(\mathbb{E}[Y|V], \mathbb{E}[Z|V]) + \mathbb{E}[\mathrm{Cov}(Y, Z|V)]$$

that holds for any random variables Y, Z, and V with finite second moments to obtain

$$R_1(s,t) = R_B(s,t) + R_W(s,t).$$

Thus, $R_W \ll R_1$ and $\mathbb{G}_W \subset \mathbb{G}_1$ due to Theorem 2.7.11. To go in the other direction, define the operator $\mathscr{L} : \mathbb{G}_1 \to \mathbb{G}_W$ by

$$\mathscr{L}(R_1(\cdot, t)) = R_W(\cdot, t)$$

for $t \in E$. Then, $\mathscr{L}$ is a one-to-one, onto and, for $h \in \mathbb{G}_W$ and $f \in \mathbb{G}_1$,

$$\langle h, f\rangle_1 = \langle h, \mathscr{L}f\rangle_W$$

because

$$\langle h, R_1(\cdot, t)\rangle_1 = h(t) = \langle h, R_W(\cdot, t)\rangle_W.$$

For $f \in \mathbb{G}_1$,

$$\begin{aligned} f(t) &= \langle R_B(\cdot, t), f\rangle_1 + \langle R_W(\cdot, t), f\rangle_1 \\ &= \langle R_B(\cdot, t), \mathscr{L}f\rangle_W + \langle R_W(\cdot, t), \mathscr{L}f\rangle_W \\ &= (\mathscr{C}_B\mathscr{L}f)(t) + (\mathscr{L}f)(t), \end{aligned}$$

which indicates that f is in the range of $\mathscr{L}$ and therefore in $\mathbb{G}_W$.

Now $\mathscr{C}_B$ induces an operator $\tilde{\mathscr{C}}_B$ on $\mathbb{G}_1$ via $(\tilde{\mathscr{C}}_B f)(t) = \langle R_B(t,\cdot), f(\cdot)\rangle_1$. One finds that $\tilde{\mathscr{C}}_B = \mathscr{C}_B\mathscr{L}$ because

$$(\mathscr{C}_B\mathscr{L}f)(t) = \langle R_B(\cdot,t), \mathscr{L}f\rangle_W = \langle R_B(\cdot,t), f\rangle_1 = (\mathscr{C}_B f)(t).$$

As $f = \mathscr{L}f + \mathscr{C}_B\mathscr{L}f$, $\mathscr{L} = I - \tilde{\mathscr{C}}_B$ and

$$\|\mathscr{L}f\|_W^2 = \langle \mathscr{L}f, f\rangle_1 = \langle (I - \tilde{\mathscr{C}}_B)f, f\rangle_1$$

for $f \in \mathbb{G}_1$.

The linear mappings Z_1, Z_W that connect $\mathbb{G}_1$ and $\mathbb{G}_W$ to $\mathbb{L}^2(X_1)$ are similarly related in that

$$Z_1(f) = Z_W(\mathscr{L}f) \tag{10.33}$$

for $f \in \mathbb{G}_1$, which follows from

$$Z_1\,(R_1(\cdot,t)) = X_1(t) = Z_W(R_W(\cdot,t)) = Z_W(\mathscr{L}(R_1(\cdot,t))).$$

For $f \in \mathbb{G}_1$, we have

$$(\mathscr{C}_{12}\mathscr{C}_{21}f)(t) = \langle R_{12}(t,\cdot), \mathscr{C}_{21}f\rangle_2 = \langle \mathscr{C}_{12}R_{12}(t,\cdot), f\rangle_1.$$

However,

$$R_{12}(\cdot,j) = \mathrm{Cov}(X_1(\cdot), X_2(j)) = \pi_j(m_j(\cdot) - m(\cdot))$$

for $j = 1, \ldots J$ and

$$\begin{aligned}\mathscr{C}_{12}R_{12}(t,\cdot) &= \langle R_{12}(\cdot,\star), R_{12}(t,\star)\rangle_2 \\ &= \sum_{j=1}^{J} \frac{R_{12}(\cdot,j)R_{12}(t,j)}{\pi_j} = R_B(t,\cdot).\end{aligned}$$

Thus, $\mathscr{C}_{12}\mathscr{C}_{12} = \tilde{\mathscr{C}}_B$.

We can now use the fact that

$$\mathscr{C}_{12}\mathscr{C}_{21}f = \tilde{\mathscr{C}}_B f = \mathscr{C}_B\mathscr{L}f$$

and $f = \mathscr{L}f + \mathscr{C}_B\mathscr{L}f$ for $f \in \mathbb{G}_1$ to write

$$\mathscr{C}_B\mathscr{L}f_{1j} = \frac{\rho_j^2}{1-\rho_j^2}\mathscr{L}f_{1j}.$$

In addition, $\mathscr{L}f_{1j} = (1-\rho_j^2)f_{1j}$ because

$$\mathscr{L}f_{1j} = (I - \tilde{\mathscr{C}}_B)f_{1j} = (I - \mathscr{C}_{12}\mathscr{C}_{21})f_{1j}$$

and

$$\|\mathscr{L} f_{1j}\|_W^2 = \left\langle \left(I - \tilde{\mathscr{C}}_B\right) f_{1j}, f_{1j} \right\rangle_1 = 1 - \rho_j^2.$$

As the h_j in $\mathbb{G}_W$ corresponding to $\ell_j = Z_W(h_j)$ satisfy $\|h_j\|_W^2 = 1$,

$$h_j = \frac{\mathscr{L} f_{1j}}{\|\mathscr{L} f_{1j}\|_W} = \frac{\mathscr{L} f_{1j}}{(1 - \rho_j^2)^{1/2}} = (1 - \rho_j^2)^{1/2} f_{1j}.$$

The proof is completed using (10.33). □

We note in passing that the results in Theorem 10.5.1 are the same as what we found for the mva case in Section 1.1. Along this same vein, we conclude the section by showing that, just as in the finite-dimensional case, classification with either Fisher's method or cca produces the same results.

Corollary 10.5.2 *The distance measures (10.30) and (10.32) are equivalent.*

Proof: The relation

$$\langle f_{1j}, m_k \rangle_1 = \langle \mathscr{L} f_{1j}, m_k \rangle_W = (1 - \rho_j^2)^{1/2} \langle h_j, m_k \rangle_W$$

has the implication that

$$\begin{aligned}\sum_{j=1}^{r} (Z_W(h_j) - \langle h_j, m_k \rangle_W)^2 &= \sum_{j=1}^{r} \frac{1}{1 - \rho_j^2} (U_{1j} - \langle f_{1j}, m_k \rangle_1)^2 \\ &= \sum_{j=1}^{r} \frac{1}{1 - \rho_j^2} (U_{1j}^c - \langle f_{1j}, m_k - m \rangle_1)^2.\end{aligned}$$

As

$$(\mathscr{C}_{21} f_{1j})(k) = \langle f_{1j}(\cdot), R_{12}(\cdot, k) \rangle_1 = \pi_k \langle f_{1j}, m_k - m \rangle_1,$$

we have

$$\begin{aligned}\langle f_{1j}, m_k - m \rangle_1 &= \pi_k^{-1} \left(\mathscr{C}_{21} f_{1j} \right)(k) \\ &= \rho_j \pi_k^{-1} f_{2j}(k).\end{aligned}$$

□

10.6 Orthogonal subspaces and partial cca

Theorem 10.1.2 represents the solution to our cca problem. There is, however, another way to obtain this result based on the orthogonal subspace decomposition detailed in Section 3.7. This approach has some advantages in that it is extensible to situations that involve more than just two random elements as will be demonstrated subsequently. To pursue this alternative development, we focus on the random element setting and introduce the Hilbert space

$$\mathbb{G}_0 = \left\{ h = (f_1, f_2) : f_i \in \mathbb{G}_i, i = 1, 2, \|h\|_0^2 = \sum_{i=1}^{2} \|f_i\|_i^2 < \infty \right\} \quad (10.34)$$

with addition of elements of $\mathbb{G}_0$ being performed component-wise. For each $h \in \mathbb{G}_0$, we then define an analog of Z_1 and Z_2 by

$$Z_0(h) = Z_1(f_1) + Z_2(f_2).$$

For $h = (f_1, f_2), h' = (f_1', f_2')$ in $\mathbb{G}_0$, the Z_0 function is

$$\begin{aligned}\mathrm{Cov}(Z_0(h), Z_0(h')) &= \mathrm{Cov}(Z_1(f_1), Z_1(f_1')) + \mathrm{Cov}(Z_2(f_2), Z_2(f_2')) \\ &\quad + \mathrm{Cov}(Z_1(f_1), Z_2(f_2')) + \mathrm{Cov}(Z_1(f_1'), Z_2(f_2)) \\ &= \langle f_1, f_1' \rangle_1 + \langle f_2, f_2' \rangle_2 + \langle f_1, \mathscr{C}_{12} f_2' \rangle_1 \\ &\quad + \langle \mathscr{C}_{21} f_1', f_2 \rangle_2. \end{aligned} \quad (10.35)$$

The Hilbert space spanned by the Z_0 process is

$$\mathbb{L}^2(Z_0) = \left\{ Z_0(h) : h \in \mathbb{G}_0, \|Z_0(h)\|_0^2 := \mathrm{Var}(Z_0(h)) < \infty \right\}. \quad (10.36)$$

Note that when $\mathrm{Var}(Z_1(f_1)) = 1 = \mathrm{Var}(Z_2(f_2))$

$$\rho^2(f_1, f_2) = \mathrm{Cov}(Z_0((f_1, 0)), Z_0((0, f_2))).$$

Thus, we have not lost track of our original problem and optimization of $\rho^2(f_1, f_2)$ over $f_i \in \mathbb{G}_i$ can be recovered through, e.g., restricted optimization of the covariance functional (10.35) over $\mathbb{G}_0$. However, the analysis becomes somewhat more tractable if we can work with a Hilbert space that is congruent to $\mathbb{L}^2(Z_0)$ so that the objective function can be expressed in terms of its norms and inner products. For this purpose, we will require

Assumption 1 *There is a constant* $B \in [0, 1)$ *such that* $|\mathrm{Corr}(Z_1(f_1), Z_2(f_2))| \leq B$ *for all* $(f_1, f_2) \in \mathbb{G}_0$. □

Assumption 1 insures that there is no linear combination of the random variables $\{\langle \chi_2, e_{2j} \rangle\}$ (with finite variance) that can exactly predict a similar linear combination of $\{\langle \chi_1, e_{1j} \rangle\}$. For the finite-dimensional multivariate analysis case, this would translate into the cross-covariance matrix for the vector representations of χ_1, χ_2 having full rank: i.e., rank equal to the smaller of the dimensions for the two random vectors. This, in turn, avoids the degenerate case where a component of one of the random vectors is just a translated and rescaled version of variables that already exist in the other. A similar interpretation can be made in the more abstract context of this chapter. Of more immediate import is the fact that

$$\|\mathscr{C}_{12}\| = \|\mathscr{C}_{21}\| \leq B \tag{10.37}$$

when Assumption 1 is in effect as can be seen by retracing the argument that was used to prove Theorem 10.1.1.

Now, for $h \in \mathbb{G}_0$, define $\mathscr{Q}h = (f_1 + \mathscr{C}_{12}f_2, f_2 + \mathscr{C}_{21}f_1)$. It will be convenient to write this in matrix form as

$$\mathscr{Q}h = \begin{bmatrix} I & \mathscr{C}_{12} \\ \mathscr{C}_{21} & I \end{bmatrix} \begin{bmatrix} f_1 \\ f_2 \end{bmatrix} \tag{10.38}$$

with the convention that the resulting vector is viewed as an element of $\mathscr{H}_0$. Observe that

$$\text{Cov}(Z_0(h), Z_0(h')) = \langle h, \mathscr{Q}h' \rangle_0.$$

Theorem 10.6.1 *Under Assumption 1, $\mathscr{Q} : \mathbb{G}_0 \mapsto \mathbb{G}_0$ is invertible with inverse defined by*

$$\mathscr{Q}^{-1}(h) = (\mathscr{C}_{11.2}^{-1}f_1 - \mathscr{C}_{12}\mathscr{C}_{22.1}^{-1}f_2, \mathscr{C}_{22.1}^{-1}f_2 - \mathscr{C}_{21}C_{11.2}^{-1}f_1) \tag{10.39}$$

for $h = (f_1, f_2) \in \mathbb{G}_0$ and $\mathscr{C}_{ii.k} = I - \mathscr{C}_{ik}\mathscr{C}_{ki} = (I - \mathscr{C}_{ik}\mathscr{C}_{ki})^*$, $i, k = 1, 2, i \neq k$.

Analogous to (10.38), (10.39) will also be expressed as

$$\mathscr{Q}^{-1}h = \begin{bmatrix} \mathscr{C}_{11.2}^{-1} & -\mathscr{C}_{12}\mathscr{C}_{22.1}^{-1} \\ -\mathscr{C}_{21}\mathscr{C}_{11.2}^{-1} & \mathscr{C}_{22.1}^{-1} \end{bmatrix} \begin{bmatrix} f_1 \\ f_2 \end{bmatrix}.$$

Proof: The form of the inverse as stated in 10.39 follows directly once we have shown all the relevant inverse operators exist. Thus, let us concentrate on the latter task.

We can write $\mathscr{Q} = I - \mathscr{T}$ with

$$\mathscr{T}h = (-\mathscr{C}_{12}f_2, -\mathscr{C}_{21}f_1) = -\begin{bmatrix} 0 & \mathscr{C}_{12} \\ \mathscr{C}_{21} & 0 \end{bmatrix} \begin{bmatrix} f_1 \\ f_2 \end{bmatrix}.$$

Then,

$$\begin{aligned} \|\mathscr{T}h\|_0^2 &= \|\mathscr{C}_{12}f_2\|_1^2 + \|\mathscr{C}_{21}f_1\|_2^2 \\ &\leq \|\mathscr{C}_{12}\|^2 \|\|f_2\|_2^2 + \|\mathscr{C}_{21}\|^2\|f_1\|_1^2 \\ &= \|\mathscr{C}_{12}\|^2[\|f_1\|_1^2 + \|f_2\|_2^2] \\ &= \|\mathscr{C}_{12}\|^2\|h\|_0^2 \\ &\leq B\|h\|_0^2 < \|h\|_0^2 \end{aligned}$$

due to (10.37) and Theorem 3.5.5 has the consequence that $I - \mathscr{T} = \mathscr{Q}$ is invertible.

To complete the proof, we need to show that $\mathscr{C}_{11.2}$ and $\mathscr{C}_{22.1}$ are invertible. This again follows from Theorem 3.5.5 because, e.g., $\mathscr{C}_{11.2} = I - \mathscr{C}_{12}\mathscr{C}_{21}$ with $\|\mathscr{C}_{21}\| = \|\mathscr{C}_{12}\| < 1$. □

Now define

$$\mathbb{H}(\mathscr{Q}) = \left\{ \tilde{h} : \tilde{h} = \mathscr{Q}\begin{bmatrix} f_1 \\ f_2 \end{bmatrix}, f_i \in \mathbb{G}_i, i = 1, 2, \|\tilde{h}\|^2_{\mathbb{H}(\mathscr{Q})} = \|\mathscr{Q}^{-1/2}\tilde{h}\|_0^2 < \infty \right\}.$$

This Hilbert space is the one that will now be the focus of our attention due to the next result.

Theorem 10.6.2 *$\mathbb{H}(\mathscr{Q})$ is congruent to $\mathbb{L}^2(\mathrm{Z}_0)$ in (10.36) under the mapping $\Psi(\tilde{h}) = \mathrm{Z}_0(\mathscr{Q}^{-1}\tilde{h})$ for $\tilde{h} \in \mathbb{H}(\mathscr{Q})$.*

Proof: Let $h, h' \in \mathbb{G}_0$ with $\tilde{h} = \mathscr{Q}h, \tilde{h}' = \mathscr{Q}h' \in \mathbb{H}(\mathscr{Q})$. Then,

$$\begin{aligned} \mathrm{Cov}\left(\mathrm{Z}_0(h), \mathrm{Z}_0(h')\right) &= \mathrm{Cov}\left(\mathrm{Z}_0(\mathscr{Q}^{-1}\tilde{h}), \mathrm{Z}_0(\mathscr{Q}^{-1}\tilde{h}')\right) \\ &= \langle h, \mathscr{Q}h' \rangle_0 \\ &= \langle \mathscr{Q}^{-1/2}\tilde{h}, \mathscr{Q}^{-1/2}\tilde{h}' \rangle_0 \\ &= \langle \tilde{h}, \tilde{h}' \rangle_{\mathbb{H}(\mathscr{Q})}. \end{aligned}$$

□

With Theorem 10.6.2 in hand, we can give our new formulation of cca. Specially, we seek elements $f_i \in \mathbb{G}_i$ of unit norm that maximize $|\mathrm{Cov}(\mathrm{Z}_1(f_1), \mathrm{Z}_2(f_2))|$. However,

$$\begin{aligned} \mathrm{Cov}(\mathrm{Z}_1(f_1), \mathrm{Z}_2(f_2)) &= \mathrm{Cov}\left(\mathrm{Z}_0\left((f_1, 0)\right), \mathrm{Z}_0\left((0, f_2)\right)\right) \\ &= \left\langle \mathscr{Q}\begin{bmatrix} f_1 \\ 0 \end{bmatrix}, \mathscr{Q}\begin{bmatrix} 0 \\ f_2 \end{bmatrix} \right\rangle_{\mathbb{H}(\mathscr{Q})}, \end{aligned}$$

which leads to the conclusion that it is equivalent to find $f_i \in \mathbb{G}_i$ to maximize the right-hand side of this last expression.

The analysis from this point is driven by the results in Section 3.7. For that purpose, we decompose $\mathbb{H}(\mathscr{Q})$ into a sum of the closed subspaces $\mathbb{M}_1$ and $\mathbb{M}_2$ with

$$\mathbb{M}_1 = \left\{ \tilde{h} \in \mathbb{H}(\mathscr{Q}) : \tilde{h} = \mathscr{Q}\begin{bmatrix} f_1 \\ 0 \end{bmatrix} = (f_1, \mathscr{C}_{21}f_1), f_1 \in \mathbb{G}_1 \right\},$$

$$\mathbb{M}_2 = \left\{ \tilde{h} \in \mathbb{H}(\mathscr{Q}) : \tilde{h} = \mathscr{Q}\begin{bmatrix} 0 \\ f_2 \end{bmatrix} = (\mathscr{C}_{12}f_2, f_2), f_2 \in \mathbb{G}_2 \right\}.$$

Regarding $\mathbb{M}_1$ and $\mathbb{M}_2$, we have the following result.

Theorem 10.6.3 $\mathbb{H}(\mathcal{Q}) = \mathbb{M}_1 + \mathbb{M}_2$ *with "+" indicating an algebraic direct sum.*

Proof: Clearly, any element of $\mathbb{G}_0$ can be written as the sum of elements in $\mathbb{M}_1$ and $\mathbb{M}_2$. We therefore need only show that $\mathbb{M}_1 \cap \mathbb{M}_2 = \{0\}$. To the contrary, suppose that there exist $f_i \in \mathbb{G}_i, i = 1, 2$, such that $(f_1, \mathscr{C}_{21} f_1) = (\mathscr{C}_{12} f_2, f_2)$. Then,

$$\mathrm{Var}(Z_1(f_1)) = \langle f_1, f_1 \rangle_1 = \langle f_1, \mathscr{C}_{12} f_2 \rangle_1$$

and

$$\mathrm{Var}(Z_2(f_2)) = \langle f_2, f_2 \rangle_2 = \langle f_2, \mathscr{C}_{21} f_1 \rangle_2 = \langle C_{12} f_2, f_1 \rangle_1.$$

However, these relations have the consequence that $|\mathrm{Corr}(Z_1(f_1), Z_2(f_2))| = 1$, which contradicts Assumption 1. □

To relate Theorem 10.6.3 to the Section 3.7 development let $\mathbb{N}_1 = \mathbb{M}_1$ and $\mathbb{N}_2 = \mathbb{M}_2 \cap \mathbb{M}_1^{\perp} = \mathbb{M}_1^{\perp}$. Then, for $\tilde{h}_1 = \mathcal{Q} \begin{bmatrix} f_1 \\ 0 \end{bmatrix} \in \mathbb{M}_1$ and $\tilde{h}_2 = \mathcal{Q} \begin{bmatrix} 0 \\ f_2 \end{bmatrix} \in \mathbb{M}_2$, the first canonical correlation satisfies

$$\begin{aligned} \rho_1^2 &= \sup_{\substack{\tilde{h}_1 \in \mathbb{M}_1, \tilde{h}_2 \in \mathbb{M}_2 \\ \|\tilde{h}_i\|_{\mathbb{H}(\mathcal{Q})} = 1, i=1,2}} \langle \tilde{h}_1, \tilde{h}_2 \rangle^2_{\mathbb{H}(\mathcal{Q})} \\ &= \sup_{\substack{\tilde{h}_1 \in \mathbb{N}_1, v \in \mathbb{N}_2 \\ \|\tilde{h}_1\|_{\mathbb{H}(\mathcal{Q})} = 1, \|v + \mathscr{T} v\|_{\mathbb{H}(\mathcal{Q})} = 1}} \langle \tilde{h}_1, \mathscr{T} v \rangle^2_{\mathbb{H}(\mathcal{Q})} \\ &\le \sup_{\substack{v \in \mathbb{N}_2 \\ \|v + \mathscr{T} v\|_{\mathbb{H}(\mathcal{Q})} = 1}} \|\mathscr{T} v\|^2_{\mathbb{H}(\mathcal{Q})} \end{aligned}$$

for $\mathscr{T} = \mathscr{P}_{\mathbb{N}_1|\mathbb{M}_2} (\mathscr{P}_{\mathbb{N}_2|\mathbb{M}_2})^{-1}$. Taking $\tilde{h}_1 = \mathscr{T} v / \|\mathscr{T} v\|_{\mathbb{H}(\mathcal{Q})}$, we see that the bound is attainable and holds with equality. Thus, we have shown that ρ_1 is obtained by maximizing $\|\mathscr{T} v\|_{\mathbb{H}(\mathcal{Q})}$ over $v \in \mathbb{N}_2$ subject to $\|\mathscr{T} v + v\|_{\mathbb{H}(\mathcal{Q})} = 1$. However,

$$\begin{aligned} \|\mathscr{T} v + v\|^2_{\mathbb{H}(\mathcal{Q})} &= \|v\|^2_{\mathbb{H}(\mathcal{Q})} + 2\langle \mathscr{T} v, v \rangle_{\mathbb{H}(\mathcal{Q})} + \|\mathscr{T} v\|^2_{\mathbb{H}(\mathcal{Q})} \\ &= \langle v, (I + \mathscr{T}^* \mathscr{T}) v \rangle_{\mathbb{H}(\mathcal{Q})} \end{aligned}$$

because $v \in \mathbb{N}_2$ is orthogonal to $\mathscr{T} v \in \mathbb{N}_1$. Thus, we maximize $\|\mathscr{T} v\|_{\mathbb{H}(\mathcal{Q})}$ subject to $\langle v, (I + \mathscr{T}^* \mathscr{T}) v \rangle_{\mathbb{H}(\mathcal{Q})} = 1$.

The operator $I + \mathscr{T}^*\mathscr{T}$ is self-adjoint, positive, Invertible, and has a self-adjoint square-root $(I + \mathscr{T}^*\mathscr{T})^{1/2}$. We can therefore work with $v' = (I + \mathscr{T}^*\mathscr{T})^{1/2}v$ and maximize

$$\|\mathscr{T}v\|_{\mathbb{H}(\mathcal{Q})} = \|\mathscr{T}(I + \mathscr{T}^*\mathscr{T})^{-1/2}v'\|_{\mathbb{H}(\mathcal{Q})}$$

subject to $v' \in \mathbb{N}_2$ and $\|v'\|^2_{\mathbb{H}(\mathcal{Q})} = 1$. If we now assume that $\mathscr{T}^*\mathscr{T}$ is compact, the maximizer is the eigenvector that corresponds to the largest eigenvalue of $(I + \mathscr{T}^*\mathscr{T})^{-1/2}\mathscr{T}^*\mathscr{T}(I + \mathscr{T}^*\mathscr{T})^{-1/2}$. Some algebra reveals that this eigenvalue problem is equivalent to finding a vector $v \in \mathbb{N}_2$ with $\|v\|^2_{\mathbb{H}(\mathcal{Q})} = 1$ such that

$$\mathscr{T}^*\mathscr{T}v = \alpha^2 v \tag{10.40}$$

in which case the corresponding canonical correlation is $\rho = \alpha/\sqrt{1+\alpha^2}$.

Let $v \in \mathbb{N}_2$ be any vector that satisfies (10.40). Its $\mathbb{M}_1$ component is $\mathscr{T}v$ and its $\mathbb{M}_2$ component is $\mathscr{T}v + v$. These correspond to the canonical variables $\Psi(\mathscr{T}v/\alpha)$ and $\Psi\left((v + \mathscr{T}v)/\sqrt{1+\alpha^2}\right)$ of the Z_1 and Z_2 spaces, respectively.

We now turn to the task of characterizing $\mathscr{T}^*\mathscr{T}$. To do so, we first need to appreciate the form of the elements of $\mathbb{N}_2 = \mathbb{M}_1^{\perp}$. In this regard, we have the following useful result.

Lemma 10.6.4 *Every element $v \in \mathbb{N}_2$ can be expressed as $v = (0, \tilde{f}_2)$ with $\tilde{f}_2 = \mathscr{C}_{22.1}f_2$ for some $f_2 \in \mathbb{G}_2$.*

Proof: Let

$$v = \mathcal{Q}\begin{bmatrix} f_1 \\ f_2 \end{bmatrix}$$

for $f_i \in \mathbb{G}_i, i = 1, 2$, be any element of $\mathbb{M}_1^{\perp}$. Then, for every $\tilde{h} = (f_1', \mathscr{C}_{21}f_1') \in \mathbb{M}_1$

$$\begin{aligned}\langle \tilde{h}, v\rangle_{\mathbb{H}(\mathcal{Q})} &= \langle \mathcal{Q}^{-1/2}\tilde{h}, \mathcal{Q}^{-1/2}v\rangle_0 \\ &= \langle (f_1', 0), v\rangle_0 \\ &= \langle f_1', f_1\rangle_1 + \langle f_1', \mathscr{C}_{12}f_2\rangle_1 \\ &= 0.\end{aligned}$$

As this is true for all $f_1' \in \mathbb{G}_1, f_1 = -\mathscr{C}_{12}f_2$. □

As a result of the previous lemma, the following theorem determines the form of $\mathscr{T}^*\mathscr{T}$ in terms of the fundamental operators $\mathscr{C}_{12}$ and $\mathscr{C}_{21}$.

Theorem 10.6.5 *For any* $v = (0, \tilde{f}_2) \in \mathbb{N}_2, \mathscr{T}^*\mathscr{T}v = (0, \mathscr{C}_{21}\mathscr{C}_{12}\mathscr{C}_{22.1}^{-1}\tilde{f}_2)$.

Proof: We will establish two results that imply the theorem statement: namely, that for $v = (0, \tilde{f}_2) \in \mathbb{N}_2$,

$$\mathscr{T}v = (\mathscr{C}_{12}\mathscr{C}_{22.1}^{-1}\tilde{f}_2, \mathscr{C}_{21}\mathscr{C}_{12}\mathscr{C}_{22.1}^{-1}\tilde{f}_2) \tag{10.41}$$

and for $\tilde{h} = (f_1, \mathscr{C}_{21}f_1) \in \mathbb{M}_1$

$$\mathscr{T}^*\tilde{h} = (0, \mathscr{C}_{21}f_1). \tag{10.42}$$

As, $\mathscr{T} = \mathscr{P}_{\mathbb{N}_1|\mathbb{M}_2}(\mathscr{P}_{\mathbb{N}_2|\mathbb{M}_2})^{-1}$, we first need to obtain explicit expressions for $\mathscr{P}_{\mathbb{N}_1|\mathbb{M}_2}$ and $(\mathscr{P}_{\mathbb{N}_2|\mathbb{M}_2})^{-1}$.

Let $\tilde{h}_1 = (f_1, \mathscr{C}_{21}f_1) \in \mathbb{M}_1 = \mathbb{N}_1$ with $\tilde{h}_2 = (\mathscr{C}_{12}f_2, f_2) \in \mathbb{M}_2$. Then,

$$\langle \mathscr{P}_{\mathbb{N}_1|\mathbb{M}_2}\tilde{h}_2, \tilde{h}_1\rangle_{\mathbb{H}(\mathscr{Q})} = \langle \tilde{h}_2, \tilde{h}_1\rangle_{\mathbb{H}(\mathscr{Q})}$$

for every $\tilde{h}_1 \in \mathbb{M}_1$. Writing $\mathscr{P}_{\mathbb{N}_1|\mathbb{M}_2}\tilde{h}_2 = (f_1', \mathscr{C}_{21}f_1')$ for some $f_1' \in \mathbb{G}_1$ leads to

$$\begin{aligned}
\langle \mathscr{P}_{\mathbb{N}_1|\mathbb{M}_2}\tilde{h}_2, \tilde{h}_1\rangle_{\mathbb{H}(\mathscr{Q})} &= \langle (f_1', \mathscr{C}_{21}f_1'), (f_1, 0)\rangle_0 = \langle f_1', f_1\rangle_1 \\
&= \langle (\mathscr{C}_{12}f_2, f_2), \tilde{h}_1\rangle_{\mathbb{H}(\mathscr{Q})} = \langle (\mathscr{C}_{12}f_2, f_2), (f_1, 0)\rangle_0 \\
&= \langle \mathscr{C}_{12}f_2, f_1\rangle_1
\end{aligned}$$

for every $f_1 \in \mathbb{G}_1$. So, $f_1' = \mathscr{C}_{12}f_2$ and

$$\mathscr{P}_{\mathbb{N}_1|\mathbb{M}_2}\tilde{h}_2 = (\mathscr{C}_{12}f_2, \mathscr{C}_{21}\mathscr{C}_{12}f_2). \tag{10.43}$$

Consequently,

$$\mathscr{P}_{\mathbb{N}_2|\mathbb{M}_2}\tilde{h}_2 = (I - \mathscr{P}_{\mathbb{N}_1|\mathbb{M}_2})\tilde{h}_2 = (0, \mathscr{C}_{22.1}f_2)$$

from which we see that

$$(\mathscr{P}_{\mathbb{N}_2|\mathbb{M}_2})^{-1}v = (\mathscr{C}_{12}\mathscr{C}_{22.1}^{-1}\tilde{f}_2, \mathscr{C}_{22.1}^{-1}\tilde{f}_2) \tag{10.44}$$

for $v = (0, \tilde{f}_2) \in \mathbb{N}_2$. In combination, (10.43) and (10.44) give us (10.41).

To verify (10.42), let $\tilde{h} = (f_1, \mathscr{C}_{21}f_1) \in \mathbb{M}_1 = \mathbb{N}_1$ and $v = (0, \tilde{f}_2) \in \mathbb{N}_2$. Then,

$$\begin{aligned}
\langle \tilde{h}, \mathscr{T}v\rangle_{\mathbb{H}(\mathscr{Q})} &= \langle \mathscr{Q}^{-1}\tilde{h}, \mathscr{T}v\rangle_0 \\
&= \langle (f_1, 0), (\mathscr{C}_{12}\mathscr{C}_{22.1}^{-1}\tilde{f}_2, \mathscr{C}_{21}\mathscr{C}_{12}\mathscr{C}_{22.1}^{-1}\tilde{f}_2)\rangle_0 \\
&= \langle f_1, \mathscr{C}_{12}\mathscr{C}_{22.1}^{-1}\tilde{f}_2\rangle_1 = \langle \mathscr{C}_{22.1}^{-1}\mathscr{C}_{21}f_1, \tilde{f}_2\rangle_2 \\
&= \left\langle \mathscr{Q}^{-1}\begin{bmatrix} 0 \\ \mathscr{C}_{21}f_1 \end{bmatrix}, v\right\rangle_0 = \left\langle \begin{bmatrix} 0 \\ \mathscr{C}_{21}f_1 \end{bmatrix}, v\right\rangle_{\mathbb{H}(\mathscr{Q})} \\
&= \langle \mathscr{T}^*\tilde{h}, v\rangle_{\mathbb{H}(\mathscr{Q})}.
\end{aligned}$$

□

Theorem 10.6.5 in conjunction with Lemma 10.6.4 entails that $\mathscr{T}^*\mathscr{T}(0,\tilde{f}_2) = (0, \mathscr{C}_{21}\mathscr{C}_{12}f_2)$ for some $f_2 \in \mathbb{G}_2$. The eigenvalue problem (10.40) is therefore equivalent to $\mathscr{C}_{21}\mathscr{C}_{12}f_2 = \alpha^2\mathscr{C}_{22.1}f_2$ or

$$\mathscr{C}_{21}\mathscr{C}_{12}f_2 = \rho^2 f_2. \tag{10.45}$$

By interchanging the roles of $\mathbb{M}_1$ and $\mathbb{M}_2$, it follows that the optimal choice for f_1 is the eigenvector corresponding to the same eigenvalue ρ^2 in (10.45) except for the operator $\mathscr{C}_{12}\mathscr{C}_{21}$. Thus, Theorem 4.3.1 now leads us to the same conclusion as Theorem 10.1.2.

So far, we have succeeded only in reproducing our previous cca result. The benefits of the orthogonal subspace approach become evident only when we are interested in more than two random elements. We will direct our attention to the simplest case of three random elements here although it should be clear that the same development can be used in a substantially more general context.

The specific, three-variable, extension of cca we will consider is the one suggested by Roy (1958) that has been termed *partial canonical correlation analysis* or pcca subsequently. The idea is that we have random vectors X_1, X_2, and X_3 and wish to study the relationship between X_2 and X_3 after removing the influence of X_1. One way to accomplish this is by examination of *partial canonical correlations* that are the ordinary canonical correlations between $X_2 - \mathscr{P}_{X_1}X_2$ and $X_3 - \mathscr{P}_{X_1}X_3$, where $\mathscr{P}_{X_1}$ denotes projection onto the linear space spanned by X_1.

The goal is to now formulate Roy's idea for three Hilbert space valued random elements. The arguments that are required to accomplish this are rather immediate extensions of those for the two random element case. Thus, we will merely highlight the important differences.

We now have three random elements $\chi_i, i = 1, 2, 3$, with covariance operators $\mathscr{K}_i$, $i = 1, 2, 3$ and cross-covariance operators $\mathscr{K}_{12}$, $\mathscr{K}_{13}$, $\mathscr{K}_{23}$. The Hilbert spaces $\mathbb{L}^2(Z_i)$ that are spanned by the $Z_i(\cdot)$ processes that are indexed by their congruent Hilbert spaces $\mathbb{G}_i = \mathbb{H}(\mathscr{K}_i), i = 1, 2, 3$, are defined as in Section 10.1. One finds, as before, that there are bounded operators $\mathscr{C}_{ij} : \mathbb{G}_j \to \mathbb{G}_i$ satisfying

$$\mathrm{Cov}(Z_i(f_i), Z_j(f_j)) = \langle f_i, \mathscr{C}_{ij}f_j\rangle_i$$

for $i, j = 1, 2, 3$ and $i \neq j$. Our previous definition of $\mathbb{G}_0$ in (10.34) now takes the form

$$\mathbb{G}_0 = \left\{ h = (f_1, f_2, f_3) : f_i \in \mathbb{G}_i, i = 1, 2, 3, \|h\|_0^2 = \sum_{i=1}^{3} \|f_i\|_i^2 < \infty \right\}$$

for which a corresponding $\mathbb{G}_0$ indexed process is

$$Z_0(h) = \sum_{i=1}^{3} Z_i(f_i)$$

that has

$$\mathrm{Cov}\left(Z_0(h), Z_0(h')\right) = \langle h, \mathscr{Q}h' \rangle_0$$

with $\mathscr{Q}$ defined by

$$\mathscr{Q}h = \begin{bmatrix} I & \mathscr{C}_{12} & \mathscr{C}_{13} \\ \mathscr{C}_{21} & I & \mathscr{C}_{23} \\ \mathscr{C}_{31} & \mathscr{C}_{32} & I \end{bmatrix} \begin{bmatrix} f_1 \\ f_2 \\ f_3 \end{bmatrix}.$$

As in Section 10.1, we need to rule out the case where certain types of exact prediction are possible. For this purpose, we require that Assumption 1 holds for both of the process pairs Z_1, Z_2 and Z_1, Z_3 as well as

Assumption 2 *There exist no* $f_2 \in \mathbb{G}_2$ or $f_3 \in \mathbb{G}_3$ *such that*

$$|\mathrm{Corr}(Z_2(f_2) - \mathscr{P}_{Z_1} Z_2(f_2), Z_3(f_3) - \mathscr{P}_{Z_1} Z_3(f_3))| = 1. \qquad \square$$

Under this condition, $\mathscr{Q}$ is invertible. Its inverse, in matrix form, is given by

$$\mathscr{Q}^{-1} = \begin{bmatrix} I + \mathscr{E}\mathscr{G}^{-1}\mathscr{F} & -\mathscr{E}\mathscr{G}^{-1} \\ -\mathscr{G}^{-1}\mathscr{F} & \mathscr{G}^{-1} \end{bmatrix} \tag{10.46}$$

for $\mathscr{E} = \begin{bmatrix} \mathscr{C}_{12} & \mathscr{C}_{13} \end{bmatrix}$, $\mathscr{F} = \begin{bmatrix} \mathscr{C}_{21} \\ \mathscr{C}_{31} \end{bmatrix}$, $\mathscr{D} = \begin{bmatrix} I & \mathscr{C}_{23} \\ \mathscr{C}_{32} & I \end{bmatrix}$ and

$$\mathscr{G} = \mathscr{D}^{1/2}(I - \mathscr{V})\mathscr{D}^{1/2}$$

with

$$\mathscr{V} = \begin{bmatrix} 0 & -\mathscr{C}_{22.1}^{-1/2}(\mathscr{C}_{23} - \mathscr{C}_{21}\mathscr{C}_{13})\mathscr{C}_{33.1}^{-1/2} \\ -\mathscr{C}_{33.1}^{-1/2}(\mathscr{C}_{32} - \mathscr{C}_{31}\mathscr{C}_{12})\mathscr{C}_{22.1}^{-1/2} & 0 \end{bmatrix}.$$

As in Theorem 10.6.2,

$$\mathbb{L}^2(Z_0) = \{Z_0(h) : h \in \mathbb{G}_0, \|Z_0(h)\|^2_{\mathbb{L}^2(Z_0)} = \mathrm{Var}(Z_0(h)) < \infty\}$$

is congruent to

$$\mathbb{H}(\mathscr{Q}) = \left\{ \tilde{h} = \mathscr{Q} \begin{bmatrix} f_1 \\ f_2 \\ f_3 \end{bmatrix} : f_i \in \mathbb{G}_i, i = 1, 2, 3, \|\tilde{h}\|^2_{\mathbb{H}(\mathscr{Q})} = \|\mathscr{Q}^{-1/2}\tilde{h}\|_0^2 < \infty \right\}$$

under the mapping $\Psi(\tilde{h}) = Z_0(\mathscr{Q}^{-1}\tilde{h})$.

One finds that the projection of $Z_2(f_2)$ onto $\mathbb{L}^2(Z_1)$ is $Z_1(\mathscr{C}_{12}f_2)$ and the projection of $Z_3(f_3)$ onto $\mathbb{L}^2(Z_1)$ is $Z_1(\mathscr{C}_{13}f_3)$. Thus, for the pcca formulation, we wish to find $f_2 \in \mathbb{G}_2$ and $f_3 \in \mathbb{G}_3$ to maximize the correlation between

$$Z_2(f_2) - Z_1(\mathscr{C}_{12}f_2) = Z_0(-\mathscr{C}_{12}f_2, f_2, 0)$$

and

$$Z_3(f_3) - Z_1(\mathscr{C}_{13}f_3) = Z_0(-\mathscr{C}_{13}f_3, 0, f_3).$$

However,

$$\begin{aligned}&\mathrm{Cov}(Z_0(-\mathscr{C}_{13}f_3, 0, f_3)), Z_0(-\mathscr{C}_{13}f_3, 0, f_3))\\ &= \left\langle \mathscr{Q}\begin{bmatrix} -\mathscr{C}_{12}f_2 \\ f_2 \\ 0 \end{bmatrix}, \mathscr{Q}\begin{bmatrix} -\mathscr{C}_{13}f_3 \\ 0 \\ f_3 \end{bmatrix} \right\rangle_{\mathbb{H}(\mathscr{Q})}.\end{aligned}$$

Thus, just as was the case for ordinary cca, we can formulate the problem of finding partial canonical correlations as one of restricted optimization in $\mathbb{H}(\mathscr{Q})$.

To proceed, we again draw on the results in Section 3.7 and write $\mathbb{H}(\mathscr{Q})$ as the direct sum of three subspaces defined by

$$\mathbb{M}_1 = \left\{ \tilde{h} \in \mathbb{H}(\mathscr{Q}) : \tilde{h} = \mathscr{Q}\begin{bmatrix} f_1 \\ 0 \\ 0 \end{bmatrix} := (f_1, \mathscr{C}_{21}f_1, \mathscr{C}_{31}f_1) \right\}$$

$$\mathbb{M}_2 = \left\{ \tilde{h} \in \mathbb{H}(\mathscr{Q}) : \tilde{h} = \mathscr{Q}\begin{bmatrix} 0 \\ f_2 \\ 0 \end{bmatrix} := (\mathscr{C}_{12}f_2, f_2, \mathscr{C}_{32}f_2) \right\},$$

$$\mathbb{M}_3 = \left\{ \tilde{h} \in \mathbb{H}(\mathscr{Q}) : \tilde{h} = \mathscr{Q}\begin{bmatrix} 0 \\ 0 \\ f_3 \end{bmatrix} := (\mathscr{C}_{13}f_3, \mathscr{C}_{23}f_3, f_3) \right\}$$

with $\mathbb{N}_1 = \mathbb{M}_1, \mathbb{N}_2 = \mathbb{M}_2 \cap \mathbb{M}_1^{\perp}$, and $\mathbb{N}_3 = \mathbb{M}_3 \cap (\mathbb{M}_1 + \mathbb{M}_2)^{\perp}$. Then, one can check that the elements of the subspaces $\tilde{\mathbb{M}}_2 = \mathscr{P}_{\mathbb{M}_1^{\perp}}\mathbb{M}_2$ and $\tilde{\mathbb{M}}_3 = \mathscr{P}_{\mathbb{M}_1^{\perp}}\mathbb{M}_3$ have the form

$$\tilde{h}_2 = \mathscr{Q}\begin{bmatrix} -\mathscr{C}_{12}f_2 \\ f_2 \\ 0 \end{bmatrix}$$

and

$$\tilde{h}_3 = \mathscr{Q}\begin{bmatrix} -\mathscr{C}_{13}f_2 \\ 0 \\ f_3 \end{bmatrix},$$

respectively, for $f_i \in \mathbb{G}_i, i = 1, 2$. Thus, the first squared partial canonical correlation is

$$\rho^2 = \sup_{\substack{\tilde{h}_i \in \tilde{\mathbb{M}}_i \\ \|\tilde{h}_i\|_{\mathbb{H}(\mathscr{Q})}=1, i=2,3}} \langle \tilde{h}_2, \tilde{h}_3 \rangle^2_{\mathbb{H}(\mathscr{Q})}.$$

However, $\tilde{\mathbb{M}}_2 = \mathbb{N}_2$ and the elements of $\tilde{\mathbb{M}}_3$ can all be expressed as

$$\tilde{h}_3 = v + \mathscr{T}v$$

for $\mathscr{T} = \mathscr{P}_{\mathbb{N}_2|\mathbb{M}_3}\left(\mathscr{P}_{\mathbb{N}_3|\mathbb{M}_3}\right)^{-1}$. Consequently,

$$\begin{aligned} \rho^2 &= \sup_{\substack{\tilde{h}_2 \in \mathbb{N}_2, v \in \mathbb{N}_3 \\ \|\tilde{h}_2\|_{\mathbb{H}(\mathscr{Q})}=1, \|v+\mathscr{T}v\|_{\mathbb{H}(\mathscr{Q})}=1}} \langle \tilde{h}_2, \mathscr{T}v \rangle^2_{\mathbb{H}(\mathscr{Q})} \\ &\le \sup_{\substack{v \in \mathbb{N}_3 \\ \|v+\mathscr{T}v\|_{\mathbb{H}(\mathscr{Q})}=1}} \|\mathscr{T}v\|^2_{\mathbb{H}(\mathscr{Q})}. \end{aligned}$$

This bound holds with equality when $\tilde{h}_2 = \mathscr{T}\tilde{h}_3/\|\mathscr{T}\tilde{h}_3\|_{\mathbb{H}(\mathscr{Q})}$, which leads to the conclusion that when $\mathscr{T}$ is compact $\rho = \alpha/\sqrt{1+\alpha^2}$ with α^2 the largest eigenvalue of $\mathscr{T}^*\mathscr{T}$. If v is an eigenvector corresponding to α^2, the partial canonical variable for the Z_2 and Z_3 spaces is $\Psi(\mathscr{T}v/\alpha)$ and $\Psi\left((v + \mathscr{T}v)/\sqrt{1+\alpha^2}\right)$.

To obtain a representation for $\mathscr{T}$ and $\mathscr{T}^*\mathscr{T}$ in terms of the $\mathscr{C}_{ij}$, first define

$$\mathscr{C}_0 = \mathscr{C}_{33.1} - (\mathscr{C}_{32} - \mathscr{C}_{31}\mathscr{C}_{12})\mathscr{C}_{22.1}^{-1}(\mathscr{C}_{23} - \mathscr{C}_{21}\mathscr{C}_{13}).$$

Then, for $v = (0, 0, \tilde{f}_3) \in \mathbb{N}_3$, we find that

$$\mathscr{T}v = (0, (\mathscr{C}_{23} - \mathscr{C}_{21}\mathscr{C}_{13})\mathscr{C}_0^{-1}\tilde{f}_3, (\mathscr{C}_{32} - \mathscr{C}_{31}\mathscr{C}_{12})\mathscr{C}_{22.1}^{-1}(\mathscr{C}_{23} - \mathscr{C}_{21}\mathscr{C}_{13})\mathscr{C}_0^{-1}\tilde{f}_3),$$

which characterizes the $\mathscr{T}$ operator. Similarly,

$$\mathscr{T}^*\mathscr{T}v = (0, 0, (\mathscr{C}_{32} - \mathscr{C}_{31}\mathscr{C}_{12})\mathscr{C}_{22.1}^{-1}(\mathscr{C}_{23} - \mathscr{C}_{21}\mathscr{C}_{13})\mathscr{C}_0^{-1}\tilde{f}_3)).$$

11

Regression

In this final chapter, we focus on the fda-specific problem of functional linear regression. There are various ways to formulate this type of regression idea and we have chosen to study one of the more common situations that has appeared in the fda literature: namely, the case of a scalar-dependent variable and functional independent variable. This leads to a special case of the functional linear model introduced in Section 6.1 which makes it natural to investigate the performance of method of regularization estimation techniques in this setting. In Section 11.1, we describe the basic regression model that we pose for study and derive a penalized least-squares estimator for the corresponding coefficient function that was originally proposed in Crambes, Kneip, and Sarda (2009). Subsequent sections examine the large sample and optimality properties of this estimator.

11.1 A functional regression model

The basic premise is that we have a probability space $(\Omega, \mathscr{F}, \mathbb{P})$ and an associated second-order stochastic process $\{X(t, \omega) : t \in [0, 1], \omega \in \Omega\}$ that is jointly measurable in t and ω with a square-integrable sample paths that can be viewed as a random element of $\mathbb{L}^2 := \mathbb{L}^2[0, 1]$. The $X(t)$ all have zero mean and the covariance kernel

$$K(s, t) = \mathbb{E}[X(s)X(t)]$$

is presumed to be continuous for all $s, t \in [0, 1]$. Then, the regression model we intend to study concerns the pairs $(X_1, Y_1), \dots, (X_n, Y_n)$ with $X_1, \dots, X_n$ independent copies of X and

$$Y_i = \langle \beta, X_i \rangle + \varepsilon_i, i = 1, \dots, n, \tag{11.1}$$

Theoretical Foundations of Functional Data Analysis, with an Introduction to Linear Operators, First Edition. Tailen Hsing and Randall Eubank.

where $\langle \cdot, \cdot \rangle$ is the $\mathbb{L}^2$ inner product, the ε_j are independent, zero mean random errors with common variance σ^2 that are independent of the X_i, and β is an unknown element of $\mathbb{W}_q := \mathbb{W}_q[0,1]$. Model (11.1) provides one possible generalization of multiple linear regression to the functional domain with β playing the role of the vector of regression coefficients. Accordingly, we will refer to it as the *coefficient function* throughout the chapter.

To connect (11.1) with the developments in Chapter 6, take $\mathbb{Y} = \mathbb{R}^n$ with

$$\|y\|_{\mathbb{Y}}^2 = \frac{1}{n} \sum_{i=1}^{n} y_i^2$$

for $y = (y_1, \ldots, y_n)^T \in \mathbb{Y}$. Then, take $\mathbb{H} = \mathbb{W}_q$ and define $\mathscr{T} \in \mathfrak{B}(\mathbb{H}, \mathbb{Y})$ by $\mathscr{T}g = (\mathscr{T}_1 g, \ldots, \mathscr{T}_n g)^T$ for

$$\mathscr{T}_i g = \langle X_i, g \rangle, \quad 1 \le i \le n, \tag{11.2}$$

when $g \in \mathbb{W}_q$. From the Cauchy–Schwarz inequality,

$$|\mathscr{T}_i g| \le \|X_i\| \|g\| \le \|X_i\| \|g\|_{\mathbb{W}_q},$$

where

$$\|g\|_{\mathbb{W}_q}^2 = \|g\|^2 + \|g^{(q)}\|^2$$

for $\|\cdot\|$ the $\mathbb{L}^2$ norm. Thus, $\mathscr{T}$ is, in fact, a bounded linear mapping from $\mathbb{H}$ to $\mathbb{Y}$. Note that our definition of $\mathscr{T}$ tacitly assumes that the X_i are fully observed. Section 11.4 relaxes that condition and allows the sample paths to only be sampled at discrete time points.

If we define

$$Y = (Y_1, \ldots, Y_n)^T$$

and

$$\varepsilon = (\varepsilon_1, \ldots, \varepsilon_n)^T,$$

(11.1) becomes

$$Y = \mathscr{T}\beta + \varepsilon$$

which now fits into the framework of the functional linear model (6.2). This suggests using a penalized least-squares estimator for β obtained by minimization of criterion (6.7). For our specific case, this entails minimization with respect to $g \in \mathbb{W}_q$ of

$$\|Y - \mathscr{T}g\|_{\mathbb{Y}}^2 + \eta \langle g, g \rangle_{\mathbb{H}} = n^{-1} \sum_{i=1}^{n} (Y_i - \mathscr{T}_i g)^2 + \eta \|g\|_{\mathbb{W}_q}^2.$$

Upon making the identification $\mathscr{W} = I$ in (6.7), an application of Theorem 6.2.1 produces

$$\beta_\eta = \mathscr{G}(\eta)^{-1}\mathscr{T}^* Y, \tag{11.3}$$

with

$$\mathscr{G}(\eta) = \mathscr{T}^*\mathscr{T} + \eta I \tag{11.4}$$

as our estimator of the coefficient function β.

The following theorem provides a first step toward understanding the squared error properties of β_η.

Theorem 11.1.1 *Let $\mathbb{E}_\varepsilon$ represent expectation with respect to ε. Then,*

$$\|\mathbb{E}_\varepsilon \mathscr{T}(\beta_\eta - \beta)\|_{\mathbb{Y}}^2 \leq \eta \|\beta\|_{\mathbb{W}_q}^2,$$

$$\mathbb{E}_\varepsilon \|\mathscr{T}(\beta_\eta - \mathbb{E}_\varepsilon \beta_\eta)\|_{\mathbb{Y}}^2 = \frac{\sigma^2}{n}\text{trace } \mathscr{T}\mathscr{G}(\eta)^{-1}\mathscr{T}^*\mathscr{T}\mathscr{G}(\eta)^{-1}\mathscr{T}^*$$

and

$$\mathbb{E}_\varepsilon \|\beta_\eta\|_{\mathbb{W}_q}^2 \leq 2\|\beta\|_{\mathbb{W}_q}^2 + \frac{2\sigma^2}{n}\text{trace } \mathscr{T}\mathscr{G}(\eta)^{-2}\mathscr{T}^*. \tag{11.5}$$

Proof: The first two bounds follow from Theorem 6.3.1. For the third note that under our model

$$\beta_\eta = \mathscr{G}(\eta)^{-1}\mathscr{T}^*\mathscr{T}\beta + \mathscr{G}(\eta)^{-1}\mathscr{T}^*\varepsilon.$$

Hence,

$$\|\beta_\eta\|_{\mathbb{W}_q}^2 \leq 2\|\mathscr{G}(\eta)^{-1}\mathscr{T}^*\mathscr{T}\beta\|_{\mathbb{W}_q}^2 + 2\|\mathscr{G}(\eta)^{-1}\mathscr{T}^*\varepsilon\|_{\mathbb{W}_q}^2$$

and we can analyze the two terms in this last expression separately.

First

$$\begin{aligned}\|\mathscr{G}(\eta)^{-1}\mathscr{T}^*\mathscr{T}\beta\|_{\mathbb{W}_q}^2 &\leq \|\mathscr{G}(\eta)^{-1}\mathscr{T}^*\mathscr{T}\|^2\|\beta\|_{\mathbb{W}_q}^2 \\ &= \|\mathscr{G}(\eta)^{-1/2}\mathscr{T}^*\mathscr{T}\mathscr{G}(\eta)^{-1/2}\|^2\|\beta\|_{\mathbb{W}_q}^2.\end{aligned}$$

Now, $\mathscr{T}^*\mathscr{T}$ is compact which means that Theorem 4.8.1 can be applied to conclude that the eigenvalues of $\mathscr{G}(\eta)^{-1/2}\mathscr{T}^*\mathscr{T}\mathscr{G}(\eta)^{-1/2}$ are at most one. Next, as the ε_j are uncorrelated,

$$\mathbb{E}_\varepsilon \|\mathscr{G}(\eta)^{-1}\mathscr{T}^*\varepsilon\|_{\mathbb{W}_q}^2 = \mathbb{E}_\varepsilon \langle \varepsilon, \mathscr{T}\mathscr{G}(\eta)^{-2}\mathscr{T}^*\varepsilon\rangle_{\mathbb{Y}} = \frac{\sigma^2}{n}\text{trace } \mathscr{T}\mathscr{G}(\eta)^{-2}\mathscr{T}^*.$$

□

The theorem quantifies the effect of only a portion of the random components that are present in β_η: namely, those that arise from the additive random errors in the model. It remains to assess the contribution from sampling the X process. We address this issue from a large sample perspective in the following section.

11.2 Asymptotic theory

One way to assess the performance of β_η as an estimator of β is through the squared error (range) loss

$$L(\eta) = \|\mathscr{T}(\beta_\eta - \beta)\|_{\mathbb{Y}}^2$$

whose probabilistic behavior is influenced by that of both X and ε. However,

$$L(\eta) = O_p(\mathbb{E}_\varepsilon L(\eta)) \tag{11.6}$$

and, from Theorem 11.1.1,

$$\begin{aligned} \mathbb{E}_\varepsilon L(\eta) &= \|\mathbb{E}_\varepsilon \mathscr{T}(\beta_\eta - \beta)\|_{\mathbb{Y}}^2 + \mathbb{E}_\varepsilon \|\mathscr{T}(\beta_\eta - \mathbb{E}_\varepsilon \beta_\eta)\|_{\mathbb{Y}}^2 \\ &\leq \eta \|\beta\|_{\mathbb{W}_q}^2 + \mathbb{E}_\varepsilon \|\mathscr{T}(\beta_\eta - \mathbb{E}_\varepsilon \beta_\eta)\|_{\mathbb{Y}}^2. \end{aligned} \tag{11.7}$$

The first term in the last expression is independent of X, and, as a result, we can direct our efforts toward assessing the magnitude of the second entry.

Recall that $\mathbb{W}_q$ is a reproducing kernel Hilbert space with rk, denoted here by R, that is defined as in (2.46) and the sample covariance X kernel is given by

$$K_n(s,t) = n^{-1} \sum_{i=1}^{n} X_i(s) X_i(t).$$

Using these two functions, we can obtain a useful characterization for $\mathscr{T}^* \mathscr{T}$.

Theorem 11.2.1 *Let $\mathscr{R}$ and $\mathscr{K}_n$ be the $\mathbb{L}^2$ integral operators with kernels R and K_n, respectively. Then,*

$$(\mathscr{T}^* Y)(\cdot) = n^{-1} \sum_{i=1}^{n} Y_i \int_0^1 X_i(t) R(\cdot, t) dt \tag{11.8}$$

and, for $g \in \mathbb{W}_q$,

$$(\mathscr{T}^* \mathscr{T} g)(\cdot) = (\mathscr{R} \mathscr{K}_n g)(\cdot). \tag{11.9}$$

Proof: First, by the reproducing property,

$$\begin{aligned} (\mathscr{T}^* Y)(s) &= \langle \mathscr{T}^* Y, R(s, \cdot) \rangle_{\mathbb{W}_q} \\ &= \langle Y, \mathscr{T} R(s, \cdot) \rangle_{\mathbb{Y}}. \end{aligned}$$

To prove (11.9), one simply concludes from (11.8) that

$$(\mathcal{T}^*\mathcal{T}g)(\cdot) = n^{-1}\sum_{i=1}^{n}\int_0^1 X_i(s)g(s)ds\int_0^1 X_i(t)R(\cdot,t)dtds$$
$$= \int_0^1 R(\cdot,t)\left[\int_0^1 K_n(s,t)g(s)ds\right]dt.$$

□

It will be convenient to use expansions based on the $\mathbb{W}_q$ functions $\{e_j\}_{j=1}^{\infty}$ described in Theorem 2.8.3 that satisfy

$$\int_0^1 e_i(t)e_j(t)dt = \delta_{ij}$$

and

$$\int_0^1 e_i^{(q)}(t)e_j^{(q)}(t)dt = \gamma_j\delta_{ij},$$

where $\gamma_1 = \cdots = \gamma_q = 0$ and $C_1 j^{2q} \le \gamma_{(j+q)} \le C_2 j^{2q}$ for $C_1, C_2 \in (0,\infty)$. The e_j provide a CONS for $\mathbb{L}^2$ while the functions $\nu_j^{-1/2}e_j$ with $\nu_j = 1+\gamma_j$ represent a CONS for $\mathbb{W}_q$. Using these functions, we can, for example, write

$$\mathcal{K}_n = \sum_{j=1}^{\infty}\sum_{k=1}^{\infty} r_{jk}e_j \otimes e_k, \tag{11.10}$$

with $\otimes$ indicating $\mathbb{L}^2$ tensor product and

$$r_{jk} = n^{-1}\sum_{i=1}^{n}\langle X_i, e_j\rangle\langle X_i, e_k\rangle.$$

Now write $g \in \mathbb{W}_q$ as $g = \sum_{j=1}^{\infty} g_j\nu_j^{-1/2}e_j$ and apply Theorem 11.2.1 to obtain

$$\mathcal{T}^*\mathcal{T}g = \sum_{j=1}^{\infty}\sum_{k=1}^{\infty} r_{jk}\nu_j^{-1/2}g_j\mathcal{R}e_k. \tag{11.11}$$

This leads us to the following conclusion.

Theorem 11.2.2 *Let $g_1, g_2 \in \mathbb{W}_q$ have the representations $g_i = \sum_{j=1}^{\infty} \nu_j^{-1/2}g_{ij}e_j$ for square summable coefficient sequences $\{g_{ij}\}$. Then,*

$$\langle \mathcal{T}^*\mathcal{T}g_1, g_2\rangle_{\mathbb{W}_q} = \sum_{j=1}^{\infty}\sum_{k=1}^{\infty} r_{jk}\nu_j^{-1/2}\nu_k^{-1/2}g_{1j}g_{2k}.$$

Proof: The result follows from (11.11) once we realize that

$$\begin{aligned}
\langle \mathscr{R}e_k, e_j\rangle_{\mathbb{W}_q} &= \int_0^1 (\mathscr{R}e_k)(s)e_j(s)ds + \int_0^1 \frac{d^q}{ds^q}(\mathscr{R}e_k)(s)e_j^{(q)}(s)ds \\
&= \int_0^1 e_k(t)\left(\int_0^1 \left[R(s,t)e_j(s) + \frac{\partial^q}{\partial s^q}R(s,t)e_j^{(q)}(s)\right] ds\right) dt \\
&= \int_0^1 e_k(t)\langle R(\cdot,t), e_j\rangle_{\mathbb{W}_q} dt = \int_0^1 e_k(t)e_j(t).
\end{aligned}$$

□

Theorem 2.8.4 has the consequence that the mapping from $\mathbb{W}_q$ to the sequence space ℓ^2 of Example 2.2.12 defined by

$$g = \sum_{j=1}^{\infty} \nu_j^{-1/2} g_j e_j \mapsto (g_1, g_2, \ldots) \tag{11.12}$$

is an isometric isomorphism. Theorem 11.2.2 then provides the ℓ^2 incarnation of $\mathscr{T}^*\mathscr{T}$ as

$$\mathscr{L} = \sum_{j=1}^{\infty}\sum_{k=1}^{\infty} r_{jk}\nu_j^{-1/2}\nu_k^{-1/2}\phi_j \otimes_{\ell^2} \phi_k,$$

where ϕ_j is the element of ℓ^2 with all zero components except for a one as its jth entry. Observe that we can express $\mathscr{L}$ as the composition

$$\mathscr{L} = \mathscr{D}\mathscr{M}\mathscr{D}, \tag{11.13}$$

where

$$\mathscr{D} = \sum_{j=1}^{\infty} \nu_j^{-1/2}\phi_j \otimes_{\ell^2} \phi_j$$

and

$$\mathscr{M} = \sum_{j=1}^{\infty}\sum_{k=1}^{\infty} r_{jk}\phi_j \otimes_{\ell^2} \phi_k.$$

Let $\lambda_{1n} \geq \lambda_{2n} \geq \cdots$ be the eigenvalues of the $\mathbb{L}^2$ integral operator $\mathscr{K}_n$, which are also the eigenvalues of $\mathscr{M}$. Similarly, take $\lambda_1 \geq \lambda_2 \geq \cdots$ to be the eigenvalues of the $\mathbb{L}^2$ integral operator $\mathscr{K}$ induced by the X covariance kernel. Then, for each k, define

$$V_{kn} = \sum_{j>k} \lambda_{jn}$$

and

$$V_k = \sum_{j>k} \lambda_j.$$

Theorem 11.2.3 *For all $j, k \geq 1$,*

$$\text{trace } \mathscr{G}(\eta)^{-1}\mathscr{T}^*\mathscr{T} \leq k + j + \frac{V_{kn}}{\eta \nu_{j+1}}.$$

Proof: The isometric isomorphism in (11.12) entails that

$$\text{trace } \mathscr{G}(\eta)^{-1}\mathscr{T}^*\mathscr{T} = \text{trace } (\mathscr{L} + \eta I)^{-1}\mathscr{L}.$$

Let $\mathscr{P}$ be the projection operator onto the span of the first k eigenfunctions of $\mathscr{M}$ and set $\mathscr{Q} = I - \mathscr{P}$. Then,

$$\begin{aligned}\text{trace } (\mathscr{L} + \eta I)^{-1}\mathscr{L} = &\text{trace } (\mathscr{L} + \eta I)^{-1}\mathscr{D}\mathscr{M}\mathscr{P}\mathscr{D} \\ &+\text{trace } (\mathscr{L} + \eta I)^{-1}\mathscr{D}\mathscr{M}\mathscr{Q}\mathscr{D}.\end{aligned}$$

Note that $\mathscr{M}$ and $\mathscr{P}$ (and, hence, $\mathscr{Q}$) commute. In addition, as $\mathscr{L} \geq \mathscr{D}\mathscr{M}\mathscr{P}\mathscr{D}$ and $\mathscr{L} \geq \mathscr{D}\mathscr{M}\mathscr{Q}\mathscr{D}$, we have

$$\begin{aligned}\text{trace } (\mathscr{L} + \eta I)^{-1}\mathscr{L} \leq &\text{trace } (\mathscr{D}\mathscr{M}\mathscr{P}\mathscr{D} + \eta I)^{-1}\mathscr{D}\mathscr{M}\mathscr{P}\mathscr{D} \\ &+\text{trace } (\mathscr{D}\mathscr{M}\mathscr{Q}\mathscr{D} + \eta I)^{-1}\mathscr{D}\mathscr{M}\mathscr{Q}\mathscr{D}. \quad (11.14)\end{aligned}$$

Observe that

$$(\mathscr{D}\mathscr{M}\mathscr{P}\mathscr{D} + \eta I)^{-1}\mathscr{D}\mathscr{M}\mathscr{P}\mathscr{D} = I - \eta(\mathscr{D}\mathscr{M}\mathscr{P}\mathscr{D} + \eta I)^{-1}.$$

The fact that $\mathscr{M}$ and $\mathscr{P}$ commute now has the consequence that $(\mathscr{D}\mathscr{M}\mathscr{P}\mathscr{D} + \eta I)^{-1}\mathscr{D}\mathscr{M}\mathscr{P}\mathscr{D}$ is self-adjoint. In addition,

$$\|(\mathscr{D}\mathscr{M}\mathscr{P}\mathscr{D} + \eta I)^{-1}\mathscr{D}\mathscr{M}\mathscr{P}\mathscr{D}\| \leq \|\mathscr{D}\mathscr{M}\mathscr{P}\mathscr{D}\|/\|\mathscr{D}\mathscr{M}\mathscr{P}\mathscr{D} + \eta I\| \leq 1$$

so that

$$\text{trace } (\mathscr{D}\mathscr{M}\mathscr{P}\mathscr{D} + \eta I)^{-1}\mathscr{D}\mathscr{M}\mathscr{P}\mathscr{D} \leq \dim\ (Im(\mathscr{P})) = k. \quad (11.15)$$

Similarly, $(\mathscr{D}\mathscr{M}\mathscr{Q}\mathscr{D} + \eta I)^{-1}\mathscr{D}\mathscr{M}\mathscr{Q}\mathscr{D}$ is self-adjoint with norm bounded by 1 and

$$(\mathscr{D}\mathscr{M}\mathscr{Q}\mathscr{D} + \eta I)^{-1}\mathscr{D}\mathscr{M}\mathscr{Q}\mathscr{D} \leq \eta^{-1}\mathscr{D}\mathscr{M}\mathscr{Q}\mathscr{D}.$$

Thus, for any j,

$$\begin{aligned}\text{trace}(\mathscr{D}\mathscr{M}\mathscr{Q}\mathscr{D}+\eta I)^{-1}\mathscr{D}\mathscr{M}\mathscr{Q}\mathscr{D} &= \sum_{i=1}^{\infty}\langle\phi_i,(\mathscr{D}\mathscr{M}\mathscr{Q}\mathscr{D}+\eta I)^{-1}\mathscr{D}\mathscr{M}\mathscr{Q}\mathscr{D}\phi_i\rangle_{\ell^2} \\ &\le j+\frac{1}{\eta}\sum_{i>j}\langle\phi_i,\mathscr{D}\mathscr{M}\mathscr{Q}\mathscr{D}\phi_i\rangle_{\ell^2} \\ &\le j+\frac{1}{\eta\nu_{j+1}}\sum_{i>j}\langle\phi_i,\mathscr{M}\mathscr{Q}\phi_i\rangle_{\ell^2} \\ &\le j+\frac{1}{\eta\nu_{j+1}}\text{trace}\,\mathscr{M}\mathscr{Q}. \end{aligned} \tag{11.16}$$

The conclusion of the theorem is now a direct consequence of (11.15) and (11.16). □

The following result provides a tool we can use for working with the V_{kn}.

Lemma 11.2.4 *For any sequence* k_n, $V_{k_n n}=O_p(V_{k_n})$.

Proof: Let $\mathscr{P}_n$ be the projection operator onto the span of the first k sample eigenfunctions from $\mathscr{K}_n$ and, similarly, let $\mathscr{P}$ be the projection operator for the first k eigenfunctions of $\mathscr{K}$. Then, V_{kn} and V_k can be expressed as

$$V_{kn}=\text{trace}\,\mathscr{K}_n(I-\mathscr{P}_n)$$

and

$$V_k=\text{trace}\,\mathscr{K}(I-\mathscr{P}).$$

Theorem 4.4.7 gives

$$\text{trace}\,\mathscr{K}_n\mathscr{P}_n\ge\text{trace}\,\mathscr{K}_n\mathscr{P}$$

and, hence,

$$\text{trace}\,\mathscr{K}_n(I-\mathscr{P}_n)\le\text{trace}\,\mathscr{K}_n(I-\mathscr{P}).$$

Now, by Chebeshev's inequality and the fact that

$$\mathbb{E}[\text{trace}\,\mathscr{K}_n(I-\mathscr{P})]=\text{trace}\,\mathscr{K}(I-\mathscr{P}),$$

we obtain

$$\begin{aligned}\mathbb{P}(\text{trace}\,\mathscr{K}_n(I-\mathscr{P}_n)>a) &\le \mathbb{P}(\text{trace}\,\mathscr{K}_n(I-\mathscr{P})>a) \\ &\le \frac{\text{trace}\,\mathscr{K}(I-\mathscr{P})}{a}.\end{aligned}$$

□

When used in conjunction with Theorems 11.1.1 and 11.2.3, the previous lemma allows us to establish

Theorem 11.2.5 *Assume that $V_k = O(k^{-\alpha})$ for some $\alpha > 0$. Then, the choice of $\eta = n^{-\frac{2q+\alpha+1}{2q+\alpha+2}}$ yields*

$$\mathbb{E}_\varepsilon \|\mathscr{T}(\beta_\eta - \beta)\|_{\mathbb{Y}}^2 = O_p\left(n^{-\frac{2q+\alpha+1}{2q+\alpha+2}}\right) \tag{11.17}$$

and

$$\mathbb{E}_\varepsilon \|\beta_\eta - \beta\|_{\mathbb{W}_q}^2 = O_p(1). \tag{11.18}$$

It may be helpful to remember here that expression (11.17) is still a stochastic function of $X_1, \ldots, X_n$, which explains the form of the bound.

Proof: By Lemma 11.2.4, we conclude that $V_{kn} = O_p(k^{-\alpha})$. Theorem 11.2.3, with $k = j = \left[n^{\frac{1}{2q+\alpha+2}}\right]$, then gives

$$\text{trace } \mathscr{G}(\eta)^{-1}\mathscr{T}^*\mathscr{T} = O_p\left(n^{\frac{1}{2q+\alpha+2}}\right). \tag{11.19}$$

Note that $\mathscr{G}(\eta)^{-1}\mathscr{T}^*\mathscr{T}$ is self-adjoint and bounded by I. Thus,

$$\text{trace } (\mathscr{G}(\eta)^{-1}\mathscr{T}^*\mathscr{T})^2 \leq \text{trace } \mathscr{G}(\eta)^{-1}\mathscr{T}^*\mathscr{T}$$

and (11.17) follows from Theorems 11.1.1.

From (11.5), we have

$$\mathbb{E}_\varepsilon \|\beta_\eta\|_{\mathbb{W}_q}^2 \leq 2\|\beta\|_{\mathbb{W}_q}^2 + \frac{2\sigma^2}{n}\text{trace } \mathscr{G}(\eta)^{-2}\mathscr{T}^*\mathscr{T}.$$

In view of the fact that

$$\text{trace } \mathscr{G}(\eta)^{-2}\mathscr{T}^*\mathscr{T} \leq \eta^{-1}\text{trace } \mathscr{G}(\eta)^{-1}\mathscr{T}^*\mathscr{T}$$

and (11.19),

$$\frac{1}{n}\text{trace } \mathscr{G}(\eta)^{-2}\mathscr{T}^*\mathscr{T} = \frac{1}{n\eta}O_p\left(n^{\frac{1}{2q+\alpha+2}}\right) = O_p(1).$$

Thus, (11.18) is established. □

A case of some interest occurs when X is the Brownian motion process with covariance kernel $K(s,t) = \min(s,t)$ for $s, t \in [0,1]$. In that instance, we know from Example 4.6.3 that $\lambda_j = 4/((2j-1)\pi)^2$. An integral estimate then shows that $V_k = O(k^{-1})$. Thus, $\alpha = 1$ in the corollary and

$$\|\mathscr{T}(\beta_\eta - \beta)\|_{\mathbb{Y}}^2 = O_p\left(n^{-\frac{2q+2}{2q+3}}\right).$$

Now let us examine the prediction properties of our estimator. For that purpose, let X be a new, independent, observation and consider

$$\mathbb{E}_{\varepsilon,X}(\langle X, \beta_\eta\rangle - \langle X, \beta\rangle)^2,$$

where $\mathbb{E}_{\varepsilon,X}$ indicates expectation with respect to both X and ε.

Note that for any $g \in \mathbb{L}^2$

$$\|\mathscr{T} g\|_{\mathbb{Y}}^2 = \langle g, \mathscr{K}_n g\rangle$$

and, from Theorem 7.2.5,

$$\mathbb{E}_X\langle X, g\rangle^2 = \langle g, \mathscr{K} g\rangle.$$

Thus,

$$\mathbb{E}_\varepsilon\|\mathscr{T}(\beta_\eta - \beta)\|_{\mathbb{Y}}^2 = \mathbb{E}_\varepsilon\langle \beta_\eta - \beta, \mathscr{K}_n(\beta_\eta - \beta)\rangle$$

and

$$\begin{aligned}\mathbb{E}_{\varepsilon,X}(\langle X, \beta_\eta\rangle - \langle X, \beta\rangle)^2 &= \mathbb{E}_\varepsilon\langle \beta_\eta - \beta, \mathscr{K}(\beta_\eta - \beta)\rangle \\ &= \mathbb{E}_\varepsilon\langle \beta_\eta - \beta, (\mathscr{K} - \mathscr{K}_n)(\beta_\eta - \beta)\rangle \\ &\quad + \mathbb{E}_\varepsilon\|\mathscr{T}(\beta_\eta - \beta)\|_{\mathbb{Y}}^2.\end{aligned} \tag{11.20}$$

The second term in the last expression was already treated in Theorem 11.2.5. It therefore suffices to consider the first term.

Let

$$\mathscr{K} = \sum_{j=1}^{\infty} \lambda_j \psi_j \otimes \psi_j,$$

be the eigen-expansion for $\mathscr{K}$ and define the scores

$$Z_{ij} = \langle X_i, \psi_j\rangle$$

for $i = 1, \ldots, n$. Their associated empirical covariance is

$$\tau_{jk} = \frac{1}{\sqrt{n\lambda_j\lambda_k}}\sum_{i=1}^{n}(Z_{ij}Z_{ik} - \lambda_j\delta_{jk})$$

and

$$\mathscr{K}_n = \sum_{j=1}^{\infty}\sum_{k=1}^{\infty}\left(n^{-1}\sum_{i=1}^{n} Z_{ij}Z_{ik}\right)\psi_j \otimes \psi_k$$

$$= n^{-1/2}\sum_{j=1}^{\infty}\sum_{k=1}^{\infty}\sqrt{\lambda_j\lambda_k}\tau_{jk}\psi_j \otimes \psi_k + \mathscr{K}. \tag{11.21}$$

Lemma 11.2.6 *Suppose that there exists a fixed $C < \infty$ such that*

$$\mathrm{Var}(Z_{ij}Z_{ik}) \le C\lambda_j\lambda_k \tag{11.22}$$

for all j, k. Then, for any $g \in \mathbb{L}^2$ and any $k \ge 1$,

$$\langle g, (\mathscr{K} - \mathscr{K}_n)g\rangle$$

$$\le O_p(n^{-1})\|g\|^2 \sum_{r=1}^{k} V_r + O_p(n^{-1/2})\|g\|\|\mathscr{T}g\|_{\mathbb{Y}}\left(\sum_{r=1}^{k} V_r\right)^{1/2} \tag{11.23}$$

$$+ O_p(n^{-3/4})\|g\|^2\left(V_k\sum_{r=1}^{k} V_r\right)^{1/2} + O_p(n^{-1/2})\|g\|^2 V_k,$$

where the O_p terms are functions of $X_1, \ldots, X_n$ only.

Proof: From (11.21),

$$\langle g, (\mathscr{K}_n - \mathscr{K})g\rangle = \frac{1}{\sqrt{n}}\sum_{j=1}^{\infty}\sum_{k=1}^{\infty} g_j g_k \sqrt{\lambda_j\lambda_k}\tau_{jk},$$

where $g_j = \langle g, e_j\rangle$. Then,

$$|\langle g, (\mathscr{K}_n - \mathscr{K})g\rangle|$$

$$\le \frac{2}{\sqrt{n}}\sum_{r=1}^{k}\sum_{s=r+1}^{\infty} |g_r g_s\sqrt{\lambda_r\lambda_s}\tau_{rs}| + \frac{2}{\sqrt{n}}\sum_{r=k+1}^{\infty}\sum_{s=r+1}^{\infty} |g_r g_s\sqrt{\lambda_r\lambda_s}\tau_{rs}|$$

$$\le \frac{2}{\sqrt{n}}\left(\sum_{r=1}^{k}\sum_{s=r+1}^{\infty}\lambda_r g_r^2 g_s^2\right)^{1/2}\left(\sum_{r=1}^{k}\sum_{s=r+1}^{\infty}\lambda_s\tau_{rs}^2\right)^{1/2}$$

$$+\frac{2}{\sqrt{n}}\left(\sum_{r=k+1}^{\infty}\sum_{s=r+1}^{\infty} g_r^2 g_s^2\right)^{1/2}\left(\sum_{r=k+1}^{\infty}\sum_{s=r+1}^{\infty}\lambda_r\lambda_s\tau_{rs}^2\right)^{1/2}.$$

We will proceed by obtaining bounds for the terms in this last expression.

Now, $\sum_{r=1}^{\infty} g_r^2 \leq \|g\|^2$ and $\sum_{r=1}^{\infty} \lambda_r g_r^2 = \langle g, \mathscr{K} g\rangle$. Thus,

$$\sum_{r=1}^{k} \sum_{s=r+1}^{\infty} \lambda_r g_r^2 g_s^2 \leq \langle g, \mathscr{K} g\rangle \|g\|^2$$

and

$$\sum_{r=k+1}^{\infty} \sum_{s=r+1}^{\infty} g_r^2 g_s^2 \leq \|g\|^4.$$

Using assumption (11.22), $\mathbb{E}\tau_{rs}^2 \leq C$ for all r, s so that

$$\sum_{r=1}^{k} \sum_{s=r+1}^{\infty} \lambda_s \tau_{rs}^2 = O_p(1) \sum_{r=1}^{k} V_r$$

and

$$\sum_{r=k+1}^{\infty} \sum_{s=r+1}^{\infty} \lambda_r \lambda_s \tau_{rs}^2 = O_p(V_k^2).$$

Upon combining all our bounds, we obtain

$$\begin{aligned} \langle g, \mathscr{K} - \mathscr{K}_n)g\rangle &\leq \langle g, \mathscr{K}_n)g\rangle \\ &\quad + O_p(n^{-1/2})\|g\|\langle g, \mathscr{K} g\rangle^{1/2}\left(\sum_{r=1}^{k} V_r\right)^{1/2} \\ &\quad + O_p(n^{-1/2})\|g\|^2 V_k. \end{aligned} \tag{11.24}$$

Now, if $x_0^2 \leq bx_0 + c$ and $x_0 \geq 0$ then it must be that x_0 is less than or equal to the positive root of the quadratic equation $x^2 - bx - c = 0$: namely,

$$0 \leq x_0 \leq \frac{b + \sqrt{b^2 + 4c}}{2} \leq \left(\frac{1}{2} + \frac{1}{\sqrt{2}}\right) b + \sqrt{2c}.$$

Applying this to (11.24), we get

$$\begin{aligned} \langle g, \mathscr{K} g\rangle^{1/2} &\leq O_p(n^{-1/2})\|g\|\left(\sum_{r=1}^{k} V_r\right)^{1/2} + O_p\left(\langle g, \mathscr{K}_n g\rangle^{1/2}\right) \\ &\quad + O_p(n^{-1/4})\|g\| V_k^{1/2}. \end{aligned}$$

Upon substituting this back in (11.24), we obtain (11.23). □

Our main result concerning prediction can now be stated as follows.

Theorem 11.2.7 *Assume that (11.22) holds and $V_k = O(k^{-\alpha})$ for some $\alpha > 0$. Then, if $\eta = n^{-\frac{2q+\alpha+1}{2q+\alpha+2}}$,*

$$\mathbb{E}_{\varepsilon,X}(\langle X, \beta_\eta\rangle - \langle X, \beta\rangle)^2 = O_p\left(n^{-\frac{2q+\alpha+1}{2q+\alpha+2}} + n^{-\frac{\alpha+1}{2}}\right). \tag{11.25}$$

Proof: By (11.20) and (11.23), it is sufficient to consider

$$A_1 = O_p(n^{-1})\left(\sum_{r=1}^{k} V_r\right) \quad \mathbb{E}_\varepsilon\|\beta_\eta - \beta\|^2,$$

$$A_2 = O_p(n^{-1/2})\left(\sum_{r=1}^{k} V_r\right)^{1/2} \{\mathbb{E}_\varepsilon(\|\beta_\eta - \beta\|^2)\mathbb{E}_\varepsilon(\|\mathscr{T}(\beta_\eta - \beta)\|_{\mathbb{Y}}^2)\}^{1/2},$$

$$A_3 = O_p(n^{-3/4})\left(V_k\sum_{r=1}^{k} V_r\right)^{1/2} \mathbb{E}_\varepsilon\|\beta_\eta - \beta\|^2,$$

$$A_4 = O_p(n^{-1/2})V_k\,\mathbb{E}_\varepsilon\|\beta_\eta - \beta\|^2.$$

Bounds for both $\mathbb{E}_\varepsilon\|\beta_\eta - \beta\|^2$ and $\mathbb{E}_\varepsilon\|\mathscr{T}(\beta_\eta - \beta)\|_{\mathbb{Y}}^2$ were obtained in Theorem 11.2.5 so that only routine calculations are required to establish the stated rates. For instance, if $\alpha > 1$ so that $\sum_{r=1}^{\infty} V_r < \infty$, then, with $k = [n^{1/2}]$,

$$A_1 = O_p(n^{-1}) = o(n^{-\frac{2q+\alpha+1}{2q+\alpha+2}}),$$

$$A_2 = O_p(n^{-\frac{1}{2}-\frac{1}{2}\frac{2q+\alpha+1}{2q+\alpha+2}}) = o(n^{-\frac{2q+\alpha+1}{2q+\alpha+2}}),$$

$$A_3 = O_p(n^{-\frac{3+\alpha}{4}}) = o(n^{-1}) = o(n^{-\frac{2q+\alpha+1}{2q+\alpha+2}}),$$

$$A_4 = O_p(n^{-\frac{\alpha+1}{2}}).$$

Bounds for $\alpha \le 1$ can be similarly established also using the choice $k = [n^{1/2}]$. □

The value of α determines the dominant term in (11.25). Provided $\alpha \ge 1$,

$$\mathbb{E}_{\varepsilon,X}(\langle X, \beta_\eta\rangle - \langle X, \beta\rangle)^2 = O_p\left(n^{-\frac{2q+\alpha+1}{2q+\alpha+2}}\right).$$

As mentioned earlier, this condition holds when X is a Brownian motion process.

Theorem 11.2.7 provides information about the large sample prediction ability of β_η. However, in order to fully appreciate its message, we need to understand what is optimal in terms of convergence rates for this problem. That is the subject of Section 11.3.

11.3 Minimax optimality

In this section, we obtain results that speak to the rate optimality of our penalized least-squares estimator of β. We do so in the context of prediction and, as before, assume that data $(X_i, Y_i), i = 1, \dots, n$, have been obtained from the regression model (11.1). Given some estimator $\tilde{\beta}$ of the coefficient function obtained from this data and a new independent observation X, we can predict the value of $\langle X, \beta\rangle$ by that of $\langle X, \tilde{\beta}\rangle$. We will now derive the minimax convergence rate of $\langle X, \tilde{\beta}\rangle$ to $\langle X, \beta\rangle$ over all choices for $\tilde{\beta} \in \mathbb{L}^2$.

Each functional regression model of the form (11.1) can be specified by a triple $\theta = (\beta, p_X, p_\varepsilon)$ with β the coefficient function and p_X, p_ε the probability distributions for X and ε. We will consider how our estimator performs over a class of models as determined by

$$\Theta = \{\theta = (\beta, p_X, p_\varepsilon) : \|\beta\|_{\mathbb{W}_q} \le 1, X \in \mathbb{L}^2, V_k \le k^{-\alpha}\}.$$

Then, given any specific choice for an estimator $\tilde{\beta}$ of the coefficient function, the worst that can happen is if the real model is $\theta_0 = (\beta_0, p_{0X}, p_{0\varepsilon})$ for which

$$\mathbb{E}_{\theta_0}\big(\langle X, \tilde{\beta}\rangle - \langle X, \beta_0\rangle\big)^2 = \sup_{\theta\in\Theta} \mathbb{E}_\theta\big(\langle X, \tilde{\beta}\rangle - \langle X, \beta\rangle\big)^2 \tag{11.26}$$

with $\mathbb{E}_\theta$ indicating expectation under the model corresponding to θ. A minimax estimator is a choice for $\tilde{\beta}$ that makes (11.26) as small as possible. In lieu of determining an explicit form for such an estimator, we will consider the rate of convergence of the infimum of (11.26) to zero as the sample size grows large. Then, any estimator that attains this rate is minimax optimal in an asymptotic or rate of convergence sense.

Theorem 11.3.1 *Assume that $\alpha \ge 1$. Then, for some constant $C \in (0, \infty)$,*

$$\liminf_{n\to\infty} \inf_{\tilde{\beta}\in\mathbb{L}^2} \sup_{\theta\in\Theta} \mathbb{E}_\theta\big(\langle X, \tilde{\beta}\rangle - \langle X, \beta\rangle\big)^2 \ge Cn^{-\frac{2q+\alpha+1}{2q+\alpha+2}}.$$

Proof: Let $k_n = \left[n^{\frac{1}{2q+\alpha+2}}\right]$ and consider a θ for which p_ε is a standard normal distribution,

$$\beta = \sum_{j=k_n+1}^{2k_n} k_n^{-1/2} \delta_j \nu_j^{-1/2} e_j$$

and

$$X = \sum_{j=k_n+1}^{2k_n} \xi_j \nu_j^{-\frac{\alpha+1}{4q}} e_j,$$

where $\delta_j = 0$ or 1, the e_j's are, again, as in Theorem 2.8.4 and the ξ_j are independent uniform random variables on $[-\sqrt{3}, \sqrt{3}]$. As $\{v_j^{-1/2} e_j\}_{j=1}^{\infty}$ is a CONS for $\mathbb{W}_q$, this specification ensures that $\|\beta\|_{\mathbb{W}_q} \le 1$. If we denote the collection of all such θ by Θ_0, it is clearly true that

$$\sup_{\theta\in\Theta} \mathbb{E}_\theta\big(\langle X, \tilde{\beta}\rangle - \langle X, \beta\rangle\big)^2 \ge \sup_{\theta\in\Theta_0} \mathbb{E}_\theta\big(\langle X, \tilde{\beta}\rangle - \langle X, \beta\rangle\big)^2. \tag{11.27}$$

Now, let $\tilde{\beta} \in \mathbb{L}^2$ be an arbitrary estimator of β and write

$$\tilde{\beta} = \sum_{j=1}^{\infty} k_n^{-1/2} v_j^{-1/2} \tilde{\delta}_j e_j$$

for some coefficient sequence $\{\tilde{\delta}_j\}$. Then, for any model $\theta \in \Theta_0$,

$$\langle X, \tilde{\beta} - \beta\rangle = \sum_{j=k_n+1}^{2k_n} k_n^{-1/2} v_j^{-\frac{2q+\alpha+1}{4q}} (\tilde{\delta}_j - \delta_j)\xi_j,$$

and it follows that

$$\mathbb{E}_\theta\langle X, \tilde{\beta} - \beta\rangle^2 = \sum_{j=k_n+1}^{2k_n} k_n^{-1} v_j^{-\frac{2q+\alpha+1}{2q}} \mathbb{E}_\theta(\tilde{\delta}_j - \delta_j)^2. \tag{11.28}$$

Combining (11.27) and (11.28) gives

$$\begin{aligned}
\sup_{\theta\in\Theta} \mathbb{E}_\theta\langle X, \tilde{\beta} - \beta\rangle^2 &\ge \frac{1}{2^{k_n}} \sum_{\theta\in\Theta_0} \sum_{j=k_n+1}^{2k_n} k_n^{-1} v_j^{-\frac{2q+\alpha+1}{2q}} \mathbb{E}_\theta(\tilde{\delta}_j - \delta_j)^2 \\
&= \sum_{j=k_n+1}^{2k_n} k_n^{-1} v_j^{-\frac{2q+\alpha+1}{2q}} \frac{1}{2^{k_n}} \sum_{\theta\in\Theta_0} \mathbb{E}_\theta(\tilde{\delta}_j - \theta_j)^2.
\end{aligned}$$

For each $j = k_n + 1, \ldots, 2k_n$ and $\theta \in \Theta_0$, let $\theta_{j0} = \theta$ and take θ_{j1} to be the same as θ except for flipping the value of δ_j: i.e., $\delta_{j1} = 1 - \delta_{j0}$. By symmetry,

$$\begin{aligned}
&\sum_{j=k_n+1}^{2k_n} k_n^{-1} v_j^{-\frac{2q+\alpha+1}{2q}} \frac{1}{2^{k_n}} \sum_{\theta\in\Theta_0} \mathbb{E}_\theta(\tilde{\delta}_j - \delta_j)^2 \\
&= \sum_{j=k_n+1}^{2k_n} k_n^{-1} v_j^{-\frac{2q+\alpha+1}{2q}} \frac{1}{2^{k_n}} \sum_{\theta\in\Theta_0} \frac{1}{2}\left[\mathbb{E}_{\theta_{j0}}(\tilde{\delta}_j - \delta_j)^2 + \mathbb{E}_{\theta_{j1}}(\tilde{\delta}_j - \delta_j)^2\right] \\
&\ge v_{2k_n}^{-\frac{2q+\alpha+1}{2q}} \inf_{k_n<j\le 2k_n} \frac{1}{2^{k_n}} \sum_{\theta\in\Theta_0} \frac{1}{2}\left[\mathbb{E}_{\theta_{j0}}(\tilde{\delta}_j - \delta_j)^2 + \mathbb{E}_{\theta_{j1}}(\tilde{\delta}_j - \delta_j)^2\right].
\end{aligned}$$

Thus, we conclude that, for some $C > 0$,

$$\sup_{\theta\in\Theta} \mathbb{E}_\theta\langle X, \tilde{\beta} - \beta\rangle^2 \geq C n^{-\frac{2q+\alpha+1}{2q+\alpha+2}} \inf_{k_n<j\leq 2k_n} \inf_{\theta\in\Theta_0} \max_{\theta\in\{\theta_{j0},\theta_{j1}\}} \mathbb{E}_\theta(\tilde{\delta}_j - \delta_j)^2.$$

The proof is completed by an application of Lemma 11.3.2. □

Lemma 11.3.2

$$\liminf_{n\to\infty} \inf_{k_n<j\leq 2k_n} \inf_{\theta\in\Theta_0} \inf_{\tilde{\delta}_j} \max_{\theta\in\{\theta_{j0},\theta_{j1}\}} \mathbb{E}_\theta(\tilde{\delta}_j - \delta_j)^2 > 0.$$

Proof: Fix j, $\tilde{\delta}_j, \theta$ and, without loss of generality, assume that $\delta_{j0} = 0$ and $\delta_{j1} = 1$. Then, define

$$\hat{\delta}_j = I(|\tilde{\delta}_j - \delta_{j0}| > |\tilde{\delta}_j - \delta_{j1}|)$$

and observe that

$$1 = |\delta_{j0} - \delta_{j1}| \leq |\tilde{\delta}_j - \delta_{j0}| + |\tilde{\delta}_j - \delta_{j1}| \leq \begin{cases} 2|\tilde{\delta}_j - \delta_{j1}|, & \hat{\delta}_j = 0, \\ 2|\tilde{\delta}_j - \delta_{j0}|, & \hat{\delta}_j = 1. \end{cases}$$

Thus,

$$\mathbb{P}_{\theta_{j0}}(|\tilde{\delta}_j - \delta_j| > 1/2) \geq \mathbb{P}_{\theta_{j0}}(\hat{\delta}_j = 1)$$

and

$$\mathbb{P}_{\theta_{j1}}(|\tilde{\delta}_j - \delta_j| > 1/2) \geq \mathbb{P}_{\theta_{j1}}(\hat{\delta}_j = 0).$$

So,

$$\begin{aligned} \max_{\theta\in\{\theta_{j0},\theta_{j1}\}} \mathbb{P}_\theta(|\tilde{\theta}_j - \theta_j| > 1/2) &\geq \max\{\mathbb{P}_{\theta_{j0}}(\hat{\delta}_j = 1), \mathbb{P}_{\theta_{j1}}(\hat{\delta}_j = 0)\} \\ &\geq \frac{1}{2}\{\mathbb{P}_{\theta_{j0}}(\hat{\delta}_j = 1) + \mathbb{P}_{\theta_{j1}}(\hat{\delta}_j = 0)\}. \end{aligned}$$

Now, let $L_{\theta_{j0}}/L_{\theta_{j1}}$ be the likelihood ratio for the models corresponding to θ_{j0} and θ_{j1}. By the Neyman–Pearson Lemma,

$$\mathbb{P}_{\theta_{j0}}(\hat{\delta}_j = 1) + \mathbb{P}_{\theta_{j1}}(\hat{\delta}_j = 0) \geq \mathbb{P}_{\theta_{j0}}(L_{\theta_{j0}}/L_{\theta_{j1}} \leq 1) + \mathbb{P}_{\theta_{j1}}(L_{\theta_{j0}}/L_{\theta_{j1}} \geq 1)$$

and, hence,

$$\begin{aligned} &\max_{\theta\in\{\theta_{j0},\theta_{j1}\}} \mathbb{P}_\theta(|\tilde{\delta}_j - \delta_j| > 1/2) \\ &\geq \frac{1}{2}\{\mathbb{P}_{\theta_{j0}}(L_{\theta_{j0}}/L_{\theta_{j1}} \leq 1) + \mathbb{P}_{\theta_{j1}}(L_{\theta_{j0}}/L_{\theta_{j1}} \geq 1)\} \\ &\geq \left\{4\mathbb{E}_{\theta_{j0}}(L^2_{\theta_{j1}}/L^2_{\theta_{j0}})\right\}^{-1}. \end{aligned}$$

Thus,

$$\max_{\theta \in \{\theta_{j0}, \theta_{j1}\}} \mathbb{E}_\theta(\tilde{\delta}_j - \delta_j)^2 \geq \frac{1}{4} \max_{\theta \in \{\theta_{j0}, \theta_{j1}\}} \mathbb{P}_\theta(|\tilde{\delta}_j - \delta_j(\theta)| > 1/2)$$

$$\geq \{16 \mathbb{E}_{\theta_{j0}}(L^2_{\theta_{j1}}/L^2_{\theta_{j0}})\}^{-1}.$$

As the ε_i are iid standard normals, it is straightforward to show that

$$\mathbb{E}_{\theta_{j0}}(L^2_{\theta_{j1}}/L^2_{\theta_{j0}}) = \left(\mathbb{E}_{\theta_0}\left[\exp\left\{\left(V_n^{-1/2} \nu_j^{-\frac{2q+\alpha+1}{4q}} \xi_j\right)^2\right\}\right]\right)^n$$

$$= (1 + O(n^{-1}))^n = O(1).$$

The result follows from this as the constants in the derivation do not depend on j, $\tilde{\delta}_j$, or θ. □

Let us conclude by relating Theorem 11.3.1 to the developments at the end of the previous section. Specifically, when using this result in combination with Theorem 11.2.7, we see that a rate optimal choice for η ensures that $\langle X, \beta_\eta \rangle$ attains the minimax optimal rate of squared error convergence as a predictor of $\langle X, \beta \rangle$.

11.4 Discretely sampled data

Until now, the assumption has been that the X_i are observed in their entirety. To conclude this chapter, we briefly examine the case where the X data can only be realized at some discrete set of sampling points.

Suppose that each X_i is observed at time ordinate $t_{i1}, \ldots, t_{iJ_i}$. Let I_{ij} be an interval containing t_{ij} such that $I_{i1}, \ldots, I_{iJ_i}$ form a partition of $[0, 1]$. Then, we approximate X_i by

$$\tilde{X}_i(t) = \sum_{j=1}^{J_i} X_i(t_{ij}) I(t \in I_{ij}).$$

This, in turn, produces approximations to the $\mathscr{T}_i$ and $\mathscr{T}$ that are given by

$$\tilde{\mathscr{T}}_i g = \langle \tilde{X}_i, g \rangle$$

and

$$\tilde{\mathscr{T}} g = \left(\tilde{\mathscr{T}}_1 g, \ldots, \tilde{\mathscr{T}}_n g\right)^T$$

for $g \in \mathbb{W}_q$. The resulting estimator of β is obtained by minimizing

$$\|Y - \tilde{\mathscr{T}} g\|^2_{\mathbb{Y}} + \eta \|g\|^2_{\mathbb{W}_q} \tag{11.29}$$

over $g \in \mathbb{W}_q$; i.e., we estimate β by

$$\tilde{\beta}_\eta = \tilde{G}(\eta)^{-1}\tilde{\mathscr{T}}^* Y,$$

where $\tilde{G}(\eta) = \tilde{\mathscr{T}}^*\tilde{\mathscr{T}} + \eta I$.

Our goal is to obtain an analog of Theorem 11.2.6. To do so, we proceed along the same lines as the developments in Section 11.2. First, take $\tilde{\mathscr{K}}_n$ to be the $\mathbb{L}^2$ integral operator corresponding to the (discretized) sample covariance kernel

$$\tilde{K}(s,t) = n^{-1}\sum_{i=1}^{n} \tilde{X}_i(s)\tilde{X}_i(t).$$

Next, define

$$\tilde{\mathscr{K}} = \mathbb{E}\tilde{\mathscr{K}}_n$$

with associated eigenvalues $\tilde{\lambda}_1 \geq \tilde{\lambda}_2 \geq \cdots$ and $\tilde{V}_k = \sum_{j>k}\tilde{\lambda}_j$. Similarly, define scores by

$$\tilde{Z}_{ik} = \langle \tilde{X}_i, \tilde{e}_k \rangle,$$

where $\tilde{e}_s$ is the eigenfunction for $\tilde{\mathscr{K}}$ that corresponds to the eigenvalue $\tilde{\lambda}_k$.

Theorem 11.4.1 *Assume that* $\tilde{V}_k = O(k^{-\alpha})$ *for some* $\alpha > 0$ *and that there exists a fixed* $C < \infty$ *such that*

$$\mathrm{Var}(\tilde{Z}_{ir}\tilde{Z}_{is}) \leq C\tilde{\lambda}_r\tilde{\lambda}_s \tag{11.30}$$

for all r, s. *Then, if* $\eta = n^{-\frac{2q+\alpha+1}{2q+\alpha+2}}$,

$$\mathbb{E}_{\varepsilon,X}(\langle X, \tilde{\beta}_\eta\rangle - \langle X, \beta\rangle)^2 = O_p\left(n^{-\frac{2q+\alpha+1}{2q+\alpha+2}} + n^{-\frac{\alpha+1}{2}} + \frac{1}{n}\sum_{i=1}^{n}\mathbb{E}\|X_i - \tilde{X}_i\|^2\right).$$

Proof: First, we show that

$$\mathbb{E}_\varepsilon\|\tilde{\mathscr{T}}(\tilde{\beta}_\eta - \beta)\|_{\mathbb{Y}}^2 = O_p\left(n^{-\frac{2q+\alpha+1}{2q+\alpha+2}} + \frac{1}{n}\sum_{i=1}^{n}\mathbb{E}\|X_i - \tilde{X}_i\|^2\right). \tag{11.31}$$

Toward this goal, observe that

$$\begin{aligned}\mathbb{E}_\varepsilon\|\tilde{\mathscr{T}}(\tilde{\beta}_\eta - \beta)\|_{\mathbb{Y}}^2 &= \mathbb{E}_\varepsilon\|\tilde{\mathscr{T}}(\tilde{\beta}_\eta - \mathbb{E}_\varepsilon\tilde{\beta}_\eta + \mathbb{E}_\varepsilon\tilde{\beta}_\eta - \beta)\|_{\mathbb{Y}}^2 \\ &\leq 2\mathbb{E}_\varepsilon\|\tilde{\mathscr{T}}(\tilde{\beta}_\eta - \mathbb{E}_\varepsilon\tilde{\beta}_\eta)\|^2 + 2\|\tilde{\mathscr{T}}\mathbb{E}_\varepsilon(\tilde{\beta}_\eta - \beta)\|_{\mathbb{Y}}^2. \end{aligned} \tag{11.32}$$

The first term on the right of (11.32) is a variance expression and computations similar to what we used for the completely observed case can be used to show that

$$\mathbb{E}_\varepsilon \|\tilde{\mathscr{T}}(\tilde{\beta}_\eta - \mathbb{E}_\varepsilon \tilde{\beta}_\eta)\|^2 = \frac{\sigma^2}{n} \text{trace } (\tilde{\mathscr{G}}(\eta)^{-1} \tilde{\mathscr{T}}^* \tilde{\mathscr{T}})^2$$
$$= O_p \left(n^{-\frac{2q+\alpha+1}{2q+\alpha+2}} \right).$$

The second term corresponds to squared bias and can be handled as follows. First observe that $\mathbb{E}_\varepsilon \tilde{\beta}_\eta = \tilde{G}(\eta)^{-1} \tilde{\mathscr{T}}^* \mathscr{T} \beta$ is the minimizer of

$$\|\mathscr{T}\beta - \tilde{\mathscr{T}} g\|_{\mathbb{Y}}^2 + \eta \|g\|_{\mathbb{W}_q}^2 .$$

Thus,

$$\|\mathscr{T}\beta - \tilde{\mathscr{T}} \mathbb{E}_\varepsilon \tilde{\beta}_\eta\|_{\mathbb{Y}}^2 + \eta \|\mathbb{E}_\varepsilon \tilde{\beta}_\eta\|_{\mathbb{W}_q}^2 \leq \|\mathscr{T}\beta - \tilde{\mathscr{T}}\beta\|_{\mathbb{Y}}^2 + \eta \|\beta\|_{\mathbb{W}_q}^2 . \tag{11.33}$$

As

$$\|\tilde{\mathscr{T}} \mathbb{E}_\varepsilon (\tilde{\beta}_\eta - \beta)\|_{\mathbb{Y}}^2 \leq 2\|\tilde{\mathscr{T}}\beta - \mathscr{T}\beta\|_{\mathbb{Y}}^2 + 2\|\mathscr{T}\beta - \tilde{\mathscr{T}} \mathbb{E}_\varepsilon \tilde{\beta}_\eta\|_{\mathbb{Y}}^2,$$

using (11.33), we arrive at the inequality

$$\|\tilde{\mathscr{T}} \mathbb{E}_\varepsilon (\tilde{\beta}_\eta - \beta)\|_{\mathbb{Y}}^2 \leq 3\|\mathscr{T}\beta - \tilde{\mathscr{T}}\beta\|_{\mathbb{Y}}^2 + 2\eta \|\beta\|_{\mathbb{W}_q}^2 . \tag{11.34}$$

By the Cauchy–Schwarz inequality,

$$\|\mathscr{T}\beta - \tilde{\mathscr{T}}\beta\|_{\mathbb{Y}}^2 \leq \|\beta\|^2 \frac{1}{n} \sum_{i=1}^n \|X_i - \tilde{X}_i\|^2$$
$$= O_p \left(\frac{1}{n} \sum_{i=1}^n \mathbb{E}\|X_i - \tilde{X}_i\|^2 \right),$$

and, from (11.34),

$$\|\tilde{\mathscr{T}} \mathbb{E}_\varepsilon (\tilde{\beta}_\eta - \beta)\|_{\mathbb{Y}}^2 = O_p \left(\eta + \frac{1}{n} \sum_{i=1}^n \mathbb{E}\|X_i - \tilde{X}_i\|^2 \right). \tag{11.35}$$

Then, (11.31) follows from (11.32) and (11.35).

Next, let $X_1', \ldots, X_n'$ be iid and have the same distribution as X_1 and let U be a discrete uniform random variable with possible values $1, \ldots, n$. Assume that $X_1, \ldots, X_n, X_1', \ldots, X_n', U$ are all independent and define

$$\tilde{X}_i'(t) = \sum_{j=1}^{J_i} X_i'(t_{ij}) I(t \in I_{ij}),$$

and

$$\tilde{X} = \sum_{i=1}^{n} I(U = i)\tilde{X}'_i.$$

Note that the covariance operator of $\tilde{X}$ is $\tilde{\mathscr{K}}$. Using (11.31) and following the lines of the proofs for Lemma 11.2.6 and Theorem 11.2.7, we see that

$$\mathbb{E}_{\varepsilon,\tilde{X}}\left(\langle \tilde{X}, \tilde{\beta}_\eta\rangle - \langle \tilde{X}, \beta\rangle\right)^2 = O_p\left(n^{-\frac{2q+\alpha+1}{2q+\alpha+2}} + n^{-\frac{\alpha+1}{2}} + \frac{1}{n}\sum_{i=1}^{n} \mathbb{E}\|X_i - \tilde{X}_i\|^2\right).$$

Of course, the object of interest is $\mathbb{E}_{\varepsilon,X}\left(\langle X, \tilde{\beta}_\eta\rangle - \langle X, \beta\rangle\right)^2$. However, we can equivalently consider X' defined by

$$X' = \sum_{i=1}^{n} I(U = i)X'_i,$$

which clearly has the same distribution as X_1. As

$$\langle X', \tilde{\beta}_\eta - \beta\rangle = \langle \tilde{X}, \tilde{\beta}_\eta - \beta\rangle + \langle X' - \tilde{X}, \tilde{\beta}_\eta - \beta\rangle,$$

we have

$$(\langle X', \tilde{\beta}_\eta\rangle - \langle X', \beta\rangle)^2 \le 2\langle \tilde{X}, \tilde{\beta}_\eta - \beta\rangle^2 + 2\langle X' - \tilde{X}, \tilde{\beta}_\eta - \beta\rangle^2.$$

Thus,

$$\begin{aligned}
&\mathbb{E}_{\varepsilon,X'}\langle X', \tilde{\beta}_\eta - \beta\rangle^2 \\
&\quad = \mathbb{E}_{\varepsilon,X'_1,\ldots,X'_n,U}\langle X', \tilde{\beta}_\eta - \beta\rangle^2 \\
&\quad \le 2\mathbb{E}_{\varepsilon,X'_1,\ldots,X'_n,U}\langle \tilde{X}, \tilde{\beta}_\eta - \beta\rangle^2 + 2\mathbb{E}_{\varepsilon,X'_1,\ldots,X'_n,U}\langle X' - \tilde{X}, \tilde{\beta}_\eta - \beta\rangle^2 \\
&\quad = 2\mathbb{E}_{\varepsilon,\tilde{X}}\langle \tilde{X}, \tilde{\beta}_\eta - \beta\rangle^2 + 2\mathbb{E}_{\varepsilon,X',\tilde{X}}\langle X' - \tilde{X}, \tilde{\beta}_\eta - \beta\rangle^2.
\end{aligned}$$

We already have a bound for the first term on the right-hand side of this last expression. For the second term, apply the Cauchy–Schwarz inequality to obtain

$$\mathbb{E}_{\varepsilon,X',\tilde{X}}\langle X' - \tilde{X}, \tilde{\beta}_\eta - \beta\rangle^2 \le \mathbb{E}_\varepsilon\|\tilde{\beta}_\eta - \beta\|^2\mathbb{E}\|X' - \tilde{X}\|^2.$$

As in the proof of Theorem 11.2.7, the first term on the right-hand side is $O_p(1)$. The second term is equal to $n^{-1}\sum_{i=1}^{n} \mathbb{E}\|X_i - \tilde{X}_i\|^2$ that completes the proof. □

Suppose, for example, that the X process has the covariance properties of Brownian motion and that the sampling occurs at a common set of points $t_j = (j-1)/J, j = 1, \ldots, J+1$. Then, one may check that

$$n^{-1}\sum_{i=1}^{n} \mathbb{E}\|X_i - \tilde{X}_i\|^2 = \mathbb{E}\|X_1 - \tilde{X}_1\|^2 = \int_0^1 tdt - J^{-1}\sum_{j=1}^{J} t_j \le J^{-1}.$$

Thus, as one might expect, for a fine sampling grid with, e.g., $J \gg n$, the quadrature error entailed by approximating the X_i by the $\tilde{X}_i$ has a negligible influence on the performance of the estimator in this particular case.

Note that β_η and $\tilde{\beta}_\eta$ are not natural splines. To compute these estimators, one can use the fact that $\mathbb{W}_q$ is an RKHS. Let us focus on $\tilde{\beta}_\eta$. Take ξ_i be the representer of the functional $g \mapsto \langle \tilde{X}_i, g\rangle, g \in \mathbb{W}_q$: namely, $\langle \xi_i, g\rangle_{\mathbb{W}_q} = \langle \tilde{X}_i, g\rangle$. If R denotes the rk for $\mathbb{W}_q$, then by the reproducing property,

$$\xi_i(t) = \langle \xi_i, R(\cdot, t)\rangle_{\mathbb{W}_q} = \langle \tilde{X}_i, R(\cdot, t)\rangle.$$

Now define the matrix

$$\mathcal{U} = \left\{ \int_0^1 \int_0^1 R(s,t)\tilde{X}_i(s)\tilde{X}_j(t)dsdt \right\}_{i,j=1:n}.$$

An application of Theorem 6.4.1 then reveals that the estimator can be expressed as

$$\tilde{\beta}_\eta = \sum_{i=1}^{n} b_i \xi_i$$

with

$$b = (\mathcal{U}^T\mathcal{U} + \eta\mathcal{U})^{-1}\mathcal{U}^T Y = (\mathcal{U} + \eta I)^{-1} Y$$

as $\mathcal{U}$ is symmetric.

References

Anderson T 2003 *An Introduction to Multivariate Statistical Analysis*. Wiley, New York, NY.

Aronszajn N 1950 Theory of reproducing kernels. *Trans. Am. Math. Soc.* **68**, 337–404.

Baker C 1970 Mutual information for Gaussian processes. *SIAM J. Appl. Math.* **19**, 451–458.

Baker C 1973 Joint measures and cross-covariance operators. *Trans. Am. Math. Soc.* **186**, 273–289.

Basilevsky A 1994 *Statistical Factor Analysis and Related Methods: Theory and Applications*. Wiley, New York, NY.

Bass R 2011 *Stochastic Processes*. Cambridge University Press, Cambridge.

Berlinet A and Thomas-Agnan C 2004 *Reproducing Kernel Hilbert Spaces in Probability and Statistics*. Kluwer Academic Publishers, Boston, MA.

Bickel P and Levina E 2004 Some theory for Fisher's linear discriminant function, 'naive bayes', and some alternatives when there are many more variables than observations. *Bernoulli* **10**, 989–1010.

Billingsley P 1995 *Probability and Measure, Third Edition*. Wiley, New York, NY.

Billingsley P 1999 *Convergence of Probability Measures*. Wiley, New York, NY.

Birkhoff G 1908 Boundary value and expansion problems of ordinary linear differential equations. *Trans. Am. Math. Soc.* **9**, 373–395.

Bochner S 1933 Integration von funktionen, deren werte die elemente eines vektorraumes sind. *Fundam. Math.* **20**, 262–276.

Cai T and Yuan M 2010 Nonparametric covariance function estimation for functional and longitudinal data. manuscript.

Cai T and Yuan M 2011 Optimal estimation of the mean function based on discretely sampled functional data: phase transition. *Ann. Stat.* **39**, 2330–2355.

Crambes C, Kneip A and Sarda P 2009 Smoothing spline estimators for functional linear regression. *Ann. Stat.* **37**, 35–72.

Dauxious J, Pousse A and Romain Y 1982 Asymptotic theory for the principal component analysis of a vector random function: some applications to statistical inference. *J. Multivariate Anal.* **12**, 136–154.

Theoretical Foundations of Functional Data Analysis, with an Introduction to Linear Operators, First Edition. Tailen Hsing and Randall Eubank.

de Acosta A 1970 Existence and convergence of probability measures in Banach spaces. *Trans. Am. Math. Soc.* **132**, 273–298.

de Boor C 1978 *A Practical Guide to Splines*. Springer, New York.

Diestel J and Uhl J 1977 *Vector Measures*. American Mathematical Society, Providence, RI.

Driscoll M 1973 The reproducing kernel Hilbert space structure of sample paths of Gaussian processes. *Z. Wahrscheinlichkeitstheorie verw. Geb.* **26**, 309–316.

Dunford N and Schwarz J 1988 *Linear Operators Part I: General Theory*. Wiley-Interscience, New York, NY.

Durrett R 1996 *Probability: Theory and Examples, The Wadsworth & Brooks/Cole Statistics/Probability Series*. Duxbury Press, ISBN: 9780534243180, LCCN: 95022544, http://books.google.com/books?id=kkc_AQAAIAAJ.

Engl H, Hanke M and Neubauer A 2000 *Regularization of Inverse Problems*. Kluwer Academic Publishers, Norwell, MA.

Etemadi N 1983 On the laws of large numbers for nonnegative random variables. *J. Multivariate Anal.* **13**, 187–193.

Eubank R and Hsing T 2007 Canonical correlation for stochastic processes. *Stochastic Processes Appl.* **118**, 1634–1661.

Fan J and Gijbels I 1996 *Local Polynomial Modeling and Its Applications*. Chapman and Hall, New York, NY.

Gittins R 1985 *Canonical Analysis: A Review with Applications in Ecology*. Springer, New York, NY.

Grigorieff R D 1991 A note on von Neumann's trace inequality. *Math. Nachr.* **151**, 327–328.

Hall P and Hosseini-Nasab M 2005 On properties of functional principal components analysis. *J. R. Stat. Soc. Ser. B* **68**, 109–126.

Hall P and Hosseini-Nasab M 2009 Theory for high-order bounds in functional principal components analysis. *Math. Proc. Cambridge Philos. Soc.* **146**, 225–256.

Hansen P 1988 Computation of the singular value expansion. *Computing* **40**, 185–199.

He G, Muller H and Wang J 2003 Functional canonical correlation analysis for square integrable stochastic processes. *J. Multivariate Anal.* **85**, 54–77.

Hotelling H 1936 Relations between two sets of variates. *Biometrika* **28**, 321–377.

Izenman A 2008 *Modern Multivariate Statistical Techniques: Regression, Classification and Manifold Learning*. Springer, New York, NY.

Johnson R and Wichern D 2007 *Applied Multivariate Statistical Analysis*, Sixth Edition. Prentice Hall, Upper Saddle River, NJ.

Jolliffe L 2004 *Principal Components Analysis*, Second Edition. Springer, New York, NY.

Kato T 1995 *Perturbation Theory for Linear Operators*. Springer, New York, NY.

Kshirsagar A 1972 *Multivariate Analysis*. Marcel-Dekker, New York, NY.

Landau H and Shepp L 1970 On the supremum of a Gaussian process. *Sankhyā Ser. A* **32**, 369–378.

Ledoux M and Talagrand M 2013 *Probability in Banach Spaces: Isoperimetry and Processes*. Springer, Berlin.

Lin Y 2000 Tensor product space ANOVA models. *Ann. Stat.* **28**, 734–755.

Li Y and Hsing T 2010 Uniform convergence rates for nonparametric regression and principal component analysis of functional/longitudinal data. *Ann. Stat.* **38**, 3321–3351.

Luenberger D 1969 *Optimization by Vector Space Methods*. Wiley, New York, NY.

Lukić M and Beder J 2001 Stochastic processes with sample paths in reproducing kernel Hilbert spaces. *Trans. Am. Math. Soc.* **353**, 3945–3969.

Mas A 2006 A sufficient condition for the CLT in the space of nuclear operators - application to covariance of random functions. *Stat. Probab. Lett.* **76**, 1503–1509.

Moore H 1916 On properly positive Hermitian matrices. *Bull. Am. Math. Soc.* **23**, 66–67.

Muirhead R and Waternaux C 1980 Asymptotic distributions in canonical correlation analysis and other multivariate procedures for nonnormal populations. *Biometrika* **67**, 31–43.

Nychka D and Cox D 1989 Convergence rates for regularized solutions of integral equalities from discrete noisy data. *Ann. Stat.* **17**, 556–572.

Ortega J and Rheinboldt W 1970 *Iterative Solution of Nonlinear Equations in Several Variables*. Academic Press, New York, NY.

Parzen E 1961 An approach to time series analysis. *Ann. Math. Stat.* **32**, 951–989.

Parzen E 1970 Statistical inference on time series by RKHS method. In *12th Annual Biennial Seminar Canadian Mathematical Congress Proceedings* (ed. Pyke R), pp. 1–37, Canadian Mathematical Congress, Montreal.

Ramsay J and Silverman B 2005 *Functional Data Analysis*, Second Edition. Springer, New York.

Rao C 1955 Estimation and tests of significance in factor analysis. *Psychometrika* **20**, 93–111.

Reed M and Simon D 1980 *Functional Analysis*. Academic Press, Salt Lake City, UT.

Resnick S 1999 *A Probability Path*. Birkhäuser, Boston, MA.

Rice J and Silverman B 1991 Estimating the mean and covariance structure nonparametrically when the data are curves. *J. R. Stat. Soc. Ser. B* **53**, 233–243.

Riesz F and Sz.-Nagy B 1990 *Functional Analysis*. Dover Publications, New York, NY.

Roy S 1958 *Some Aspects of Multivariate Analysis*. Wiley, New York, NY.

Royden H and Fitzpatrick P 2010 *Real Analysis*, Fourth Edition. Pearson, Upper Saddle River, NJ.

Rudin W 1991 *Functional Analysis*. McGraw-Hill, Boston, MA.

Rynne B and Youngson M 2001 *Linear Functional Analysis*. Springer, New York, NY.

Salaff S 1968 Regular boundary conditions for ordinary differential operators. *Trans. Am. Math. Soc.* **134**, 355–373.

Schumaker L 1981 *Spline Functions: Basic Theory*. Wiley, New York, NY.

Shin H 2008 An extension of Fisher's discriminant analysis for stochastic processes. *J. Multivariate Anal.* **99**, 1191–1216.

Stewart G 1993 On the early history of the singular value decomposition. *SIAM Rev.* **35**, 551–566.

Stone C 1982 Optimal global rates of convergence for nonparametric regression. *Ann. Stat.* **10**, 1040–1053.

Stone M 1926 A comparison of the series of Fourier and Birkhoff. *Trans. Am. Math. Soc.* **28**, 695–761.

Sunder V 1988 N subspaces. *Can. J. Math.* **XL**, 38–54.

Thompson R and Freede L 1971 On the eigenvalues of sums of Hermitian matrices. *Linear Algebra Appl.* **4**, 369–376.

Utreras FI 1983 Natural splines functions: their associated eigenvalue problem. *Numer. Math.* **63**, 107–117.

Utreras F 1988 Boundary effects on convergence rates for Tikhonov regularization. *J. Approx. Theory* **54**, 235–249.

Yosida K 1971 *Functional Analysis*. Springer-Verlag, New York, NY.

Index

Theoretical Foundations of Functional Data Analysis, with an Introduction to Linear Operators, First Edition. Tailen Hsing and Randall Eubank.

Notation Index

Theoretical Foundations of Functional Data Analysis, with an Introduction to Linear Operators,
First Edition. Tailen Hsing and Randall Eubank.

WILEY SERIES IN PROBABILITY AND STATISTICS

ESTABLISHED BY WALTER A. SHEWHART AND SAMUEL S. WILKS

The ***Wiley Series in Probability and Statistics*** is well established and authoritative. It covers many topics of current research interest in both pure and applied statistics and probability theory. Written by leading statisticians and institutions, the titles span both state-of-the-art developments in the field and classical methods.

Reflecting the wide range of current research in statistics, the series encompasses applied, methodological and theoretical statistics, ranging from applications and new techniques made possible by advances in computerized practice to rigorous treatment of theoretical approaches.

This series provides essential and invaluable reading for all statisticians, whether in academia, industry, government, or research.

† ABRAHAM and LEDOLTER · Statistical Methods for Forecasting

AGRESTI · Analysis of Ordinal Categorical Data, *Second Edition*

AGRESTI · An Introduction to Categorical Data Analysis, *Second Edition*

AGRESTI · Categorical Data Analysis, *Third Edition*

ALSTON, MENGERSEN and PETTITT (editors) · Case Studies in Bayesian Statistical Modelling and Analysis

ALTMAN, GILL, and McDONALD · Numerical Issues in Statistical Computing for the Social Scientist

AMARATUNGA and CABRERA · Exploration and Analysis of DNA Microarray and Protein Array Data

AMARATUNGA, CABRERA, and SHKEDY · Exploration and Analysis of DNA Microarray and Other High-Dimensional Data, *Second Edition*

ANDĚL · Mathematics of Chance

ANDERSON · An Introduction to Multivariate Statistical Analysis, *Third Edition*

* ANDERSON · The Statistical Analysis of Time Series

ANDERSON, AUQUIER, HAUCK, OAKES, VANDAELE, and WEISBERG · Statistical Methods for Comparative Studies

ANDERSON and LOYNES · The Teaching of Practical Statistics

ARMITAGE and DAVID (editors) · Advances in Biometry

ARNOLD, BALAKRISHNAN, and NAGARAJA · Records

* ARTHANARI and DODGE · Mathematical Programming in Statistics

AUGUSTIN, COOLEN, DE COOMAN and TROFFAES (editors) · Introduction to Imprecise Probabilities

* BAILEY · The Elements of Stochastic Processes with Applications to the Natural Sciences

† Now available in a lower priced paperback edition in the Wiley–Interscience Paperback Series.
* Now available in a lower priced paperback edition in the Wiley Classics Library.

BAJORSKI · Statistics for Imaging, Optics, and Photonics

BALAKRISHNAN and KOUTRAS · Runs and Scans with Applications

BALAKRISHNAN and NG · Precedence-Type Tests and Applications

BARNETT · Comparative Statistical Inference, *Third Edition*

BARNETT · Environmental Statistics

BARNETT and LEWIS · Outliers in Statistical Data, *Third Edition*

BARTHOLOMEW, KNOTT, and MOUSTAKI · Latent Variable Models and Factor Analysis: A Unified Approach, *Third Edition*

BARTOSZYNSKI and NIEWIADOMSKA-BUGAJ · Probability and Statistical Inference, *Second Edition*

BASILEVSKY · Statistical Factor Analysis and Related Methods: Theory and Applications

BATES and WATTS · Nonlinear Regression Analysis and Its Applications

BECHHOFER, SANTNER, and GOLDSMAN · Design and Analysis of Experiments for Statistical Selection, Screening, and Multiple Comparisons

BEH and LOMBARDO · Correspondence Analysis: Theory, Practice and New Strategies

BEIRLANT, GOEGEBEUR, SEGERS, TEUGELS, and DE WAAL · Statistics of Extremes: Theory and Applications

BELSLEY Conditioning Diagnostics: Collinearity and Weak Data in Regression

† BELSLEY, KUH, and WELSCH · Regression Diagnostics: Identifying Influential Data and Sources of Collinearity

BENDAT and PIERSOL · Random Data: Analysis and Measurement Procedures, *Fourth Edition*

BERNARDO and SMITH · Bayesian Theory

BHAT and MILLER · Elements of Applied Stochastic Processes, *Third Edition*

BHATTACHARYA and WAYMIRE · Stochastic Processes with Applications

BIEMER, GROVES, LYBERG, MATHIOWETZ, and SUDMAN · Measurement Errors in Surveys

BILLINGSLEY · Convergence of Probability Measures, *Second Edition*

BILLINGSLEY · Probability and Measure, *Anniversary Edition*

BIRKES and DODGE · Alternative Methods of Regression

BISGAARD and KULAHCI · Time Series Analysis and Forecasting by Example

BISWAS, DATTA, FINE, and SEGAL · Statistical Advances in the Biomedical Sciences: Clinical Trials, Epidemiology, Survival Analysis, and Bioinformatics

BLISCHKE and MURTHY (editors) · Case Studies in Reliability and Maintenance

BLISCHKE and MURTHY · Reliability: Modeling, Prediction, and Optimization

BLOOMFIELD · Fourier Analysis of Time Series: An Introduction, *Second Edition*

BOLLEN · Structural Equations with Latent Variables

BOLLEN and CURRAN · Latent Curve Models: A Structural Equation Perspective

BONNINI, CORAIN, MAROZZI and SALMASO · Nonparametric Hypothesis Testing: Rank and Permutation Methods with Applications in R

BOROVKOV · Ergodicity and Stability of Stochastic Processes

† Now available in a lower priced paperback edition in the Wiley–Interscience Paperback Series.

BOSQ and BLANKE · Inference and Prediction in Large Dimensions

BOULEAU · Numerical Methods for Stochastic Processes

* BOX and TIAO · Bayesian Inference in Statistical Analysis

BOX · Improving Almost Anything, *Revised Edition*

* BOX and DRAPER · Evolutionary Operation: A Statistical Method for Process Improvement

BOX and DRAPER · Response Surfaces, Mixtures, and Ridge Analyses, *Second Edition*

BOX, HUNTER, and HUNTER · Statistics for Experimenters: Design, Innovation, and Discovery, *Second Editon*

BOX, JENKINS, and REINSEL · Time Series Analysis: Forcasting and Control, *Fourth Edition*

BOX, LUCEÑO, and PANIAGUA-QUIÑONES · Statistical Control by Monitoring and Adjustment, *Second Edition*

* BROWN and HOLLANDER · Statistics: A Biomedical Introduction

CAIROLI and DALANG · Sequential Stochastic Optimization

CASTILLO, HADI, BALAKRISHNAN, and SARABIA · Extreme Value and Related Models with Applications in Engineering and Science

CHAN · Time Series: Applications to Finance with R and S-Plus^, *Second Edition*

CHARALAMBIDES · Combinatorial Methods in Discrete Distributions

CHATTERJEE and HADI · Regression Analysis by Example, *Fourth Edition*

CHATTERJEE and HADI · Sensitivity Analysis in Linear Regression

CHEN · The Fitness of Information: Quantitative Assessments of Critical Evidence

CHERNICK · Bootstrap Methods: A Guide for Practitioners and Researchers, *Second Edition*

CHERNICK and FRIIS · Introductory Biostatistics for the Health Sciences

CHILÈS and DELFINER · Geostatistics: Modeling Spatial Uncertainty, *Second Edition*

CHIU, STOYAN, KENDALL and MECKE · Stochastic Geometry and Its Applications, *Third Edition*

CHOW and LIU · Design and Analysis of Clinical Trials: Concepts and Methodologies, *Third Edition*

CLARKE · Linear Models: The Theory and Application of Analysis of Variance

CLARKE and DISNEY · Probability and Random Processes: A First Course with Applications, *Second Edition*

* COCHRAN and COX · Experimental Designs, *Second Edition*

COLLINS and LANZA · Latent Class and Latent Transition Analysis: With Applications in the Social, Behavioral, and Health Sciences

CONGDON · Applied Bayesian Modelling, *Second Edition*

CONGDON · Bayesian Models for Categorical Data

CONGDON · Bayesian Statistical Modelling, *Second Edition*

CONOVER · Practical Nonparametric Statistics, *Third Edition*

COOK · Regression Graphics

COOK and WEISBERG · An Introduction to Regression Graphics

* Now available in a lower priced paperback edition in the Wiley Classics Library.

COOK and WEISBERG · Applied Regression Including Computing and Graphics

CORNELL · A Primer on Experiments with Mixtures

CORNELL · Experiments with Mixtures, Designs, Models, and the Analysis of Mixture Data, *Third Edition*

COX · A Handbook of Introductory Statistical Methods

CRESSIE · Statistics for Spatial Data, *Revised Edition*

CRESSIE and WIKLE · Statistics for Spatio-Temporal Data

CSÖRGÖ and HORVÁTH · Limit Theorems in Change Point Analysis

DAGPUNAR · Simulation and Monte Carlo: With Applications in Finance and MCMC

DANIEL · Applications of Statistics to Industrial Experimentation

DANIEL · Biostatistics: A Foundation for Analysis in the Health Sciences, *Eighth Edition*

* DANIEL · Fitting Equations to Data: Computer Analysis of Multifactor Data, *Second Edition*

DASU and JOHNSON · Exploratory Data Mining and Data Cleaning

DAVID and NAGARAJA · Order Statistics, *Third Edition*

DAVINO, FURNO and VISTOCCO · Quantile Regression: Theory and Applications

* DEGROOT, FIENBERG, and KADANE · Statistics and the Law

DEL CASTILLO · Statistical Process Adjustment for Quality Control

DEMARIS · Regression with Social Data: Modeling Continuous and Limited Response Variables

DEMIDENKO · Mixed Models: Theory and Applications with R, *Second Edition*

DENISON, HOLMES, MALLICK and SMITH · Bayesian Methods for Nonlinear Classification and Regression

DETTE and STUDDEN · The Theory of Canonical Moments with Applications in Statistics, Probability, and Analysis

DEY and MUKERJEE · Fractional Factorial Plans

DILLON and GOLDSTEIN · Multivariate Analysis: Methods and Applications

* DODGE and ROMIG · Sampling Inspection Tables, *Second Edition*

* DOOB · Stochastic Processes

DOWDY, WEARDEN, and CHILKO · Statistics for Research, *Third Edition*

DRAPER and SMITH · Applied Regression Analysis, *Third Edition*

DRYDEN and MARDIA · Statistical Shape Analysis

DUDEWICZ and MISHRA · Modern Mathematical Statistics

DUNN and CLARK · Basic Statistics: A Primer for the Biomedical Sciences, *Fourth Edition*

DUPUIS and ELLIS · A Weak Convergence Approach to the Theory of Large Deviations

EDLER and KITSOS · Recent Advances in Quantitative Methods in Cancer and Human Health Risk Assessment

* ELANDT-JOHNSON and JOHNSON · Survival Models and Data Analysis

ENDERS · Applied Econometric Time Series, Third Edition

† ETHIER and KURTZ · Markov Processes: Characterization and Convergence

* Now available in a lower priced paperback edition in the Wiley Classics Library.

† Now available in a lower priced paperback edition in the Wiley–Interscience Paperback Series.

EVANS, HASTINGS, and PEACOCK · Statistical Distributions, *Third Edition*

EVERITT, LANDAU, LEESE, and STAHL · Cluster Analysis, *Fifth Edition*

FEDERER and KING · Variations on Split Plot and Split Block Experiment Designs

FELLER · An Introduction to Probability Theory and Its Applications, Volume I, *Third Edition,* Revised; Volume II, *Second Edition*

FITZMAURICE, LAIRD, and WARE · Applied Longitudinal Analysis, *Second Edition*

* FLEISS · The Design and Analysis of Clinical Experiments

FLEISS · Statistical Methods for Rates and Proportions, Third Edition

† FLEMING and HARRINGTON · Counting Processes and Survival Analysis

FUJIKOSHI, ULYANOV, and SHIMIZU · Multivariate Statistics: High-Dimensional and Large-Sample Approximations

FULLER · Introduction to Statistical Time Series, Second Edition

† FULLER · Measurement Error Models

GALLANT · Nonlinear Statistical Models

GEISSER · Modes of Parametric Statistical Inference

GELMAN and MENG · Applied Bayesian Modeling and Causal Inference from ncomplete-Data Perspectives

GEWEKE · Contemporary Bayesian Econometrics and Statistics

GHOSH, MUKHOPADHYAY, and SEN · Sequential Estimation

GIESBRECHT and GUMPERTZ · Planning, Construction, and Statistical Analysis of Comparative Experiments

GIFI · Nonlinear Multivariate Analysis

GIVENS and HOETING · Computational Statistics

GLASSERMAN and YAO · Monotone Structure in Discrete-Event Systems

GNANADESIKAN · Methods for Statistical Data Analysis of Multivariate Observations, *Second Edition*

GOLDSTEIN · Multilevel Statistical Models, *Fourth Edition*

GOLDSTEIN and LEWIS · Assessment: Problems, Development, and Statistical Issues

GOLDSTEIN and WOOFF · Bayes Linear Statistics

GRAHAM · Markov Chains: Analytic and Monte Carlo Computations

GREENWOOD and NIKULIN · A Guide to Chi-Squared Testing

GROSS, SHORTLE, THOMPSON, and HARRIS · Fundamentals of Queueing Theory, *Fourth Edition*

GROSS, SHORTLE, THOMPSON, and HARRIS · Solutions Manual to Accompany Fundamentals of Queueing Theory, *Fourth Edition*

* HAHN and SHAPIRO · Statistical Models in Engineering

HAHN and MEEKER · Statistical Intervals: A Guide for Practitioners

HALD · A History of Probability and Statistics and their Applications Before 1750

† HAMPEL · Robust Statistics: The Approach Based on Influence Functions

HARTUNG, KNAPP, and SINHA · Statistical Meta-Analysis with Applications

HEIBERGER · Computation for the Analysis of Designed Experiments

* Now available in a lower priced paperback edition in the Wiley Classics Library.

† Now available in a lower priced paperback edition in the Wiley–Interscience Paperback Series.

HEDAYAT and SINHA · Design and Inference in Finite Population Sampling

HEDEKER and GIBBONS · Longitudinal Data Analysis

HELLER · MACSYMA for Statisticians

HERITIER, CANTONI, COPT, and VICTORIA-FESER · Robust Methods in Biostatistics

HINKELMANN and KEMPTHORNE · Design and Analysis of Experiments, Volume 1: Introduction to Experimental Design, *Second Edition*

HINKELMANN and KEMPTHORNE · Design and Analysis of Experiments, Volume 2: Advanced Experimental Design

HINKELMANN (editor) · Design and Analysis of Experiments, Volume 3: Special Designs and Applications

HOAGLIN, MOSTELLER, and TUKEY · Fundamentals of Exploratory Analysis of Variance

* HOAGLIN, MOSTELLER, and TUKEY · Exploring Data Tables, Trends and Shapes

* HOAGLIN, MOSTELLER, and TUKEY · Understanding Robust and Exploratory Data Analysis

HOCHBERG and TAMHANE · Multiple Comparison Procedures

HOCKING · Methods and Applications of Linear Models: Regression and the Analysis of Variance, *Third Edition*

HOEL · Introduction to Mathematical Statistics, *Fifth Edition*

HOGG and KLUGMAN · Loss Distributions

HOLLANDER, WOLFE, and CHICKEN · Nonparametric Statistical Methods, *Third Edition*

HOSMER and LEMESHOW · Applied Logistic Regression, *Second Edition*

HOSMER, LEMESHOW, and MAY · Applied Survival Analysis: Regression Modeling of Time-to-Event Data, *Second Edition*

HUBER · Data Analysis: What Can Be Learned From the Past 50 Years

HUBER · Robust Statistics

† HUBER and RONCHETTI · Robust Statistics, *Second Edition*

HUBERTY · Applied Discriminant Analysis, *Second Edition*

HUBERTY and OLEJNIK · Applied MANOVA and Discriminant Analysis, *Second Edition*

HUITEMA · The Analysis of Covariance and Alternatives: Statistical Methods for Experiments, Quasi-Experiments, and Single-Case Studies, *Second Edition*

HUNT and KENNEDY · Financial Derivatives in Theory and Practice, *Revised Edition*

HURD and MIAMEE · Periodically Correlated Random Sequences: Spectral Theory and Practice

HUSKOVA, BERAN, and DUPAC · Collected Works of Jaroslav Hajek – with Commentary

HUZURBAZAR · Flowgraph Models for Multistate Time-to-Event Data

JACKMAN · Bayesian Analysis for the Social Sciences

† JACKSON · A User's Guide to Principle Components

* Now available in a lower priced paperback edition in the Wiley Classics Library.
† Now available in a lower priced paperback edition in the Wiley–Interscience Paperback Series.

JOHN · Statistical Methods in Engineering and Quality Assurance

JOHNSON · Multivariate Statistical Simulation

JOHNSON and BALAKRISHNAN · Advances in the Theory and Practice of Statistics: A Volume in Honor of Samuel Kotz

JOHNSON, KEMP, and KOTZ · Univariate Discrete Distributions, *Third Edition*

JOHNSON and KOTZ (editors) · Leading Personalities in Statistical Sciences: From the Seventeenth Century to the Present

JOHNSON, KOTZ, and BALAKRISHNAN · Continuous Univariate Distributions, Volume 1, *Second Edition*

JOHNSON, KOTZ, and BALAKRISHNAN · Continuous Univariate Distributions, Volume 2, *Second Edition*

JOHNSON, KOTZ, and BALAKRISHNAN · Discrete Multivariate Distributions

JUDGE, GRIFFITHS, HILL, LÜTKEPOHL, and LEE · The Theory and Practice of Econometrics, *Second Edition*

JUREK and MASON · Operator-Limit Distributions in Probability Theory

KADANE · Bayesian Methods and Ethics in a Clinical Trial Design

KADANE AND SCHUM · A Probabilistic Analysis of the Sacco and Vanzetti Evidence

KALBFLEISCH and PRENTICE · The Statistical Analysis of Failure Time Data, *Second Edition*

KARIYA and KURATA · Generalized Least Squares

KASS and VOS · Geometrical Foundations of Asymptotic Inference

† KAUFMAN and ROUSSEEUW · Finding Groups in Data: An Introduction to Cluster Analysis

KEDEM and FOKIANOS · Regression Models for Time Series Analysis

KENDALL, BARDEN, CARNE, and LE · Shape and Shape Theory

KHURI · Advanced Calculus with Applications in Statistics, *Second Edition*

KHURI, MATHEW, and SINHA · Statistical Tests for Mixed Linear Models

* KISH · Statistical Design for Research

KLEIBER and KOTZ · Statistical Size Distributions in Economics and Actuarial Sciences

KLEMELÄ · Smoothing of Multivariate Data: Density Estimation and Visualization

KLUGMAN, PANJER, and WILLMOT · Loss Models: From Data to Decisions, *Third Edition*

KLUGMAN, PANJER, and WILLMOT · Loss Models: Further Topics

KLUGMAN, PANJER, and WILLMOT · Solutions Manual to Accompany Loss Models: From Data to Decisions, *Third Edition*

KOSKI and NOBLE · Bayesian Networks: An Introduction

KOTZ, BALAKRISHNAN, and JOHNSON · Continuous Multivariate Distributions, Volume 1, *Second Edition*

KOTZ and JOHNSON (editors) · Encyclopedia of Statistical Sciences: Volumes 1 to 9 with Index

† Now available in a lower priced paperback edition in the Wiley–Interscience Paperback Series.

* Now available in a lower priced paperback edition in the Wiley Classics Library.

KOTZ and JOHNSON (editors) · Encyclopedia of Statistical Sciences: Supplement Volume

KOTZ, READ, and BANKS (editors) · Encyclopedia of Statistical Sciences: Update Volume 1

KOTZ, READ, and BANKS (editors) · Encyclopedia of Statistical Sciences: Update Volume 2

KOWALSKI and TU · Modern Applied U-Statistics

KRISHNAMOORTHY and MATHEW · Statistical Tolerance Regions: Theory, Applications, and Computation

KROESE, TAIMRE, and BOTEV · Handbook of Monte Carlo Methods

KROONENBERG · Applied Multiway Data Analysis

KULINSKAYA, MORGENTHALER, and STAUDTE · Meta Analysis: A Guide to Calibrating and Combining Statistical Evidence

KULKARNI and HARMAN · An Elementary Introduction to Statistical Learning Theory

KUROWICKA and COOKE · Uncertainty Analysis with High Dimensional Dependence Modelling

KVAM and VIDAKOVIC · Nonparametric Statistics with Applications to Science and Engineering

LACHIN · Biostatistical Methods: The Assessment of Relative Risks, *Second Edition*

LAD · Operational Subjective Statistical Methods: A Mathematical, Philosophical, and Historical Introduction

LAMPERTI · Probability: A Survey of the Mathematical Theory, *Second Edition*

LAWLESS · Statistical Models and Methods for Lifetime Data, *Second Edition*

LAWSON · Statistical Methods in Spatial Epidemiology, *Second Edition*

LE · Applied Categorical Data Analysis, *Second Edition*

LE · Applied Survival Analysis

LEE · Structural Equation Modeling: A Bayesian Approach

LEE and WANG · Statistical Methods for Survival Data Analysis, *Fourth Edition*

LePAGE and BILLARD · Exploring the Limits of Bootstrap

LESSLER and KALSBEEK · Nonsampling Errors in Surveys

LEYLAND and GOLDSTEIN (editors) · Multilevel Modelling of Health Statistics

LIAO · Statistical Group Comparison

LIN · Introductory Stochastic Analysis for Finance and Insurance

LINDLEY · Understanding Uncertainty, *Revised Edition*

LITTLE and RUBIN · Statistical Analysis with Missing Data, *Second Edition*

LLOYD · The Statistical Analysis of Categorical Data

LOWEN and TEICH · Fractal-Based Point Processes

MAGNUS and NEUDECKER · Matrix Differential Calculus with Applications in Statistics and Econometrics, *Revised Edition*

MALLER and ZHOU · Survival Analysis with Long Term Survivors

MARCHETTE · Random Graphs for Statistical Pattern Recognition

MARDIA and JUPP · Directional Statistics

MARKOVICH · Nonparametric Analysis of Univariate Heavy-Tailed Data: Research and Practice

MARONNA, MARTIN and YOHAI · Robust Statistics: Theory and Methods

MASON, GUNST, and HESS · Statistical Design and Analysis of Experiments with Applications to Engineering and Science, *Second Edition*

McCULLOCH, SEARLE, and NEUHAUS · Generalized, Linear, and Mixed Models, *Second Edition*

McFADDEN · Management of Data in Clinical Trials, *Second Edition*

* McLACHLAN · Discriminant Analysis and Statistical Pattern Recognition

McLACHLAN, DO, and AMBROISE · Analyzing Microarray Gene Expression Data

McLACHLAN and KRISHNAN · The EM Algorithm and Extensions, *Second Edition*

McLACHLAN and PEEL · Finite Mixture Models

McNEIL · Epidemiological Research Methods

MEEKER and ESCOBAR · Statistical Methods for Reliability Data

MEERSCHAERT and SCHEFFLER · Limit Distributions for Sums of Independent Random Vectors: Heavy Tails in Theory and Practice

MENGERSEN, ROBERT, and TITTERINGTON · Mixtures: Estimation and Applications

MICKEY, DUNN, and CLARK · Applied Statistics: Analysis of Variance and Regression, *Third Edition*

* MILLER · Survival Analysis, *Second Edition*

MONTGOMERY, JENNINGS, and KULAHCI · Introduction to Time Series Analysis and Forecasting

MONTGOMERY, PECK, and VINING · Introduction to Linear Regression Analysis, *Fifth Edition*

MORGENTHALER and TUKEY · Configural Polysampling: A Route to Practical Robustness

MUIRHEAD · Aspects of Multivariate Statistical Theory

MULLER and STOYAN · Comparison Methods for Stochastic Models and Risks

MURTHY, XIE, and JIANG · Weibull Models

MYERS, MONTGOMERY, and ANDERSON-COOK · Response Surface Methodology: Process and Product Optimization Using Designed Experiments, *Third Edition*

MYERS, MONTGOMERY, VINING, and ROBINSON · Generalized Linear Models. With Applications in Engineering and the Sciences, *Second Edition*

NATVIG · Multistate Systems Reliability Theory With Applications

† NELSON · Accelerated Testing, Statistical Models, Test Plans, and Data Analyses

† NELSON · Applied Life Data Analysis

NEWMAN · Biostatistical Methods in Epidemiology

NG, TAIN, and TANG · Dirichlet Theory: Theory, Methods and Applications

OKABE, BOOTS, SUGIHARA, and CHIU · Spatial Tesselations: Concepts and Applications of Voronoi Diagrams, *Second Edition*

OLIVER and SMITH · Influence Diagrams, Belief Nets and Decision Analysis

* Now available in a lower priced paperback edition in the Wiley Classics Library.

† Now available in a lower priced paperback edition in the Wiley–Interscience Paperback Series.

PALTA · Quantitative Methods in Population Health: Extensions of Ordinary Regressions
PANJER · Operational Risk: Modeling and Analytics
PANKRATZ · Forecasting with Dynamic Regression Models
PANKRATZ · Forecasting with Univariate Box-Jenkins Models: Concepts and Cases
PARDOUX · Markov Processes and Applications: Algorithms, Networks, Genome and Finance
PARMIGIANI and INOUE · Decision Theory: Principles and Approaches
* PARZEN · Modern Probability Theory and Its Applications
PEÑA, TIAO, and TSAY · A Course in Time Series Analysis
PESARIN and SALMASO · Permutation Tests for Complex Data: Applications and Software
PIANTADOSI · Clinical Trials: A Methodologic Perspective, *Second Edition*
POURAHMADI · Foundations of Time Series Analysis and Prediction Theory
POURAHMADI · High-Dimensional Covariance Estimation
POWELL · Approximate Dynamic Programming: Solving the Curses of Dimensionality, *Second Edition*
POWELL and RYZHOV · Optimal Learning
PRESS · Subjective and Objective Bayesian Statistics, *Second Edition*
PRESS and TANUR · The Subjectivity of Scientists and the Bayesian Approach
PURI, VILAPLANA, and WERTZ · New Perspectives in Theoretical and Applied Statistics
† PUTERMAN · Markov Decision Processes: Discrete Stochastic Dynamic Programming
QIU · Image Processing and Jump Regression Analysis
* RAO · Linear Statistical Inference and Its Applications, *Second Edition*
RAO · Statistical Inference for Fractional Diffusion Processes
RAUSAND and HØYLAND · System Reliability Theory: Models, Statistical Methods, and Applications, *Second Edition*
RAYNER, THAS, and BEST · Smooth Tests of Goodnes of Fit: Using R, *Second Edition*
RENCHER and SCHAALJE · Linear Models in Statistics, *Second Edition*
RENCHER and CHRISTENSEN · Methods of Multivariate Analysis, *Third Edition*
RENCHER · Multivariate Statistical Inference with Applications
RIGDON and BASU · Statistical Methods for the Reliability of Repairable Systems
* RIPLEY · Spatial Statistics
* RIPLEY · Stochastic Simulation
ROHATGI and SALEH · An Introduction to Probability and Statistics, *Second Edition*
ROLSKI, SCHMIDLI, SCHMIDT, and TEUGELS · Stochastic Processes for Insurance and Finance
ROSENBERGER and LACHIN · Randomization in Clinical Trials: Theory and Practice
ROSSI, ALLENBY, and MCCULLOCH · Bayesian Statistics and Marketing
† ROUSSEEUW and LEROY · Robust Regression and Outlier Detection

* Now available in a lower priced paperback edition in the Wiley Classics Library.
† Now available in a lower priced paperback edition in the Wiley–Interscience Paperback Series.

ROYSTON and SAUERBREI · Multivariate Model Building: A Pragmatic Approach to Regression Analysis Based on Fractional Polynomials for Modeling Continuous Variables

* RUBIN · Multiple Imputation for Nonresponse in Surveys

RUBINSTEIN and KROESE · Simulation and the Monte Carlo Method, *Second Edition*

RUBINSTEIN and MELAMED · Modern Simulation and Modeling

RUBINSTEIN, RIDDER, and VAISMAN · Fast Sequential Monte Carlo Methods for Counting and Optimization

RYAN · Modern Engineering Statistics

RYAN · Modern Experimental Design

RYAN · Modern Regression Methods, *Second Edition*

RYAN · Sample Size Determination and Power

RYAN · Statistical Methods for Quality Improvement, *Third Edition*

SALEH · Theory of Preliminary Test and Stein-Type Estimation with Applications

SALTELLI, CHAN, and SCOTT (editors) · Sensitivity Analysis

SCHERER · Batch Effects and Noise in Microarray Experiments: Sources and Solutions

* SCHEFFE · The Analysis of Variance

SCHIMEK · Smoothing and Regression: Approaches, Computation, and Application

SCHOTT · Matrix Analysis for Statistics, *Second Edition*

SCHOUTENS · Levy Processes in Finance: Pricing Financial Derivatives

SCOTT · Multivariate Density Estimation: Theory, Practice, and Visualization

* SEARLE · Linear Models

† SEARLE · Linear Models for Unbalanced Data

† SEARLE · Matrix Algebra Useful for Statistics

† SEARLE, CASELLA, and McCULLOCH · Variance Components

SEARLE and WILLETT · Matrix Algebra for Applied Economics

SEBER · A Matrix Handbook For Statisticians

† SEBER · Multivariate Observations

SEBER and LEE · Linear Regression Analysis, Second Edition

† SEBER and WILD · Nonlinear Regression

SENNOTT · Stochastic Dynamic Programming and the Control of Queueing Systems

* SERFLING · Approximation Theorems of Mathematical Statistics

SHAFER and VOVK · Probability and Finance: It's Only a Game!

SHERMAN · Spatial Statistics and Spatio-Temporal Data: Covariance Functions and Directional Properties

SILVAPULLE and SEN · Constrained Statistical Inference: Inequality, Order, and Shape Restrictions

SINGPURWALLA · Reliability and Risk: A Bayesian Perspective

SMALL and MCLEISH · Hilbert Space Methods in Probability and Statistical Inference

SRIVASTAVA · Methods of Multivariate Statistics

* Now available in a lower priced paperback edition in the Wiley Classics Library.

† Now available in a lower priced paperback edition in the Wiley–Interscience Paperback Series.

STAPLETON · Linear Statistical Models, *Second Edition*

STAPLETON · Models for Probability and Statistical Inference: Theory and Applications

STAUDTE and SHEATHER · Robust Estimation and Testing

STOYAN · Counterexamples in Probability, *Second Edition*

STOYAN and STOYAN · Fractals, Random Shapes and Point Fields: Methods of Geometrical Statistics

STREET and BURGESS · The Construction of Optimal Stated Choice Experiments: Theory and Methods

STYAN · The Collected Papers of T. W. Anderson: 1943–1985

SUTTON, ABRAMS, JONES, SHELDON, and SONG · Methods for Meta-Analysis in Medical Research

TAKEZAWA · Introduction to Nonparametric Regression

TAMHANE · Statistical Analysis of Designed Experiments: Theory and Applications

TANAKA · Time Series Analysis: Nonstationary and Noninvertible Distribution Theory

THOMPSON · Empirical Model Building: Data, Models, and Reality, *Second Edition*

THOMPSON · Sampling, *Third Edition*

THOMPSON · Simulation: A Modeler's Approach

THOMPSON and SEBER · Adaptive Sampling

THOMPSON, WILLIAMS, and FINDLAY · Models for Investors in Real World Markets

TIERNEY · LISP-STAT: An Object-Oriented Environment for Statistical Computing and Dynamic Graphics

TROFFAES and DE COOMAN · Lower Previsions

TSAY · Analysis of Financial Time Series, *Third Edition*

TSAY · An Introduction to Analysis of Financial Data with R

TSAY · Multivariate Time Series Analysis: With R and Financial Applications

UPTON and FINGLETON · Spatial Data Analysis by Example, Volume II: Categorical and Directional Data

† VAN BELLE · Statistical Rules of Thumb, *Second Edition*

VAN BELLE, FISHER, HEAGERTY, and LUMLEY · Biostatistics: A Methodology for the Health Sciences, *Second Edition*

VESTRUP · The Theory of Measures and Integration

VIDAKOVIC · Statistical Modeling by Wavelets

VIERTL · Statistical Methods for Fuzzy Data

VINOD and REAGLE · Preparing for the Worst: Incorporating Downside Risk in Stock Market Investments

WALLER and GOTWAY · Applied Spatial Statistics for Public Health Data

WEISBERG · Applied Linear Regression, *Fourth Edition*

WEISBERG · Bias and Causation: Models and Judgment for Valid Comparisons

WELSH · Aspects of Statistical Inference

WESTFALL and YOUNG · Resampling-Based Multiple Testing: Examples and Methods for p-Value Adjustment

† Now available in a lower priced paperback edition in the Wiley–Interscience Paperback Series.

* WHITTAKER · Graphical Models in Applied Multivariate Statistics

WINKER · Optimization Heuristics in Economics: Applications of Threshold Accepting

WOODWORTH · Biostatistics: A Bayesian Introduction

WOOLSON and CLARKE · Statistical Methods for the Analysis of Biomedical Data, *Second Edition*

WU and HAMADA · Experiments: Planning, Analysis, and Parameter Design Optimization, *Second Edition*

WU and ZHANG · Nonparametric Regression Methods for Longitudinal Data Analysis

YAKIR · Extremes in Random Fields

YIN · Clinical Trial Design: Bayesian and Frequentist Adaptive Methods

YOUNG, VALERO-MORA, and FRIENDLY · Visual Statistics: Seeing Data with Dynamic Interactive Graphics

ZACKS · Examples and Problems in Mathematical Statistics

ZACKS · Stage-Wise Adaptive Designs

* ZELLNER · An Introduction to Bayesian Inference in Econometrics

ZELTERMAN · Discrete Distributions – Applications in the Health Sciences

ZHOU, OBUCHOWSKI, and MCCLISH · Statistical Methods in Diagnostic Medicine, *Second Edition*

* Now available in a lower priced paperback edition in the Wiley Classics Library.

Printed and bound by CPI Group (UK) Ltd, Croydon, CR0 4YY